T0249852

CONCEPTUAL DENSITY FUNCTIONAL THEORY AND ITS APPLICATION IN THE CHEMICAL DOMAIN

CONCEPTUAL DENSITY FUNCTIONAL THEORY AND ITS APPLICATION IN THE CHEMICAL DOMAIN

Edited by

Nazmul Islam, PhD
Savaş Kaya, PhD

APPLE
ACADEMIC
PRESS

Apple Academic Press Inc.	Apple Academic Press Inc.
3333 Mistwell Crescent	9 Spinnaker Way
Oakville, ON L6L 0A2 Canada	Waretown, NJ 08758 USA

© 2018 by Apple Academic Press, Inc.

First issued in paperback 2021

Exclusive worldwide distribution by CRC Press, a member of Taylor & Francis Group

No claim to original U.S. Government works

ISBN 13: 978-1-77-463532-2 (pbk)
ISBN 13: 978-1-77-188665-9 (hbk)

All rights reserved. No part of this work may be reprinted or reproduced or utilized in any form or by any electric, mechanical or other means, now known or hereafter invented, including photocopying and recording, or in any information storage or retrieval system, without permission in writing from the publisher or its distributor, except in the case of brief excerpts or quotations for use in reviews or critical articles.

This book contains information obtained from authentic and highly regarded sources. Reprinted material is quoted with permission and sources are indicated. Copyright for individual articles remains with the authors as indicated. A wide variety of references are listed. Reasonable efforts have been made to publish reliable data and information, but the authors, editors, and the publisher cannot assume responsibility for the validity of all materials or the consequences of their use. The authors, editors, and the publisher have attempted to trace the copyright holders of all material reproduced in this publication and apologize to copyright holders if permission to publish in this form has not been obtained. If any copyright material has not been acknowledged, please write and let us know so we may rectify in any future reprint.

Trademark Notice: Registered trademark of products or corporate names are used only for explanation and identification without intent to infringe.

Library and Archives Canada Cataloguing in Publication

Conceptual density functional theory and its application in the chemical domain / edited by Nazmul Islam, PhD, Savaş Kaya, PhD.

Includes bibliographical references and index.
Issued in print and electronic formats.
ISBN 978-1-77188-665-9 (hardcover).--ISBN 978-0-203-71139-2 (PDF)
1. Density functionals. 2. Quantum chemistry. I. Islam, Nazmul, editor II. Kaya, Savaş, editor

QD462.6.D45C66 2018   541'.28   C2018-902161-6 C2018-902162-4

CIP data on file with US Library of Congress

Apple Academic Press also publishes its books in a variety of electronic formats. Some content that appears in print may not be available in electronic format. For information about Apple Academic Press products, visit our website at **www.appleacademicpress.com** and the CRC Press website at **www.crcpress.com**

CONTENTS

ABOUT THE EDITORS

Nazmul Islam, PhD
Vice Principal, Government Engineering College, Ramgarh,
Jharkhand–825101, India

Nazmul Islam, PhD, is now working as a vice principal at the Government Engineering College, Ramgarh (India). He has published more than 70 research papers in several prestigious peer-reviewed journals and has written many book chapters and research books. In addition, he is the editor-in-chief of *The SciTech, Journal of Science and Technology*; *The SciTech, International Journal of Engineering Sciences*; and the *Signpost Open Access Journal of Theoretical Sciences*. He also serves as a member on the editorial boards of several journals, such Journal of *Chemical Engineering and Materials Science, International Journal of Physical Sciences, Polymers Research Journal*, and *International Journal of Current Physical Sciences*. He is also as editor of several books: *Handbook of Nano Science and Nano Technology, Theoretical and Computational Research in 21st Century*, and more.

Dr. Islam has been is selected for inclusion in the 2011 and 2012 (11th and 12th) editions of *Who's Who in Science and Engineering, Who's Who in America, Scientist's Reference Directory*—One, *World Series in Science, International Biographical Centre*, Great Britain. He has also been selected as an 2000 outstanding intellectual of the 21st century–2011 and as the member of the IBC (Top 100 Scientists 2012) by the International Biographical Centre, Cambridge, England.

Dr. Islam's research interests are in theoretical chemistry, particularly quantum chemistry, conceptual density functional theory (CDFT), periodicity, SAR, QSAR/QSPR study, drug design, HMO theory, biological function of chemical compounds, quantum biology, nanochemistry, and more.

Savaş Kaya, PhD
Lecturer, Imranli Vocational School, Cumhuriyet University, Turkey

Savaş Kaya, PhD, is a lecturer at the Imranli Vocational School, Cumhuriyet University, Turkey. His research is mainly focused on density functional theory, theoretical chemistry, and physical inorganic chemistry. He has authored many theoretical papers related to quantum chemical parameters, corrosion science, and chemical equalization principles. He completed his MSc degree in the field of theoretical inorganic chemistry at the Cumhuriyet University Science Institute and then went on to pursue his PhD education under the guidance of Professor Dr. Cemal Kaya.

LIST OF CONTRIBUTORS

Paul W. Ayers
Department of Chemistry and Chemical Biology, McMaster University, Hamilton, Ontario, L8S 4M1, Canada

Priyabrata Banerjee
Surface Engineering and Tribology Group, CSIR-Central Mechanical Engineering Research Institute, Mahatma Gandhi Avenue, Durgapur 713209, West Bengal, India / Academy of Scientific and Innovative Research (AcSIR), CSIR-CMERI Campus, West Bengal, Durgapur 713209, India

Steven K. Burger
Department of Chemistry and Chemical Biology, McMaster University, Hamilton, Ontario L8S 4M1, Canada

Ramon Carbó-Dorca
Institut de Química Computacional i Catàlisi, Universitat de Girona, Girona, Spain

Rogelio Cuevas-Saavedra
Department of Chemistry, The University of Western Ontario, London, Ontario, N6A 5B7, Canada / Department of Chemistry and Chemical Biology, McMaster University, Hamilton, Ontario, L8S 4M1, Canada

Walaa Fares
Department of Chemistry, Al Azhar University-Gaza, Gaza City, Palestine

Dulal C. Ghosh
Department of Chemistry, University of Kalyani, Kalyani, 741235, India

Farnaz Heidar-Zadeh
Department of Chemistry and Chemical Biology, McMaster University, Hamilton, Ontario, Canada / Department of Inorganic and Physical Chemistry, Ghent University, Krijgslaan 281 (S3), 9000 Gent, Belgium / Center for Molecular Modeling, Ghent University, Technologiepark 903, 9052 Zwijnaarde, Belgium

Nazmul Islam
Theoretical and Computational Chemistry Research Laboratory, Government Engineering College, Ramgarh, Jharkhand–825101, India

Cemal Kaya
Cumhuriyet University, Faculty of Science, Department of Chemistry, Sivas, 58140, Turkey

Savaş Kaya
Cumhuriyet University, Faculty of Science, Department of Chemistry, Sivas, 58140, Turkey

Yuli Liu
Department of Chemistry and Chemical Biology, McMaster University, Hamilton, Ontario L8S 4M1, Canada

Ramón Alain Miranda-Quintan
Laboratory of Computational and Theoretical Chemistry, Faculty of Chemistry, University of Havana, Havana, Cuba / Department of Chemistry and Chemical Biology; McMaster University; Hamilton, Ontario, Canada, E-mail: rmiranda@fq.uh.cu, ramirandaq@gmail.com

Ahmed A. K. Mohammed
Department of Chemistry and Chemical Biology, McMaster University, Hamilton, Ontario L8S 4M1, Canada

Roman F. Nalewajski
Department of Theoretical Chemistry, Jagiellonian University, Gronostajowa 2, 30-387 Cracow, Poland

Ime Bassey Obot
Centre of Research Excellence in Corrosion, Research Institute, King Fahd University of Petroleum and Minerals, Dhahran 31261, Kingdom of Saudi Arabia

Nataly Rabi
Department of Chemistry and Chemical Biology, McMaster University, Hamilton, Ontario, L8S 4M1, Canada

Zaki S. Safi
Department of Chemistry, Al Azhar University-Gaza, Gaza City, Palestine

Sourav Kr. Saha
Surface Engineering and Tribology Group, CSIR-Central Mechanical Engineering Research Institute, Mahatma Gandhi Avenue, Durgapur 713209, West Bengal, India / Academy of Scientific and Innovative Research (AcSIR), CSIR-CMERI Campus, West Bengal, Durgapur 713209, India

LIST OF ABBREVIATIONS

2-HBP	N,N'-bis(2-hydroxybenzaldehyde)-1,3-propandiimine
2-PCT	2-pyridinecarboxaldehyde thiosemicarbazone
3-HBP	N,N'-bis(3-hydroxybenzaldehyde)-1,3-propandiimine
4-HBP	N,N'-bis(4-hydroxybenzaldehyde)-1,3-propandiimine
4-PCT	4-pyridinecarboxaldehyde thiosemicarbazone
A	electron affinity
AIM	atoms-in-molecules
API	American Petroleum Institute
ATA	anisalicylal-[5-(p-methyl)-phenyl-4-amino-(1,2,4-triazolyl)-2-thiol]-acylhydrazone
B3LYP	hybrid density functional theory
BCP	bond critical point
BDTC	4-(4-bromophenyl)-N'-(2,4-dimethoxybenzylidene) thiazole-2-carbo-hydrazide
BHF	Born–Haber–Fajans
BHTC	4-(4-bromophenyl)-N'-(4-hydroxybenzylidene) thiazole-2 carbohydrazide
BMTC	4-(4-bromophenyl)-N'-(4-methoxybenzylidene)thiazole-2-carbohydrazide
BSSE	basis set superposition error
CDFT	conceptual density functional theory
CG	contragradience
CIT	classical information theory
COMPASS	condensed-phase optimized molecular potentials for atomistic simulation studies
CORE-C	Center of Research Excellence in Corrosion
CSA	charge sensitivity analysis
CT	charge transfer
DFT	density functional theory
DHDFs	doubly hybrid density functionals
DMs	density matrices

DPE	deprotonation enthalpy
EEP	electronegativity equalization principle
EIS	electrochemical impedance spectroscopy
EIS	Electrochemical impedance spectroscopy
ELF	electron localization function
EO	equidensity orbitals
FMM	fast marching method
FMOs	frontier molecular orbitals
FTSC	furoin thiosemicarbazone
GB	gas phase basicities
GC	grand-canonical
GDP	gross domestic product
GGA	generalized-gradient approximation
GIAO	gauge-independent atomic orbital
GSM	growing string algorithm
HBTT	3-[(2-hydroxy-benzylidene)-amino]-2-thioxo-thiazolidin-4-one
HCl	hydrochloric acid
HF	Hartree–Fock
HK	Hohenberg and Kohn
HOMO	highest occupied molecular orbital
HSAB	hard/soft acid/base
HZM	Harriman-Zumbach-Maschke
I	ionization energy
IEF-PCM	integral equation formalism polarizable continuum model
INHB	N'-(phenylmethylene) isonicotinohydrazide
INHC	N'-(3-phenylallylidene) isonicotinohydrazide
INHF	N'-(furan-2-ylmethylene) isonicotinohydrazide
INHS	N'-(2-hydroxybenzylidene) isonicotinohydrazide
IRC	intrinsic reaction coordinate
IRI	intrinsic reactivity index
KACST	King Abdulaziz City for Science and Technology
KFUPM	King Fahd University of Petroleum & Minerals
KS	Kohn–Sham
LCS	low carbon steel
LDA	local density approximation

LPBE	Linearized Poisson–Boltzmann equation
LRD	local reactivity descriptor
LSDA	local spin-density approximation
LUMO	lowest unoccupied molecular orbital
MD	molecular dynamics
MEP	minimum energy path
MEP	molecular electrostatic potential
MHP	maximum hardness principle
MI	3-(4-(4-methoxybenzylideneamino)phenylimino)indolin-2-one
MM	molecular mechanics
MMQT	3-((5-methylthiazol-2-ylimino)methyl) quinoline-2-thiol
MO	molecular orbital
MPP	minimum polarizability principle
MTMP	2-((5-mercapto-1,3,4-thiadiazol-2-ylimino)methyl)phenol
NAO	natural atomic orbital
NBO	natural bonding orbital
NMR	nuclear magnetic resonance
NSTIP	National Science Technology Plan
NT	Newton trajectory
NVAO	natural valence atomic orbital energies
OCP	open circuit potential
OCT	orbital communication theory
OMTKY3	turkey ovomucoid third domain
PA	proton affinity
PCM	polarized continuum method
PDTT	5-((E)-4-phenylbuta-1,3-dienylideneamino)-1,3,4-thiadiazole-2-thiol
PEIDs	property enhanced intramolecular distances
PES	photo electron spectroscopy
PES	potential energy surface
PEST	property encoded surface translator
PI	3-(4-(3-phenylallylideneamino)phenylimino)indolin-2-one
PMA	2-(2-{[2-(4-pyridylcabonyl)hydrazono]methyl}phenoxy)acetic acid

PMQ	3-((phenylimino)methyl)quinoline-2-thiol
PPLB	Perdew, Parr, Levy, and Balduz
PT	prototropic tautomerism
PTM	2-(phenylthio)phenyl)-1-(o-tolyl)methanimine
QIT	quantum information theory
QM	quantum mechanics
QMSI	quantum molecular similarity indices
QMSM	quantum molecular similarity measures
QSAR	quantitative structure activity relationships
QSM	quadratic string method
QSPR	quantitative structure-property relationships
SCE	saturated calomel electrode
SCF	self-consistent field
SCRF	self-consistent reaction field
SDP	steepest descent path
SE	Schrödinger equation
SEM	scanning electron microscope
SRL	separated-reactant limit
STA	salicylal-[5-(p-methyl)-phenyl-4-amino-(1,2,4-triazolyl)-2-thiol]-acylhydrazone
TEM	transmission electron microscope
TS	transition states
VBT	volume based thermodynamics
VTA	vanillin-[5-(p-methyl)-phenyl-4-amino-(1,2,4-triazolyl)-2-thiol]-acylhydrazone
WE	working electrode
XC	exchange-correlation
ZPE	zero point energy

PREFACE

The scientific and technological progress in the 20th century was definitely incredible. In this century one of the most outstanding amendment was the shrinkage of the quantum chemical world due to the introduction of a new concept—the density functional theory (DFT). Till then, the evaluation of theoretical concepts has been strongly influenced by density functional theory. The chemical concepts such as chemical potential, chemical hardness, electronegativity, electrophilicity, and nucleophilicity have important applications in topics like prediction of reaction mechanisms and the analysis of chemical reactions.

DFT is a very popular mainstay of quantum mechanics. It depends upon the electron density (a three variable function) as a basic variable and is developed from the efforts of scientists in search of a simpler method alternative to wave function formalism. The DFT is not only a simpler quantum chemical method based upon a three variable density function but also it has proved its amazing power to put some very important qualitative chemical concepts like electronegativity and global hardness on sound quantum chemical foundation. The DFT is shown to provide a useful balance between accuracy and computational cost.

The chemical reactivity theory can provide important insights into the nature of atoms and molecule. The ultimate aim of the reactivity theory is to provide the answer to the fundamental questions like *Why some molecules are more stable and other more reactive?*, *Is it possible to predict which atomic site is the most susceptible to undergo either a nucleophilic or an electrophilic attack?*, *Why certain atomic sites of a molecule is more reactive?*, etc. Theoretical scientists had the ambition to rationalize those experimental facts with the help of chemical theories. The most fruitful and promising framework so far is probably the DFT. The quest for the theoretical basis of the hard-soft acid base behavior has created such a surge of fundamental research in chemistry that it gave birth of a new branch of density functional based theoretical science known as conceptual density functional theory (CDFT). The framework and some recent developments of CDFT has been facilitated by the works of conceptual density functional scientists.

In this book, some new developments based on CDFT and some applications of CDFT are discussed. In addition, we have discussed some applications in corrosion and conductivity and synthesis studies based on CDFT. The electronic structure principle such as electronegativity equalization principle, hardness equalization principle, electrophilicity equalization principle, nucleophilicity equalization principle, and studies based on these electronic structure principles are broadly explained.

In Chapter 1, Islam briefly discussed the origin and development of CDFT. In recent years some novel methodologies has been developed in the field of CDFT. These methodologies have been used to explore mutual relationships between the descriptors of CDFT namely electronegativity, hardness, etc. The mutual relationship between the electronegativity and the hardness depend on the electronic configuration of the neutral atomic species. The method of calculation of equalized molecular electronegativity, equalized molecular electrophilicity, and equalized molecular nucleophilicity based on the assumption of charge equalization during the chemical event of molecule formation and by combining electrostatic theorem and CDFT is also reviewed. In Chapter 2, Ramón Alain Miranda-Quintana presented a short introduction to the use of DFT in the study of chemical reactions. In Chapter 3, Ayers et al. computed the unconstrained local hardness using an exact equation for the unconstrained local hardness in Kohn-Sham density-functional theory. In Chapter 4, Ayers et al. analyzed the physical grounds of common approaches used to model the dependency of the ground state energy on the number of electrons in molecular systems (E vs. N models). Ayers et al. elaborated on a recent result indicating that smooth interpolation models are inconsistent with popular grand ensemble approach to open quantum systems and the authors have illustrated this discrepancy using the parabolic and exponential E vs. N models.

In Chapter 5, Kaya et al. revisited the chemical equalization principles and presented some new applications of chemical equalization principles related to hardness, electronegativity, electrophilicity, and nucleophilicity. Prevention of metallic corrosion is a very challenging job and would therefore be a mammoth task considering the enormous role of metals and their alloys in several industrial applications. To prevent solution state metallic corrosion, suitable small organic molecules as inhibitors are widely used as additive to the aggressive solution. Among the organic inhibitors, different N, O, S donor Schiff base molecules are considerably used due to its low cost starting precursor material, easy to follow synthetic route and environmental friendly manner. In Chapter 6, inhibition of metallic corrosion by N, O, S donor schiff base molecules was analyzed by Saha and Banerjee using CDFT methods. In Chapter 7, Obot et al. showed that the CDFT parameters can be used as a reliable approach to screen and select potential organic corrosion inhibitors prior to experimental validation. In Chapter 8, the equilibrium states of molecular systems, which extremize the system resultant entropy combining the *classical* (probability) and *nonclassical* (phase/current) information contributions, are explored in both the bimolecular complex R = A----B and its acidic(A) and basic(B) reactants by Nalewajski. The failures of embedded cluster models is discussed by Ayers et al. in Chapter 9. In Chapter 10, Ayers and coworkers showed that similarity-based kriging method for property prediction is more appealing than previous approaches. Finding and characterizing the pathways between reactants and products is very important for studying the mechanisms and energetics of chemical reactions, not only in the gas phase but also in complex environments like enzymes. Liu and Ayers, in Chapter 11, discussed three different methods to study chemical reaction using quantum mechanical models. Safi, in Chapter 12, calculated proton affinity, gas phase basicity, and enthalpy of some polyfunctional compounds using DFT method. In Chapter 13, Safi investigated the tautomerization process using the DFT in the neutral, protonated and deprotonated forms. In addition the effect of the interaction with the transition metal cations on the tautomerization process is also considered. In Chapter 14, Islam et al. compared the ionization energies data of the atoms computed using semi-empirical and DFT methods. In the last chapter, Heidar-Zadeh and Paul W. Ayers studied the molecular similarity from manifold learning on D2-property images.

Finally, we would like to thank the governing body of Apple Academic Press along with all authors, reviewers, and the Editorial Board Members for their invaluable contributions.

—**Dr. Nazmul Islam**
Dr. Savaş Kaya

CHAPTER 1

THE CONCEPTUAL DENSITY FUNCTIONAL THEORY: ORIGIN AND DEVELOPMENT TO STUDY ATOMIC AND MOLECULAR HARDNESS

NAZMUL ISLAM

Theoretical and Computational Chemistry Research Laboratory, Government Engineering College, Ramgarh, Jharkhand–825101, India

CONTENTS

ABSTRACT

In this chapter, we briefly discuss the origin and development of conceptual density functional theory (CDFT). We also review the method of calculating equalized molecular electronegativity, equalized molecular electrophilicity, and equalized molecular nucleophilicity based on the assumption of charge equalization during the chemical event of molecule formation and by combining the electrostatic theorem and CDFT.

1.1   INTRODUCTION

To solve the Schrödinger wave equation for many electronic systems, in the late 20's, a simple method known as density functional theory (DFT) [1] as an alternative to wave function formalism was developed. In quantum mechanics, we know that all information about a system is contained in the system's wave function, Ψ. Hence, it is known to all readers of quantum mechanics that "if you know Ψ of a system, you know everything about that system." Similarly, the DFT teaches us that all information about a system is coded in the system's wave functional. Pearson [2] has beautifully defined the term "functional" as "a functional is a recipe for turning a function in to a number, just as a function is a recipe for turning a variable in to a number." The DFT has revolutionized the theoretical study of chemical reactivity as it has a good rigid conceptual framework and fundamental concept. The basic idea of DFT is to replace the complicated N-electron wavefunction, Ψ, by the electron density, ρ, a three-variable function.

After its introduction, the DFT became the central plenum of quantum mechanics. The DFT is currently a valuable alternative for including correlations effects, without using complicated wave function methods.

1.2   THEORETICAL BACKGROUND OF CONCEPTUAL DENSITY FUNCTIONAL THEORY

The principles of DFT are conventionally developed by the attempts of theoreticians to solve the many-body Schrödinger equation

$$H\Psi = E\Psi \tag{1}$$

where H is the N-electron Hamiltonian, Ψ is the N-electron antisymmetric wave function, and E is the corresponding energy eigenvalue of H.

The first approximation may be considered as the one proposed by Hartree [3] in 1928. Hartree postulated that the N-electron wave function can be written as a simple product of N one-electron wave functions, each of which verifies a one-particle Schrödinger equation.

The DFT started with the landmark work of Hohenberg and Kohn [4] and the subsequent work of Kohn and Sham [5]. After the introduction of the Hohenberg and Kohn theorem-based DFT, the use of DFT gave a major boost to the field of computational chemistry. Hohenberg and Kohn showed in their first theorem that the ground state properties of a many-electron system are uniquely determined by an electron density that depends on only three spatial coordinates.

The Hamiltonian H has the form

$$H = T_N + V_N + U_N$$

where is T_N is the N-electron kinetic energy, V_N is the N-electron potential energy from the external field, and U_N is the electron-electron interaction energy.

The famous Hartree-Fock approximation, which is considered as the central plenum of most ab initio calculations, transformed the wavefunction Ψ into a Slater determinant of one-electron wavefunctions ψ_i

$$\left(-\frac{1}{2}\nabla^2 + v_{ext}(r) + v_H(r) + v_{x,i}^{HF}(r) \right) \psi_i(r) = \epsilon_i \, \psi_i(r) \qquad (2)$$

In the above equation, $v_{ext}(r)$, $v_H(r)$, and $v_{x,i}^{HF}$ are the external, Hartree, and non-local exchange potentials, respectively, and together are known as the effective HF operator, v_{eff}^{HF}.

$$\left(-\frac{1}{2}\nabla^2 + v_{eff}^{HF}(r) \right) \psi_i(r) = \epsilon_i \, \psi_i(r) \qquad (3)$$

where ∇^2 is the Laplacian operator and ϵ_i is the eigenvalue of electron i in spin-orbital ψ_i.

The problem with the Hartree-Fock method is that it does not account for Coulomb correlation due to the rigid form of the single determinant wave function. To take correlation into account, the so-called post-HF methods such as configuration interaction, coupled-cluster, or Møller–Plesset perturbation theory [6] must be taken into consideration. In post-HF methods, the wave function is generally represented by a linear combination of determinants accounting for correlation. Moreover, while the post-HF methods offer a systematic way to improve the accuracy of the results, they scale as fifth or even higher power with the size of the system, thus implying a considerable computational effort. Thomas [7] and Fermi [8] made a very important assumption – "Electrons are distributed uniformly in the six-dimensional spaces for the motion of an electron at the rate of two for each h^3 of volume." In 1964, Hohenberg and Kohn [4] provided the very first theorem that established the Thomas–Fermi assumption as an exact theory – the DFT.

The first Hohenberg and Kohn [4] theorem allows us to determine the external potential v_r of a given system under certain conditions by the electron density ρ_r of that system. They also demonstrated that the total energy of the system is stationary with respect to the density ρ_r. Unfortunately, the Hohenberg–Kohn theorems do not provide the exact form of the total energy functional $E[r]$. Among the different components of the total energy, the exact density functional forms of both kinetic and exchange-correlation terms remain unknown. To circumvent the problem of the kinetic part (much larger than the exchange correlation one), Kohn and Sham (KS) [7] proposed to introduce a set of fictitious one-electron wave functions ψ_i to build a Slater determinant. This leads to the KS equation

$$\left(-\frac{1}{2}\nabla^2 + v_{ext}(r) + v_H(r) + v_{x,i}^{KS}(r)\right)\psi_i(\mathbf{r}) = \epsilon_i \, \psi_i(\mathbf{r}) \tag{4}$$

In the above equation, $v_{ext}(r)$, $v_H(r)$, and $v_{x,i}^{KS}$ are the external, Hartree, and exchange potentials, respectively, and together are known as the effective KS operator, v_{eff}^{KS}.

$$(-1/2\nabla^2 + v_{eff}^{KS}(r))\psi_i(r) = \epsilon_r \, \psi_i(r) \tag{5}$$

The energy is a functional of the wave function.

$$E[\psi_i(\mathbf{r})] = \int \psi_i(\mathbf{r}) H \psi_i(\mathbf{r})^* \, dr \equiv \langle \psi_i(\mathbf{r})|H|\psi_i(\mathbf{r})^*\rangle \tag{6}$$

The variational theorem states that the energy is higher than that of the ground state unless Ψ correspond to Ψ_0. The ground state wave function and energy may be found by searching all possible wave functions for the one that minimizes the total energy. The Hartree-Fock theory consists of an ansatz for the structure of Ψ; it is assumed to be an antisymmetric product of functions, each of which depends on the coordinates of a single electron.

The density can then be obtained by squaring $\Psi_i(r)$ and integrating the coordinates of all the electrons (i.e., $i = 1$ to $i = N$)

$$\rho_r = \sum_{i=1}^{N} |\psi_i(\mathbf{r})|^2 \qquad (7)$$

Now, like Hartree's scheme, the KS scheme also failed to account for the exchange correlation $v_{x,i}^{KS}(\mathbf{r})$.

To apply the density scheme to describe atomic and molecular states and properties, Parr et al. [9] have developed an important concept known as conceptual density functional theory (CDFT). The CDFT is a subfield of DFT in which one tries to extract from the electronic density relevant concepts and principles that help to understand and predict the chemical behavior of a molecule. CDFT successfully provides easy computational algorithms for chemical concepts such as electronegativity, hardness, softness, and electrophilicity and provides the framework to use them in the domain of chemical reactivity by proving some principles such as the electronegativity equalization principle [10], the hard soft acid base (HSAB) principle [11], and the maximum hardness principle [12].

DFT methods are in general capable of generating a variety of isolated molecular properties. The commercial exploitation of organic compounds as a medicinal drug, for example, is likely to require, at some stage of its development, the determination of biological or chemical activities or properties related to the intended end use of the compound. It is therefore desirable to have at hand relatively straightforward and inexpensive procedures that enable the efficient and accurate prediction of a molecular activity or property, especially when its direct measurement by experiment is, for any reason, to be avoided if at all possible. The procedures that are conventionally used for indirect determination of activities employ molecular "descriptors" that have suitable molecular properties and physical-organic constructs obtained from both experimental and computational sources.

1.3   CDFT DESCRIPTORS AND THEIR USES

In 1968, Gyftopoulos and Hatsopoulos [13] following the statistics of ensembles and by considering free atom or ion as a thermodynamic system put forward a quantum thermodynamic definition of electronegativity. Given the electron density function, $\rho(r)$, in a chemical system (atom or molecule) and the energy functional, $E(\rho)$, the chemical potential, μ, of that system in equilibrium is defined as the derivative of the energy with respect to the number of electrons at fixed molecular geometry.

$$\mu = \left[\frac{\delta E(\rho)}{\delta \rho} \right]_v \qquad (8)$$

where v is the external potential acting on an electron due to the presence of the nucleus.

The differential definition more appropriate to the atomic system is based on the assumption that for a system of N electrons with the ground state energy, $E[N,v]$,

$$\mu = -\chi = \left(\frac{\partial E}{\partial N} \right)_{v(r)} \qquad (9)$$

where χ is the electronegativity of the chemical species.

$$\eta = \left(\frac{\partial^2 E}{\partial N^2} \right)_{v(r)} = \left(\frac{\partial \mu}{\partial N} \right)_{v(r)} \qquad (10)$$

where η is the hardness of the chemical species.

A distinct breakthrough came with the landmark work of Parr and Pearson [14] in 1983 when they derived operational and approximate definitions of chemical hardness, electronegativity, and chemical potential of any chemical species (atom, ion, or molecule) by using the method of finite difference approximation. Electronegativity, χ, and chemical hardness, η, are defined mathematically as

$$\chi = -\mu = \frac{(I + A)}{2} \tag{11}$$

$$\eta = \left(\frac{\partial^2 E}{\partial N^2}\right)_{\upsilon(r)} = \left(\frac{\partial \mu}{\partial N}\right)_{\upsilon(r)} = \frac{(I - A)}{2} \tag{12}$$

Parr et al. [15] by using an empirical relationship described by Maynard et al. [16] developed another very useful descriptor of CDFT known as electrophilicity index (ω). The mathematical definition of ω is

$$\omega = \frac{\mu^2}{2\eta} = \frac{\chi^2}{2\eta} \tag{13}$$

Nucleophilicity (ε) is physically the inverse of electrophilicity.

$$\varepsilon = 1 / \omega \tag{14}$$

We can calculate the four reactivity descriptors with the operational and approximate formula of all such descriptors in terms of equations (11–14) by using only the ionization energy and electron affinity value of the corresponding atom or molecule. But the problem is that I and A of all molecules are not known experimentally, and the accurate theoretical evaluation is very costly for sizable molecules [17, 18]. To overcome the abovementioned problems, we, following Feynman [19] and relying upon classical electrostatic theorems, have derived [17, 18] radial-dependent formulae for computing atomic χ, η, ω, and ε as follows:

Classically, the energy $E(N)$ of charging a conducting sphere of the radius, r, with the charge, q, is given by

$$E(N) = q^2 / 2r \quad (in\ C.G.S.\ Unit) \tag{15}$$

$$E(N) = q^2 / (4\pi\varepsilon_0)2r \quad (in\ S.I.\ Unit) \tag{16}$$

where $E(N)$ is in ergs, q is in electrostatic unit, and r is in cm.

Now, for an atom, the change in energy associated with the increase of q, on removal of an electron (of the charge, e), would be the ionization energy, I. Similarly, the energy generated on addition of an electron with q would be the electron affinity, A. Hence,

$$I = E(N+1) - E(N) = \left[\left\{ (q+e)^2 / 2r \right\} - \left(q^2 / 2r \right) \right] \qquad (17)$$

and

$$A = E(N) - E(N-1) = \left[\left(q^2 / 2r \right) - \left\{ (q-e)^2 / 2r \right\} \right] \qquad (18)$$

Now, putting the value of I and A into the CDFT definition of χ, η, ω, and ε, Eqs. (11, 12, 13, and 14), respectively, we offered the formulae for computing χ, η, ω, and ε as follows:

$$\chi \propto eq / r \qquad (19)$$

$$\eta \propto e^2 / 2r \qquad (20)$$

$$\omega \propto e^2 / r \qquad (21)$$

$$\varepsilon \propto r / e^2 \qquad (22)$$

1.4   COMPUTATION OF THE EQUALIZED MOLECULAR DESCRIPTORS

Let us consider the formation of a polyatomic molecule ABC… from its constituents. The polyatomic molecule is assumed to be a cluster of atoms where one atom is at the center and the other atoms are surrounding it. Let us assume that the central atom is A and the ligands surrounding the central atom are B, C… as represented in Eq. (23).

$$A + B + C + \ldots \rightarrow ABC \ldots \qquad (23)$$

Let the hardness of the molecule be η and that of the combining atoms be η_A, η_B, η_C .... η_n, respectively.

Let us further assume that the absolute atomic radii of the atoms A, B, C .... etc. are r_A, r_B, r_C, ... r_n, respectively. Rigorous investigation of the status and the physical condition of atoms in molecules has revealed that the atoms remain in a slightly modified state in the molecule [20, 21]. Because the radii of atoms in any molecule is not available and there is no report of any evaluation method of the radius of any atom being part of any molecule, we can, therefore, safely assume that the radius of the atom in a polyatomic molecule is approximately equal to its absolute radius for all approximate purposes.

Now, let us visualize the charge equalization on the process of molecule formation from its constituent atoms or fragments. While the physical process of charge transfer during the event of chemical reaction leading to bond formation takes place, the charge kernel of atoms changes and in the process, it would increase somewhere and decrease elsewhere so that the ultimate values of the charge-dependent parameters of the atomic fragments will equalize to some intermediate values common to all.

The direction of charge transfer mainly depends on the difference in electronegativity, hardness, electrophilicity, or nucleophilicity values of the atoms forming the molecule. However, for the derivation of the necessary formulae of our work, we arbitrarily assume that the central atom has a lower value and the other atoms have higher values of the charge-dependent parameters.

Now, let us assume that during the formation of the polyatomic molecule, δ is the total amount of charge transferred from the central atom A to the n number of the ligands surrounding the central atom. Although the total amount of charge, (δ), is distributed among the ligands, the amount of charge received by the individual atom is governed by the hardness of that atom.

Let B, C,... n^{th} ligands have the charge δ_1, δ_2, ... δ_n, respectively, in the molecular cluster and let

$$\delta = \delta_1 + \delta_2 + \ldots\ldots + \delta_n \qquad (24)$$

Now, after the charge transfer, the hardness of the central atom A in the polyatomic molecule becomes

$$\eta'_A = K_\eta \left(e - \delta\right)^2 / \ 2r'_A \qquad (25)$$

and the hardness of the ligands in the molecule becomes

$$\eta'_B = K_h \left\{e + \left(d_1\right)\right\}^2 / \ 2r'_B,$$
$$\eta'_C = K_h \left\{e + \left(d_2\right)\right\}^2 / \ 2r'_C, \qquad (26)$$
$$\cdots\cdots\cdots\cdots\cdots\cdots$$
$$\eta_n = K_h \left\{e + \left(d_n\right)\right\}^2 / \ 2r'_n$$

where r'_A, r'_B, r'_C ...... r'_n are the radii of atoms in the molecule, respectively. Similarly, η'_A, η'_B, η'_C, ... η'_n are the hardness of the atoms in the molecule and K_η is the proportionality constant.

Expanding Eq. (18), $(e-\delta)^2$ and neglecting the δ^2 term, we get the hardness of the central atom A as

$$\eta'_A = K_\eta \left(e^2 - 2e\delta\right) / 2r'_A \qquad (27)$$

Similarly, expanding Eq. (19) and neglecting the δ^2 terms, the formulae for hardness of atoms in the molecule are

$$\eta'_B = K_\eta (e^2 + 2e\delta_1) / \ 2r'_B, \eta'_C = K_\eta (e^2 + 2e\delta_2) / \ 2r'_C, \ldots \eta'_n$$
$$= K_\eta (e^2 + 2e\delta_n) / \ 2r'_n \qquad (28)$$

Now, invoking the hardness equalization principle after the formation of the molecule, the hardness of the individual constituents must be equalized, i.e.,

$$\eta_M = \eta'_A = \eta'_B = \eta'_C = \ldots\ldots = \eta'_n \qquad (29)$$

Eq. (22) implies

$$\eta_M = K_\eta (e^2 - 2e\delta)/\ 2r'_A = K_\eta (e^2 + 2e\delta_1)/\ 2r'_B = K_\eta (e^2 + 2e\delta_2)/\ 2r'_C$$
$$= \ldots = K_\eta (e^2 + 2e\delta_n)/\ r'_n = K_\eta \{(e^2 - 2e\delta) + (e^2 + 2e\delta_1)$$
$$+ (e^2 + 2e\delta_2) + \ldots + (e^2 + 2e\delta_n)\}\ /\ \left(2r'_A + 2r'_B + 2r'_C \ldots\ldots + 2r'_n\right)$$
$$= K_\eta \{(e^2 - 2e\delta) + ne^2 + 2e(\delta_1 + \delta_2 + \delta_3 + \ldots + \delta_n)\}$$
$$/\ \left(2r'_A + 2r'_B + 2r'_C \ldots\ldots + 2r'_n\right)$$
$$= K_\eta (e^2 - 2e\delta + ne^2 + 2e\delta)\ /\ \left(2r'_A + 2r'_B + 2r'_C \ldots\ldots + 2r'_n\right)$$

or

$$\eta_M = K_n (n+1)e^2 / 2\sum_i r'_i \tag{30}$$

Invoking the approximation that atoms retain their identity in the molecule, we can replace the r' term by the absolute radii, r, of the corresponding atom in Eq. (30).

Thus, we obtain

$$\eta_M = \frac{K_\eta\ e^2 (n+1)}{2\sum_i r_i} \tag{31}$$

where the atomic radius, r_i, is expressed in Angstrom unit and K is the constant depending on the fundamental nature of hardness.

Similarly, when we consider the electronegativity equalization, electrophilicity equalization, and nucleophilicity equalization processes, we get the molecular equalized electronegativity, electrophilicity, and nucleophilicity as

$$\chi_M = K_\chi \frac{(n+1)eq}{\sum_i r_i} \qquad \text{(in eV)} \tag{32}$$

$$\omega_M = K_\omega \frac{7.2(n+1)}{\sum_i r_i} \qquad \text{(in eV)} \tag{33}$$

and

$$\varepsilon_M = \mathrm{K}_\varepsilon \frac{\sum_i r_i}{7.2(n+1)} \qquad \text{(in eV)} \tag{34}$$

where r_i is the atomic radius in Angstrom unit.

The proportionality constant (K) of the above equations plays a significant role in the evaluation of the quantitative values of the molecular reactivity parameters. The K values can be computed using mathematical correlation of the RHS of Eqs. (31–34) with some other sets of corresponding molecular reactivity data [see Refs. 17, 18]. The choice of reference data may improve the results. Hence, the computed molecular reactivity parameters can still be extended and refined. In the last 5 years, we have carried out several attempts [17, 18] to investigate whether the assumption of charge equalization is valid in the real world, and we found that the charge equalization is a valid process during the heteronuclear molecule formation. Our previous investigations also suggest that the CDFT descriptors are useful to study the process of charge equalization during the heteronuclear molecule formation.

1.5 CONCLUSION

It is well known today that the CDFT descriptors can be successfully applied in the atomic and molecular domain. In this book, we have attempted to explore the grand application of the CDFT to study the atomic and molecular properties.

We reviewed the method of calculating equalized molecular electronegativity, equalized molecular electrophilicity, and equalized molecular nucleophilicity based on the assumption of charge equalization during the chemical event of molecule formation and by combining the electrostatic theorem and CDFT.

KEYWORDS

- **CDFT**
- **charge equalization**

- **DFT**
- **electrostatic theorem**
- **quantum mechanics**

REFERENCES

1. Parr, R. G., & Yang, W., (1989). *Density Functional Theory of Atoms and Molecules*, Oxford University Press, New York.
2. Pearson, R. G., (2005). Chemical Hardness and Density Functional Theory. *J. Chem. Sci.,* Bangalore, *117*, 369.
3. Hartree, D. R., (1928). The wave mechanics of an atom with a non-Coulomb central field. Part I. Theory and methods. *Math. Proc. Camb. Philos. Soc., 24*, 89–110.
4. Hohenberg, P., & Kohn, W., (1964). Inhomogeneous Electron Gas. *Phys. Rev., 136*, 864.
5. Kohn, W., & Sham, L. J., (1965). Self-consistent equations including exchange and correlation effects. *Phys. Rev., 140*, 1113.
6. Møller, C., & Plesset, M. S., (1934). Note on an approximation treatment for many-electron systems. *Phys. Rev., 46*, 618.
7. Thomas, L. H., (1927). The Calculation of Atomic Fields. *Proc. Cambridge. Philos. Soc., 23*, 542.
8. Fermi, E., (1928). Einestatistische Methodezur Bestimmungeiniger Eigenschaften des Atoms und ihreAnwendung auf die Theorie des periodischen Systems der Elemente.Z, *Phys., 48*, 73–79.
9. Kohn, W., Becke, A. D., & Parr, R. G., (1996). Density Functional Theory of Electronic Structure. *J. Phys. Chem., 100*, 12974–12980.
10. Pauling, L., (1960). The Nature of the Chemical Bond, vol. 3. Ithaca, NY: Cornell University Press.
11. Pearson, R. G., (1963). Hard and Soft Acids and Bases. *J. Am. Chem. Soc., 85*, 3533–3539.
12. Parr, R. G., & Chattaraj, P. K., (1991). Principle of Maximum Hardness. *J. Am. Chem. Soc., 113* 1854–1855.
13. Gyftopoulos, E. P., & Hatsopoulos, G. N., (1965). Self-Consistent-Charge Density-Functional Tight-Binding Method for Simulations of Complex Materials Properties. *Proc. Natl. Acad. Sci.,* USA, *60*, 786.
14. Parr, R. G., & Pearson, R. G., (1983). Absolute Hardness: Companion Parameter to Absolute Electronegativity. *J. Am. Chem. Soc., 105*, 7512–7516.
15. Parr, R. G., Szentpaly, L., & Liu, S., (1999). Electrophilicity Index. *J. Am. Chem. Soc., 121*, 1922–1924.
16. Maynard, A. T., Huang, M., Rice, W. G., & Covell, D. G., (1998). Reactivity of the HIV-1 nucleocapsid protein p7 zinc finger domains from the perspective of density-functional theory. *Proc. Natl. Acad. Sci.,* USA, *95*, 11578–11583.

17. Ghosh, D. C., & Islam, N. (2011). A Quest for the Algorithm for Evaluating the Molecular Hardness. *Int. J. Quant. Chem., 111,* 1931–1941.
18. Islam, N., & Ghosh, D. C., (2011). A new algorithm for the evaluation of the global hardness of polyatomic molecules. *Mol. Phys., 109,* 917–931.
19. Feynman, R. P., Leighton, R. B., & Sands, M., (1964). *The Feynman Lecture on Physics.* Boston: Addison-Wesley.
20. Ghosh, D. C., & Islam, N., (2011). Whether there is a hardness equalization principle analogous to the electronegativity equalization principle—A Quest. *Int. J. Quant. Chem., 111,* 1961–1969.
21. Islam, N., & Ghosh, D. C., (2012). On the electrophilic character of molecules through its relation with electronegativity and chemical hardness. *Int. J. Mol. Sci., 13,* 2160–2175.

CHAPTER 2

DENSITY FUNCTIONAL THEORY FOR CHEMICAL REACTIVITY

RAMÓN ALAIN MIRANDA-QUINTANA

Laboratory of Computational and Theoretical Chemistry, Faculty of Chemistry, University of Havana, Havana, Cuba

Department of Chemistry and Chemical Biology, McMaster University, Hamilton, Ontario, Canada, E-mail: rmiranda@fq.uh.cu, ramirandaq@gmail.com

CONTENTS

ABSTRACT

We present a short introduction to the use of density functional theory (DFT) in the study of chemical reactions (what is termed as conceptual DFT, chemical reactivity theory, or DFT for chemical reactivity). As the guiding theme for this endeavor, we took the notion of fractional numbers of electrons. The key role of this concept in conceptual DFT will give us a glimpse of the foundations of this theory and its relationship to "traditional" DFT. Our goal is to help the uninitiated reader gain an understanding of the successes and failures of conceptual DFT, and therefore, we focus on the fundamental theoretical aspects of the theory and strive to keep this exposition—which is by no means comprehensive—self-contained.

2.1   INTRODUCTION

The most important problem faced in quantum chemistry is (at least within the nonrelativistic and Born–Oppenheimer approximations) the solution to the electronic Schrödinger equation [1–3]:

$$\hat{H}|\Psi\rangle = E|\Psi\rangle, \tag{1}$$

where the electronic Hamiltonian, $\hat{H}$, depends parametrically on the nuclear coordinates, and it is constituted by the sum of the electronic kinetic energy, $\hat{T}_e$, the interaction energy of the electronic cloud with an external potential, $\hat{V}_{Ne}$, and the electron-electron repulsion, $\hat{V}_{ee}$:

$$\hat{H} = \hat{T}_e + \hat{V}_{Ne} + \hat{V}_{ee} \tag{2}$$

or, in a simpler and more explicit way, for a system with N electrons and M nuclei (and using atomic units):

$$\hat{H} = \sum_{i=1}^{N} -\frac{1}{2}\nabla_i^2 + \sum_{i=1}^{N}\sum_{A=1}^{M} -\frac{Z_A}{|r_i - R_A|} + \frac{1}{2}\sum_{i=1}^{N}\sum_{j=1}^{N}\frac{1}{|r_i - r_j|} \tag{3}$$

where Z_A and R_A represent the charges and coordinates of the nuclei, respectively.

In general, the overall understanding of this problem was what inspired Dirac to say that the underlying mathematical and physical laws necessary to understand the whole of chemistry are completely known. Hyperbole aside, adhering to this point of view is equivalent to reducing chemistry to a mere computation, with the sole purpose of obtaining reliable and accurate results for Eq. (1). However, we cannot forget Parr's remark that to calculate a molecule is not the same as to understand it.

In order to gain a deeper insight into the chemical behavior of atoms and molecules, obtaining accurate solution to the Schrödinger equation is a necessary and challenging problem, but by no means exhausts all the content of quantum chemistry. We also need tools to interpret the chemical information hidden in these solutions, because concepts so primordial to chemistry like the notion of an atom or chemical bonds inside a molecule are seemingly absent from Eq. (1).

There have been notable efforts to devise methods aimed to retrieve such chemical insight from the solutions of the Schrödinger equation, which focus mainly on the resulting wave function. However, these approaches face many problems, mainly due to the scaling complexity of this function (for a system of N electrons, it depends on $3N$ spatial and N spin coordinates). But perhaps more importantly, as soon as we increase the complexity of our computational approach, moving beyond the simplest molecular orbital and valence bond theories, it becomes harder to regain the chemical information. This is problematic because, in many cases, the use of such simple theories leads to qualitatively wrong results for the system under study, and as such, the interpretation of the resulting wave functions is, at the very best, controversial. This situation prompted Mulliken to say "the more accurate the calculations became, the more the concepts tended to vanish into thin air."

The problem with the wave function, underlying this situation, is that it is not an observable parameter. As such, it seems convenient to find other ways to extract the chemical information from a quantum mechanical calculation. One of the most popular alternatives is to use the electron density, $\rho(r)$ [4–7]. For a system with N electrons, described by a normalized wave function ($\langle \psi | \psi \rangle = 1$), this magnitude is defined as:

$$\rho(r) = N \int \Psi^* \left(r, s_1, r_2, s_2 ..., r_N, s_N\right) \Psi \left(r, s_1, r_2, s_2 ..., r_N, s_N\right) ds_1 dr_2 ds_2 ... dr_N ds_N$$

$$(4)$$

where s_i denotes the spin variables. From this definition, we can easily infer two properties:

$$\int \rho(r)\,dr = N \tag{5}$$

$$\forall r,\ \rho(r) \geq 0 \tag{6}$$

In other words, the electron density integrates to the number of electrons of the system and, given that it is a probability distribution, is nowhere negative. Contrary to the wave function, $\rho(r)$ only depends on three spatial coordinates, independent of N. Moreover, the electron density is an observable parameter of the system.

For these reasons, there has been an ever-increasing interest in the use of the electron density as a vehicle to rationalize chemical structures and chemical reactivity. We devote this contribution to the latter, making special emphasis on what is known as conceptual DFT (or chemical reactivity theory or DFT for chemical reactivity). We analyze the foundations of this theory, with particular interest in the notion of fractional numbers of electrons. Much of the analysis is devoted to discuss this notion, motivating its introduction, presenting a rigorous framework to work with it in a quantum mechanical sense, and showing its relevance at the time of defining chemical concepts within conceptual DFT. Given our interest that the present chapter could be used by newcomers to this field as an introduction to the machinery of conceptual DFT, we intend it to be as self-contained as possible. For this reason, we begin our exposition with a short introduction of "traditional" DFT, in which we also remark the importance of working with fractional N and present the electronic chemical potential as the common link between computational and conceptual DFT.

2.2 DENSITY FUNCTIONAL THEORY

Since the early stages of the development of quantum mechanics, it was evident that procedures based on magnitudes rather than on the wave function could provide simpler ways to study quantum phenomena. This motivated Thomas and Fermi, who proposed an heuristic model to study atomic systems, whose fundamental variable was the electron density, $\rho(r)$ [8–11]. However, the many approximations included in this model and the generally poor results (most notably, its inability to account for the formation of stable chemical bonds) [12–15] diminished the interest of the community of quantum chemists in this approach. The interest in the theory based on $\rho(r)$ was revamped thanks to the work of Hohenberg and Kohn [16], who proved that the electron density of the ground state contains all the information needed to describe a system. Presently, the works of Levy [17–19] and Lieb [76, 79, 82] are considered the most rigorous formulations of the DFT. Its solid theoretical foundation, along with the development, mainly through the Kohn–Sham formalism [20], of attractive computational tools based on it, have contributed to the ever-increasing popularity of DFT as a basic tool in different areas of modern theoretical chemistry.

2.2.1 FOUNDATIONS

It is enough to know the Hamiltonian of a system to determine its state, given that we solve Eq. (1). At this point, we may note that $\hat{H}$ is completely defined by the number of electrons (terms $\hat{T}_e$ and $\hat{V}_{ee}$) and by the external potential (term $\hat{V}_{Ne}$). According to Eq. (5), $\rho(r)$ allows us to determine N; therefore, the only thing that remains is to check whether the ground state electron density, $\rho(r)_0$, is enough to univocally define the external potential acting on the electrons.[1] This potential, as can be seen in the second term of Eq. (3), usually corresponds to the potential associated with the atomic nuclei, even though in a more general formulation, it can include other effects as long as they are local (this means depending on the spatial coordinates as, e.g., $\langle r'|\hat{V}_{Ne}|r\rangle = v(r)\delta(r-r')$. Hohenberg and Kohn (HK) [75, 78] proved that, effectively, each ground state electron density corresponds to at most one external potential. This means that $\rho(r)_0$ contains all the necessary information to build $\hat{H}$ and, therefore, to determine the state of the system. For example, it is possible to write the ground state energy, E_0, using a functional of the electron density:

$$E_0 = E_{HK}[\rho_0] = F_{HK}[\rho_0] + \int \rho(r)_0 v(r) dr \tag{7}$$

---

[1] In general, the external potential will be determined only up to an additive constant. This means that the potentials $v(r)$ and $v(r)+C$ are equivalent, given that the introduction of C only changes the zero of the energy scale.

DFT's biggest attractiveness is that the functional $F_{HK}[\rho]$ (which corresponds to the sum of the kinetic and electron-electron repulsion energies) is universal. In other words, it is independent of $v(r)$, and can, in principle, be used to study any system [74, 77–79].

HK also proved a variational principle in terms of the electron density:

$$E_0 = \min_{\rho:\rho(r)\geq 0, \int \rho(r)dr=N, \rho(r)\rightarrow v} E_{HK}[\rho] \qquad (8)$$

In this expression, we highlight that the variational principle will be valid as long as we use density functions that satisfy Eqs. (5) and (6) and that correspond to the ground state of some Hamiltonian with a local external potential ($\rho(r)\rightarrow v$). This condition, known as v-representability [77–79], is necessary given that $F_{HK}[\rho]$ is defined only over these densities:

$$F_{HK}[\rho] = E_0[\rho] - \int \rho(r)v(r)_{\rho(r)} dr \qquad (9)$$

where we have pointed out that one must use both the ground state energy, $E_0[\rho]$, and the external potential, $v(r)_{\rho(r)}$, of the system where $\rho(r)$ is the ground state. This represents the biggest drawback of the HK formulation, because we do not know the general conditions under which a given $\rho(r)$ is v-representable. This implies that the variational principle, Eq. (8), cannot be used in a consistent way.

To solve this problem, Levy proposed formulating DFT starting from the variational principle based on the wave function [17–19]:

$$E_0 = \min_{\Psi} \langle \Psi | \hat{H} | \Psi \rangle \qquad (10)$$

All the admissible wave functions, over which we try to find the minimum of Eq. (10), are normalized, well-behaved, and antisymmetric (given that we are working with fermions). The key step now is to realize that we can separate this minimization into two steps:

$$E_0 = \min_{\rho:\rho(r)\rightarrow N} \left\{ \min_{\Psi \rightarrow \rho(r)} \langle \Psi | \hat{H} | \Psi \rangle \right\} \qquad (11)$$

First, we consider the set of N-representables ($\rho(r)\rightarrow N$) electron densities; this set includes those derived from an admissible wave function according to Eq. (4). Then, for each of these densities, we find the minimum of the mean value of the Hamiltonian over all the wave functions that lead to the selected density. This procedure (known as constrained search) has the advantage that, as opposed to the case of v-representability, the conditions for N-representability are well understood. In general, $\rho(r)$ will be N-representable if it satisfies Eqs. (5), (6), and [77, 78]:

$$\int \left| \nabla \rho(r)^{\frac{1}{2}} \right|^2 dr < \infty \tag{12}$$

Levy's formulation can be presented in a form similar to the one used by HK, which helps to highlight the central role of the electron density [77, 78]:

$$E_0 = \min_{\rho:\rho(r) \to N} E_{LL}[\rho] \tag{13}$$

$$F_{LL}[\rho] = F_{LL}[\rho] + \int \rho(r)v(r)dr \tag{14}$$

$$F_{LL}[\rho] = \min_{\Psi \to \rho(r)} \left\langle \Psi \left| \hat{T}_e + \hat{V}_{ee} \right| \Psi \right\rangle \tag{15}$$

Obviously, for a v-representable $\rho(r)$:

$$E_{LL}[\rho] = E_{HK}[\rho], \ F_{LL}[\rho] = F_{HK}[\rho] \tag{16}$$

In other words, the functionals of Levy-Lieb, $E_{LL}[\rho]$ and $F_{LL}[\rho]$, are extensions of their HK counterparts over the set of N-representable $\rho(r)$'s. To simplify the notation, density functionals will be presented without sub-indices, i.e., $E[\rho]$ and $F[\rho]$, unless there is cause for confusion.

In the formulations of HK and Levy, there appear variational principles where the electron density plays a key role (Eqs. (8) and (13)). In both cases, we should be careful to use only densities that integrate to the number of electrons of the system, as given in Eq. (4), and we should also note that this number is usually a non-negative integer. This is certainly not convenient from a conceptual and practical point of view. Therefore, the variational principle is re-formulated using Lagrange multipliers. Now, the equation to solve is [77, 78]:

$$\delta\left\{E[\rho]-\mu\left[\int\rho(r)dr-N\right]\right\}=0 \tag{17}$$

In this expression, the Lagrange multiplier μ, associated with the constraint of constant N, is identified as the electronic chemical potential. The advantage of this procedure is that the test densities can now integrate to an arbitrary real (non-negative) number of electrons. In the forthcoming sections, we will discuss both about the justification for introducing fractional numbers of electrons in this formalism and about the relevance of the chemical potential.

2.2.2 KOHN–SHAM TREATMENT

In the preceding section, we presented the basic formulation of DFT; however, none of the previously discussed functionals has practical applications. For example, even if we can write the energy functional as:

$$E[\rho]=T_e[\rho]+V_{Ne}[\rho]+V_{ee}[\rho] \tag{18}$$

Only the second term can be calculated easily:

$$V_{Ne}[\rho]=\int\rho(r)v(r)dr \tag{19}$$

Of the remaining terms, the most difficult to approximate is the kinetic energy. A possible strategy to overcome this is to re-write Eq. (18) as:

$$E[\rho]=T_S[\rho]+V_{Ne}[\rho]+J[\rho]+\left(T_e[\rho]-T_S[\rho]+V_{ee}[\rho]-J[\rho]\right) \tag{20}$$

Now, the functional, $J[\rho]$, represents the interaction energy of a (classical) electron cloud with itself:

$$J[\rho]=\int\frac{\rho(r)\rho(r')}{|r-r'|}drdr' \tag{21}$$

The functional, $T_s[\rho]$, is chosen such as it contains a considerable fraction of the exact kinetic energy of the system, while having a simple closed form. The most popular choice (albeit not the only one) [21, 22] is the proposal of Kohn and Sham [20] of taking $T_s[\rho]$ as the kinetic energy of a fictitious system of noninteracting electrons with the same density as that of the real system. The wave function of this ideal system is given by a Slater determinant [33, 34]:

$$\Phi_{KS}(x_1,...,x_N) = \frac{1}{\sqrt{N!}} \begin{vmatrix} \varphi_1^{KS}(x_1) & \varphi_2^{KS}(x_1) & ... & \varphi_N^{KS}(x_1) \\ \varphi_1^{KS}(x_2) & \varphi_2^{KS}(x_2) & ... & \varphi_N^{KS}(x_2) \\ ... & ... & ... & ... \\ \varphi_1^{KS}(x_N) & \varphi_2^{KS}(x_N) & ... & \varphi_N^{KS}(x_N) \end{vmatrix} \qquad (22)$$

where χ_i represent the spatial and spin coordinates of each electron and φ_i^{KS} are the spin-orbitals of the auxiliary system, also known as Kohn–Sham (KS) orbitals. Then:

$$T_S[\rho] = \sum_{i=1}^{N} -\frac{\nabla_i^2}{2} \varphi_i^{KS}(x_i) \qquad (23)$$

The term in parentheses in Eq. (20) is identified, in analogy with the Hartree-Fock (HF) [2] treatment, with the exchange-correlation functional [74, 77]:

$$E_{xc}[\rho] = T_e[\rho] - T_S[\rho] + V_{ee}[\rho] - J[\rho] \qquad (24)$$

In this case, despite containing the nonclassical effects of the interaction between the electrons, $V_{ee}[\rho]$-$J[\rho]$, (correlation and exchange), we have the kinetic contribution, $T_e[\rho]$-$T_S[\rho]$, due to the approximate characteristic of $T_S[\rho]$.

Up to this point, we have not introduced any approximation, and the KS spin-orbitals can be obtained solving the single-particle equations [20]:

$$\left\{ -\frac{\nabla_i^2}{2} + v_{KS}(r) \right\} \varphi_i^{KS}(x_i) = \varepsilon_i^{KS} \varphi_i^{KS}(x_i) \qquad (25)$$

being:

$$v_{KS}\left(r\right) = v\left(r\right) + \int \frac{\rho\left(r'\right)}{|r-r'|}dr' + v_{xc}\left(r\right) \tag{26}$$

$$v_{xc}\left(r\right) = \frac{\delta E_{xc}\left[\rho\right]}{\delta\rho\left(r\right)} \tag{27}$$

Given that the electron density appears in the left hand side of Eq. (25) (i.e., second term in Eq. (26)) and that, by construction:

$$\rho(r) = \sum_{i=1}^{N}\left|\varphi_i^{KS}\left(r_i\right)\right|^2 \tag{28}$$

the KS equations (Eq. (25)) must be solved self-consistently. Like other standard wave function-based methods, the most usual approach is to expand the spatial part of φ_i^{KS} on the basis of approximate basis functions and then solve the resulting algebraic equations [35, 131].

What is left is to know the exchange-correlation functional, $E_{xc}[\rho]$, or its associated potential, $v_{xc}(r)$. Given the complexity of the terms included in $E_{xc}[\rho]$, its exact form is unknown. In modern DFT [74, 132], finding accurate, yet computationally feasible, approaches for this functional is considered the central problem of DFT. Among the different strategies devised to accomplish this, we have:

Local functionals: Estimate $E_{xc}[\rho]$ starting with the local density approximation (LDA) [74, 77] taking the expressions derived for the homogeneous electron gas. The Dirac exchange functional [10] and the correlation functional VWN (Vosko, Wilk, and Nusair) [23] are examples of local functionals.

Generalized gradient functionals: Add information concerning the nonlocal behavior of the density, through the use of its gradient, $\nabla\rho(r)$. In a sense, this is equivalent to expanding the functional $E_{xc}[\rho]$ in a power series around the LDA expressions, thus imposing some extra conditions known to hold for the exact functional (e.g., the sum rules for the Coulomb and Fermi holes) [4]. The Perdew, Burke, and Ernzerhof (PBE) functional [24], which does not contain any empirical parameter, belongs to this family.

Hybrid functionals: They consider a fraction of the exact (HF) exchange, calculated by using the KS orbitals [74, 135, 136]. This category belongs to the B3LYP functional [25–27], one of the most widely used functionals in DFT applications.

There are other families of functionals, some of which use meta-generalized gradients [140, 141] (which include information of the kinetic energy density) and range-separated [28] (which consider different treatments for the Coulomb interaction, regarding the distance between the electrons). All these functionals depend fundamentally on the electron density (expressed through the KS spin-orbitals). Also, even when the electron density is enough, all of them can be easily extended to include the information regarding the spin density [29–31]. Thus, we can improve the quality of the results, especially for systems with unpaired electrons.

Finally, we should mention that the KS spin-orbitals and their corresponding energies, ε_i^{KS}, are just auxiliary magnitudes that are associated with an ideal system, and as such, in principle, they do not have any physical meaning. Nonetheless, it can be proven that the energy of the highest occupied KS orbital (KS HOMO), calculated using an exact functional, is equal to the additive inverse of the ionization energy, I, of the real system [32–37]:

$$\varepsilon_i^{KS-HOMO} = -I \tag{29}$$

This equality ceases to hold if we use approximate functionals. However, it is a common practice to use the values of $\varepsilon_i^{KS-HOMO}$ and $\varepsilon_{ii}^{KS-LUMO}$ (unoccupied KS orbital of minimum energy or KS LUMO) as approximations to the ionization energy and the electron affinity, A, respectively [38].

2.3   OPEN SYSTEMS AND DENSITY MATRICES

The formulation of the variational principle in terms of the density that makes use of Lagrange multipliers, Eq. (17), implies that the density (and its functionals) must be extended to include systems with fractional numbers of electrons [4]. Even if this notion can be seen as a mere mathematical artifact, it can be rigorously justified if we consider open systems [32]. An open system can exchange particles (electrons) with its environment and therefore can have, on average, a real (non-negative) number of particles. What remains then is to discuss the formalism that allows us to describe such systems with quantum mechanics.

2.3.1   DENSITY MATRICES

In most quantum chemistry formalisms and applications, the state of the system is represented by the wave function. However, a wave function is only valid for closed systems [39]. When we work with open systems, the states must be represented by density matrices (DMs) [77, 104, 115].

The most general DM that we will be considering has the form [40–42]:

$$D = \sum_M \sum_k {}^M\omega_k \left| {}^M\Psi_k \right\rangle \left\langle {}^M\Psi_k \right| \tag{30}$$

where the wave functions $\left| {}^M\psi_k \right\rangle$ are eigenstates of the Hamiltonian and of the particle number operator, $\hat{N}$

$$\hat{H}\left| {}^M\Psi_k \right\rangle = {}^M E_k \left| {}^M\Psi_k \right\rangle \tag{31}$$

$$\hat{N}\left| {}^M\Psi_k \right\rangle = M \left| {}^M\Psi_k \right\rangle \tag{32}$$

The index, k, stands for both the accessible states and their possible degeneracy for each (integer) particle number M. The coefficients, ${}^M\omega_k$, are statistical weights, and as such, they must fulfill the following conditions:

$$\forall M, k : 0 \leq {}^M\omega_k \leq 1 \tag{33}$$

$$\sum_M \sum_k {}^M\omega_k = 1 \tag{34}$$

A less general kind of DM can be obtained if in Eq. (30), we do not include the summation over different particle numbers, namely:

$$D_M = \sum_k {}^M\omega_k \left| {}^M\Psi_k \right\rangle \left\langle {}^M\Psi_k \right| \tag{35}$$

Finally, the simplest form considers only one eigenstate of the Hamiltonian:

$$D_{M,k} = \left| {}^{M}\Psi_k \right\rangle \left\langle {}^{M}\Psi_k \right| \tag{36}$$

A state of this form is termed "pure," while those given by Eqs. (30) or (35) are called "mixed" [39]. Obviously, a pure state can be described by either a DM or a wave function. Nonetheless, it should be clear that we are working with distinct mathematical objects. While the ψ are vectors in a Hilbert space, the DMs are operators acting on this space. This must be taken into account at the time of computing the mean value of an observable $\hat{Q}$ If using wave functions we write:

$$Q = \left\langle \Psi \middle| \hat{Q} \middle| \Psi \right\rangle \tag{37}$$

the expression in terms of DMs will be:

$$Q = \mathrm{Tr}\left(D\hat{Q} \right) \tag{38}$$

where $\mathrm{Tr}(\hat{A})$ is the trace of the operator, $\hat{A}$, (it is easy to show that, for arbitrary operators, $\mathrm{Tr}(\hat{A}\,\hat{B}) = \mathrm{Tr}(\hat{B}\,\hat{A})$). For example, the electron density will be given by:

$$\rho(r) = \mathrm{Tr}\left(D\hat{\rho} \right) \tag{39}$$

In analogy with the statistical thermodynamics [4], those states with the form given in Eq. (36) will be called micro-canonical, because their energies are completely determined. Those corresponding to Eq. (35) will be canonical, because even when they have the same number of particles, each of the states of the ensemble can possess an arbitrary energy. Finally, states given in Eq. (30) will be called grand-canonical (GC), because they are the combination of pure states with different energies and numbers of particles. These are precisely the states that allow us to formalize the work with quantum open systems. For example, the average number of particles in one such state can be calculated as:

$$N = \sum_{M}\sum_{k} {}^{M}\omega_k M \tag{40}$$

from where it is evident that N can be any real number equal to or greater than zero.

The last thing remaining is to extend the domain of definition of the energy functionals, in a way that includes the densities obtained from GC DMs according to Eq. (39). This can be easily done through Levy's constrained search, where the starting point is the variational principle formulated in terms of the DMs [76, 78, 82]:

$$E(N)_0 = \min_{D:Tr(D\hat{N})=N} Tr(\hat{H}D) \tag{41}$$

The search is performed for the ground state of the system with N electrons; therefore, there is constraint of only using those DMs such that $Tr(D\hat{N})=N$ (assuming that we are working with normalized DMs, i.e., $Tr(D=1)$). Starting from this expression, we can define Lieb's functionals, $E_L[\rho]$ and $F_L[\rho]$:

$$E(N)_0 = \min_{\rho:\rho(r)\rightarrow N} E_L[\rho] \tag{42}$$

$$E_L[\rho] = F_L[\rho] + \int \rho(r)v(r)dr \tag{43}$$

$$F_L[\rho] = \min_{D\rightarrow\rho(r)} Tr\left[D\left(\hat{T}_e + \hat{V}_{ee}\right)\right] \tag{44}$$

With these functionals, we can give a precise meaning to Eq. (17), which can now be formally solved as:

$$\left.\frac{\delta E[\rho]}{\delta\rho(r)}\right|_{\rho_0} = \left.\frac{\delta F[\rho]}{\delta\rho(r)}\right|_{\rho_0} + \rho(r)_0 = \mu \tag{45}$$

This equation, where we have pointed out the fact that the derivatives are taken at the ground state density, is considered by Parr as the fundamental equation of DFT [4].

Given that Lieb's functional is also an extension of the HK functional, for the case of integer numbers of particles, the solution of Eq. (41) reduces to the corresponding Schrödinger equation. However, we can think about finding the solution of this equation when the system has $N=M+x$

particles, with M being an integer, and $0 \leq x \leq 1$. In this case, it has been proven in different ways that the energy can be obtained as [32, 40, 43–45]:

$$E(N = M + x)_0 = x^{M+1} E_0 + (1 - x)^M E_0 \qquad (46)$$

In other words, $E(N=M+x)_0$ is determined by the linear interpolation between the states with M and $M+1$ electrons. Besides, it can be shown that if those states are not degenerate, the DM that minimizes Eq. (41) has a similar form:sy

$$D(N = M + x)_0 = x \left| {}^{M+1}\Psi \right\rangle \left\langle {}^{M+1}\Psi \right| + (1 - x) \left| {}^M \Psi \right\rangle \left\langle {}^M \Psi \right| \qquad (47)$$

These results are valid under the assumption that the ground state energy is a convex function of the number of particles [4]. This result has not been proven analytically, but there is a great amount of experimental work that suggests that it is true for isolated Coulomb systems.

2.3.2 FINITE TEMPERATURE EXTENSIONS

Everything discussed so far is implicitly referred to the case where the temperature, T, is zero Kelvin. The interest to extend DFT to temperatures above absolute zero [46–51] (what is known as finite temperatures) started with Mermin [52], who proved the analogue of the HK theorems for this case.

The difference between the systems with $T=0$ and $T \neq 0$ is a qualitative one and is better reflected by the fact that the state functions that govern the spontaneity of the processes in both cases are different [4]. At $T=0$, we focus on the energy and in the states that minimize this magnitude (ground states). At finite temperature, the processes are governed by the increase in entropy, S, and the electronic systems tend to evolve toward equilibrium states. The entropy of a system in quantum mechanics is defined in terms of its DM, by using an expression known as Boltzmann-Gibbs-Shannon entropy [46–49]:

$$S(D) = -k_B \mathrm{Tr}(D \ln D) = -k_B \sum_M \sum_k {}^M \omega_k \ln {}^M \omega_k \qquad (48)$$

In this expression, k_B is Boltzmann's constant (which is often taken with the value of 1, when the temperature is given in energy units).

The equilibrium state is associated with the DM that maximizes the entropy, subject to the constraints that the average energy and number of particles of the system correspond to the observed values. This problem can be easily formulated using Lagrange multipliers [77, 158–161]:

$$\delta\left\{S(D) - \beta\left[\text{Tr}(\hat{H}D) - \mu\text{Tr}(D\hat{N})\right]\right\} = 0 \qquad (49)$$

where we have assumed that we are working with normalized DMs. The Lagrange multipliers are:

$$\beta = \frac{1}{k_B T} \qquad (50)$$

and the electronic chemical potential, μ.

It is easy to see that solving Eq. (49) is equivalent to minimizing any of the given state functions:

$$\mathcal{A} = E - TS \qquad (51)$$

$$\Omega = E - TS - \mu N \qquad (52)$$

Note that during the minimization of the GC potential (or thermodynamical potential), Ω, we can use any DM, while in the minimization of the Helmholtz's free energy, $\mathcal{A}$, we can only use those DMs that correspond to the number of particles of the system under study. Then, for the equilibrium DM, D_0, we will have [4]:

$$\mathcal{A}_0 = \mathcal{A}\left[D_0 = D(N)_T\right] = \min_{D:\text{Tr}(D\hat{N})=N} \mathcal{A}[D] \qquad (53)$$

$$\Omega_0 = \Omega\left[D_0 = D(\mu)_T\right] = \min_D \Omega[D] \qquad (54)$$

The reason for representing the equilibrium DM as a function of the number of particles in the case of $\mathcal{A}$ and of the chemical potential in the case of Ω will be clarified in the following sections. Note also that, starting from Eqs. (53) and (54), and applying the constrained search over the admissible DMs in each case, we can define density functionals that would allow us to calculate $\mathcal{A}$ and Ω.

As it was the case when we were working at $T=0$, it is not hard to obtain the expression for the DM that results from solving the corresponding variational principles. Thus, for example, if we decide to express such DM as a function of the chemical potential and temperature, we will have [49]:

$$
\begin{aligned}
D(\mu)_T &= \frac{\exp\left[-\beta\left(\hat{H} - \mu\hat{N}\right)\right]}{\mathrm{Tr}\left\{\exp\left[-\beta\left(\hat{H} - \mu\hat{N}\right)\right]\right\}} \\
&= \frac{\sum_{M,k}\left|{}^{M}\Psi_k\right\rangle\left\langle{}^{M}\Psi_k\right|\exp\left[-\beta\left({}^{M}E_k - \mu M\right)\right]}{\sum_{M,k}\exp\left[-\beta\left({}^{M}E_k - \mu M\right)\right]}
\end{aligned}
\tag{55}
$$

From this expression, the equilibrium mean values of the energy and other observables can be easily obtained using Eq. (38).

Typically, to study the properties of a system with M_0 particles, at the time of working with expressions like Eq. (55), the trace is taken only over the states with M_0, M_0+1, and M_0-1 electrons. This is because under a wide range of conditions, these are the most important states to describe the chemical behavior of the system [161, 163, 164].

2.4   CONCEPTUAL DENSITY FUNCTIONAL THEORY

Thus far, we have seen some of the formal aspects of DFT as well as the basic machinery that makes it successful from a computational point of view. But, even if this is the most recognized aspect of this theory, it is by no means the only one. Within the formulation of DFT, we may precisely define multiple chemical concepts. Besides, we can justify the use of these concepts within the framework of different theoretical principles that help us to rationalize the results of electronic structure calculations and to gain a deeper understanding about the chemical reactivity. This branch of DFT is commonly called conceptual DFT, chemical reactivity theory of DFT for chemical reactivity [53–55].

2.4.1 BASIC ASPECTS OF CONCEPTUAL DFT

As we pointed out previously, N and $v(r)$ for a system is all we need to determine its properties. Therefore, it is natural to infer that, in a similar way, every process can be described by means of the changes in the number of particles, ΔN, and external potential, $\Delta v(r)$. This implies that if we know the energy (and, from now on, we will always assume that we are working with the ground state energy) of a system with N electrons in an external potential $v(r)$, $E[N,v(r)]$, then we can determine the energy of a perturbed system, $E[N+\Delta N, v(r)+\Delta v(r)]$ [56]:

$$\Delta E = \left(\frac{\partial E}{\partial N}\right)_{v(r)} \Delta N + \int \left(\frac{\delta E}{\delta v(r)}\right)_{N} \Delta v(r) dr + \frac{1}{2}\left(\frac{\partial^2 E}{\partial N^2}\right)_{v(r)} \Delta N^2$$

$$+ \int \left(\frac{\delta^2 E}{\delta v(r)\partial N}\right) \Delta N \Delta v(r) dr$$

$$+ \frac{1}{2}\int \left(\frac{\delta^2 E}{\delta v(r)\delta v(r')}\right)_{N} \Delta v(r')\Delta v(r) drdr' + \dots \qquad (56)$$

where $\Delta E = E[N+\Delta N, v(r)+\Delta v(r)] - E[N,v(r)]$. This equation is nothing more than a perturbative Taylor series (which is usually truncated at the second order). Then, to calculate $E[N+\Delta N, v(r)+\Delta v(r)]$, it is more important to analyze if a given change in the system will be more or less favored. For example, if we focus on the study of chemical reactivity, we can interpret Eq. (56) as the change in energy as one of the reactants interacts with the other (at least in the early stage of the reaction). In this context, the coefficients of the series can represent the intrinsic response of the attacked system (it is a common practice to assume that these coefficients are independent of the surroundings). On the other hand, the variations ΔN and $\Delta v(r)$ are defined to model specific features of the attacking partner. The relative importance of these changes will determine whether the reaction is governed by the charge transfer (bigger ΔN) or by the electrostatic interactions (bigger $\Delta v(r)$). This simplifies the analysis of the process under study, where we only consider the dominant terms in Eq. (56). In the case of charge-transfer control, where the geometry of the system remains approximately constant, the truncated Taylor expansion has the form [77, 165]:

$$\Delta E = \left(\frac{\partial E}{\partial N}\right)_{v(r)} \Delta N + \frac{1}{2}\left(\frac{\partial^2 E}{\partial N^2}\right)_{v(r)} \Delta N^2 \qquad (57)$$

In any case, determining which changes are more favorable (that is, which lead to a more negative, or less positive, ΔE), we can obtain information about the chemical reaction at hand (i.e., characteristics of its mechanism, predominant products, etc.).

Finally, we can see that the presented form of conceptual DFT, starting from Eq. (56), is simply a perturbative framework [56]. Essentially, identical results can be obtained using thermodynamic [55] and variational formulations [57–59].

2.4.2 REACTIVITY DESCRIPTORS IN CONCEPTUAL DFT

The coefficients of the expansion given in Eq. (56) form the basis of chemical reactivity within conceptual DFT. They provide a way to rigorously define (e.g., reify) concepts previously proposed in a qualitative way as well as to develop a systematic framework to find new molecular reactivity descriptors. In general, these descriptors can be classified as global (independent of the spatial coordinates, i.e., $\left(\frac{\partial E}{\partial N}\right)_{v(r)}$, locals

(depending on the coordinates of one spatial point, i.e., $\left(\frac{\delta E}{\delta v(r)}\right)_{N}$), and

nonlocal (depending on the spatial coordinates of more than one point, i.e.,

$\left(\frac{\delta^2 E}{\delta v(r)\delta v(r')}\right)_{N}$ [53].

According to Eq. (56), the most relevant coefficients are those corresponding to the first-order derivatives. It is easy to obtain explicit expressions for them; for example, by using perturbation theory, we can show that (for nondegenerate states) [60, 61]:

$$\left(\frac{\delta E}{\delta v(r)}\right)_N = \rho(r) \tag{58}$$

The other first-order derivative can be obtained from the variational principle shown in Eq. (17), taking into account that the Lagrange multiplier associated with a constraint is equal to the derivative of the minimum of the optimized functional with respect to the value of the constraint [62]. In the present context, this means that:

$$\mu = \left(\frac{\partial E}{\partial N}\right)_{v(r)} = \left.\frac{\delta E}{\delta \rho(r)}\right|_{\rho_0} \tag{59}$$

In this form, it is obvious that μ gives a measure of the tendency of the electrons to abandon the system. Furthermore, it can be proven that when two systems exchange electrons, they move from the region with higher values of μ toward those with smaller values [63]. This is why μ is known as the electronic chemical potential (or just the chemical potential, as is customary within DFT) [4]. These properties also motivated Parr to define the electronegativity, χ, as the additive inverse of this descriptor [62]:

$$\chi = -\mu \tag{60}$$

Similarly, we can define the chemical (or Pearson) [21, 22] hardness, η [64]:

$$\eta = \left(\frac{\partial^2 E}{\partial N^2}\right)_{v(r)} = \left(\frac{\partial \mu}{\partial N}\right)_{v(r)} \tag{61}$$

representing a measure of the tendency of the system to suffer disproportionation (which can also be considered a way to quantify the resistance toward a change of the electronic cloud). Of the local descriptors, besides the electronic density, the most relevant are the Fukui function [65, 66], $f(r)$, and the dual descriptor, $\Delta f(r)$ [67]:

$$f(r) = \left(\frac{\delta^2 E}{\delta v(r)\partial N}\right) = \left(\frac{\partial \rho(r)}{\partial N}\right)_{v(r)} = \left(\frac{\delta \mu}{\delta v(r)}\right)_N \tag{62}$$

$$\Delta f(r) = \left(\frac{\delta^3 E}{\delta v(r) \partial^2 N} \right) = \left(\frac{\partial f(r)}{\partial N} \right)_{v(r)} = \left(\frac{\delta \eta}{\delta v(r)} \right)_N \qquad (63)$$

(note that several of these expressions are just Maxwell relations). These descriptors show how the electron density changes after the system changes its number of electrons, and as such, they serve to determine which regions are more prone to suffer an electrophilic or nucleophilic attack.

Among the nonlocal descriptors, the linear response kernel, $\chi(r, r')$, is the most commonly used [68, 69]:

$$\chi(r,r') = \left(\frac{\delta^2 E}{\delta v(r) \delta v(r')} \right)_N \qquad (64)$$

and it has been successfully used to quantify the degree of electronic delocalization in several systems.

All the definitions given in Eqs. (59) and (61)–(64) are purely formal, in the sense that the practical use of these descriptors requires them to be expressed in terms of easily computable properties. We need explicit models for the dependency of the energy with the number of particles (E vs. N models). There are different strategies to do so, perhaps the more popular being to just consider three accessible states (with M, $M + 1$, and $M - 1$ electrons) and to define E in terms of the parabola fitted through them [4]. In this way (which is reminiscent of truncating the Taylor series at the second order, Eq. (57)), the global descriptors are given by:

$$\mu = -\frac{I + A}{2} \qquad (65)$$

$$\eta = I - A \qquad (66)$$

Interestingly enough, the expression for the electronegativity derived from Eq. (65) coincides with that proposed by Mulliken [70]. Nonetheless, these are not the only possible forms. For example, if we use the linear model given in Eq. (46), we will have [32]:

$$\mu^- = -I; \; \mu^+ = -A \tag{67}$$

The supra-index + (−) indicates that the derivatives are taken "by the right" ("by the left"). This means that we are only considering variations with $\Delta N > 0$ ($\Delta N < 0$). This way of proceeding is a consequence of the non-derivability of E at integer numbers of electrons in this model.

From these equations, it is clear that the descriptors from conceptual DFT can be calculated using computational tools different from those of DFT. For example, essentially, any electronic structure method can be used to determine I and A, which in turn can be used to estimate the chemical potential and/or hardness.

Finally, it is necessary to note that not all the reactivity descriptors derived from conceptual DFT are just coefficients of Eq. (56) (or its extensions that we will consider shortly) [71–73]. For example, the electrophilicity [17, 19, 182], ω, is defined by Parr as the stabilization energy gained by a system when it reaches equilibrium with a reservoir of electrons with $\mu = 0$. It can be calculated using Eq. (57), and the final expression is a combination of μ and η:

$$\omega = \frac{\mu^2}{2\eta} \tag{68}$$

It is also interesting to determine the maximum number of electrons gained by the system during this process, ΔN_{max} [74]:

$$\Delta N_{max} = -\frac{\mu}{\eta} \tag{69}$$

Given that $\mu < 0$ and, due to the convexity of the energy, $\eta > 0$, we can conclude that the system will be stabilized by gaining electrons under these conditions.

2.4.3 GENERALIZATIONS: LEGENDRE TRANSFORMS AND SPIN COMPONENTS

In the previous section, we focused on the energy of the system as the leading state function to study chemical reactivity, taking the number of particles and external potential as independent variables. In analogy with the classical thermodynamics, these variables can be changed, and in doing so, we also change the state function [75]. For example, from the definition of the chemical potential, Eq. (59), it is clear that if we use a Legendre transform [76], we can obtain a new state function, G, that depends on μ and $v(r)$ [55]:

$$G = E - \mu N \tag{70}$$

The study of chemical reactivity using this state function is known as the "open system picture," as opposed to the formulation based on E, which is termed the "closed system picture" [55]. Even though these approaches are equivalent, it is convenient to have more than one form to study the chemical reactivity of a system. First, the description in terms of μ instead of N is more natural in those cases where the number of particles is hard or impossible to determine (the typical example being species in solution). The other advantage is that now we have a new perturbative series:

$$\Delta G = \left(\frac{\partial G}{\partial \mu} \right)_{v(r)} \Delta \mu + \int \left(\frac{\delta G}{\delta v(r)} \right)_{\mu} \Delta v(r) dr + \frac{1}{2} \left(\frac{\partial^2 G}{\partial \mu^2} \right)_{v(r)} \Delta \mu^2 +$$

$$\int \left(\frac{\delta^2 G}{\delta v(r) \partial \mu} \right) \Delta \mu \Delta v(r) dr +$$

$$\frac{1}{2} \int \left(\frac{\delta^2 G}{\delta v(r) \delta v(r')} \right)_{\mu} \Delta v(r') \Delta v(r) dr dr' + \dots \tag{71}$$

Therefore, we have a new set of reactivity descriptors, such as the softness, S, and local softness, $s(r)$ (both are measures of the electronic polarizability) [53]:

$$S = \left(\frac{\partial^2 G}{\partial \mu^2} \right)_{v(r)} = \left(\frac{\partial N}{\partial \mu} \right)_{v(r)} = \frac{1}{\eta} \tag{72}$$

$$s(r) = \left(\frac{\delta^2 G}{\delta v(r) \partial \mu} \right) = \left(\frac{\partial \rho(r)}{\partial \mu} \right)_{v(r)} = S f(r) \tag{73}$$

As a proof of the link between both representations, we can see the relations linking their descriptors.

It is important to note that the perturbative series analyzed thus far (Eqs. (56) and (71)) assumes that $T = 0$. At finite temperatures, the analogues of E and G are the Helmholtz free energy, Eq. (51), and the GC potential, Eq. (52), respectively [46–49]. It is clear now why the equilibrium DM is a function of N when we work with $\mathcal{A}$ or a function of μ when we work with Ω. This change of state functions is fundamental; however, the mathematical structure of the theory remains unaltered. For example, at $T \neq 0$, the chemical potential [77], the hardness, and the softness are respectively given by:

$$\mu = \left(\frac{\partial \mathcal{A}}{\partial N} \right)_{v(r)} \tag{74}$$

$$\eta = \left(\frac{\partial^2 \mathcal{A}}{\partial N^2} \right)_{v(r)} \tag{75}$$

$$S = \left(\frac{\partial^2 \Omega}{\partial \mu^2} \right)_{v(r)} = \frac{1}{\eta} \tag{76}$$

We can also generalize conceptual DFT such that the different spin components are explicitly included [78–84]. This can be done by using two different representations: one where the fundamental variables are the number of spin alpha, N_α, and the number of spin beta, N_β electrons; and other where the fundamental variables are the number of electrons, N, and the number of unpaired electrons (or spin number), N_S:

$$N_S = N_\alpha - N_\beta \tag{77}$$

These representations are equivalent, being possible to move from the equations of one to the other by using simple linear transformations [85]. Besides, all the equations and identities valid in the spin-free case can be generalized to the spin-resolved case by using a simple vector notation [78]. However, we now have more descriptors, given the number of independent variables. For example, we will have two chemical potentials in each representation:

$$\mu_N = \left(\frac{\partial E}{\partial N} \right)_{N_S, v(r)} , \mu_S = \left(\frac{\partial E}{\partial N_S} \right)_{N, v(r)} \tag{78}$$

$$\mu_\alpha = \left(\frac{\partial E}{\partial N_\alpha} \right)_{N_\beta, v(r)} , \mu_\beta = \left(\frac{\partial E}{\partial N_\beta} \right)_{N_\alpha, v(r)} \tag{79}$$

Although much of the current literature on the spin-resolved conceptual DFT strives to obtain the working expressions for the descriptors, there is a lack of a general framework to do so [79–81] as well as a detailed analysis of the consistency of the most popular approaches.

2.4.4 REACTIVITY PRINCIPLES

When working with conceptual DFT, the descriptors are often used in different reactivity principles [53]. Among them, we find Sanderson's electronegativity equalization principle [86, 87]. It states that when a molecule is formed, the electronegativities of the constituent atoms are equalized; thus, the final system has only one global electronegativity, constant over all space. This result is a direct consequence of the definition of electronegativity within conceptual DFT [Eqs. (59) and (60)], given that it is, by construction, a global quantity [4]. Note that this is only valid for $\rho(r)_0$; for other electron densities, we will have:

$$\mu(r) = \frac{\delta E}{\delta \rho(r)} \tag{80}$$

This means that the electronegativity will have a local dependency, with different values for different points over the molecule. This reflects the fact that when we reach the equilibrium state (and only in this state), there is no net charge transfer within the molecular fragments. Apart from its conceptual importance, this principle is used to determine atomic charges and in studies that combine quantum mechanics and molecular mechanics [88–90].

Another well-known reactivity principle is the hard/soft acid/base (HSAB) principle [91, 92]. By analyzing a great amount of experimental data, Pearson qualitatively defined the hardness (and its counterpart, the softness) and stated that acids and bases have more affinity for their counterparts of similar hardness (softness). Years later, Parr and Pearson [64] precisely defined the concept of hardness (Eq. (61)), a definition that paved the way for a rigorous proof of this principle [93–96]. Pearson's principle can also be formulated in a local way, which allows determining the regions within a molecule that will have more affinity for those of an attacking partner, given the relative hardness/softness of their respective fragments [53, 97]. This local formulation, even though criticized at times [98–100], has served to explain the regioselectivity patterns observed in different families of organic reactions [101–103].

Although both Sanderson and HSAB principles can be rigorously proven within conceptual DFT, they were originally stated before the conception of this theory and were profusely used by the chemical community. But, with the development of conceptual DFT, it has been possible to discover new reactivity principles that shed some light on the behavior of the chemical reactions and to highlight the importance of less popular approaches. Among these cases, probably the most popular is the maximum hardness principle (MHP) [104–109], which states that "there seems to be a rule of nature that molecules arrange themselves to be as hard as possible" [92]. This principle, even when it is valid in less general conditions than those of the HSAB principle, has proven useful for rationalizing reaction paths [110, 111]. In a similar way, it has been shown that in certain occasions, a system will be more stable if it decreases its electrophilicity. This is known as the minimum electrophilicity principle [112–114], and it is useful in the same conditions where the MHP holds. All these principles usually complement one other and provide valuable information when studying different chemical processes.

2.5   CONCLUDING REMARKS

In this contribution, we tried to provide a short introduction to some of the basic theoretical aspects of conceptual DFT. Given the importance of the notion of fractional numbers of electrons for this theory, we took this as our guiding topic. We emphasized that fractional numbers of electrons are necessary for the consistent definition of conceptual DFT descriptors and for the derivation of their practical working expressions, and also as a fundamental part of the formulation of the variational principles within standard computational DFT. As such, it also serves to connect these two branches, with the most evident link being the electronic chemical potential. We concentrated more on the broad features of conceptual DFT than on particular applications, thus hoping to give the reader a concise survey, where the general structure is put on display in a way that highlights the common features between its many possible realizations and corresponding generalizations.

ACKNOWLEDGMENTS

The author thanks Paul W. Ayers and Taewon David Kim for useful discussions and their help to improve the manuscript.

KEYWORDS

- **DFT**
- **CDFT**
- **chemical reactivity**
- **density matrices**
- **chemical potential**
- **chemical hardness**

REFERENCES

1. Levine, I. N., (2009). *Quantum Chemistry*; 6th ed., Pearson Prentice Hall: Upper Saddle River, N.J.
2. Szabo, A., & Ostlund, N. S., (1989). *Modern Quantum Chemistry: Introduction to Advanced Electronic Structure Theory*, Dover: Mineola, NY.
3. Helgaker, T., Jørgensen, P., & Olsen, J., (2000). *Modern Electronic Structure Theory*, Wiley: Chichester.
4. Parr, R. G., & Yang, W., (1989). *Density-Functional Theory of Atoms and Molecules*, Oxford UP: New York.

5. Koch, W., & Holthausen, M. C., (2001). Wiley-VCH, New York, 2nd volume.
6. Dreizler, R. M., & Gross, E. K. U., (1990). *Density Functional Theory: An Approach to the Quantum Many-Body Problem*, Springer-Verlag, Berlin.
7. Eschrig, H., (2003). *The Fundamentals of Density Functional Theory*, Eagle: Leipzig.
8. Thomas, L. H., (1927). *Proc. Camb. Phil. Soc., 23*, 542.
9. Fermi, E., (1928). *Z .Phys., 48*, 73.
10. Dirac, P. A. M., (1930). *Proc. Cambridge Phil. Soc., 26*, 376.
11. Lieb, E. H., (1981). *Rev. Mod. Phys., 53*, 603.
12. Teller, E., (1962). *Rev. Mod. Phys., 34*, 627.
13. Lieb, E. H., & Simon, B., (1973). *Phys. Rev. Lett., 31*, 681.
14. Lieb, E. H., & Simon, B., (1977). *Advances in Mathematics, 23*, 22.
15. Balazs, N. L., (1967). *Phys. Rev., 156*, 42.
16. Hohenberg, P., & Kohn, W., (1964). *Phys. Rev., 136*, B864.
17. Levy, M., (1982). *Phys. Rev. A., 26*, 1200.
18. Levy, M., (1983). *Lecture Notes in Physics, 187*, 9.
19. Levy, M., & Perdew, J. P., (1985). *NATO ASI Series, Series B, 123*, 11.
20. Kohn, W., & Sham, L. J., (1965). *Phys. Rev., 140*, A1133.
21. Levy, M., & Ouyang, H., (1988). *Phys. Rev. A., 38*, 625.
22. Holas, A., & March, N. H., (1991). *Phys. Rev. A., 44*, 5521.
23. Vosko, S. H., Wilk, L., & Nusair, M., (1980). *Can. J. Phys., 58*, 1200.
24. Perdew, J. P., Burke, K., & Ernzerhof, M., (1996). *Phys. Rev. Lett., 77*, 3865.
25. Becke, A. D., (1993). *J. Chem. Phys., 98*, 5648.
26. Lee, C., Yang, W., & Parr, R. G., (1988). *Phys. Rev. B., 37*, 785.
27. Stephens, P. J., Devlin, F. J., Chabalowski, C. F., & Frisch, M., (1994). *Joural of Physical Chemistry, 98*, 11623.
28. Cramer, C. J., & Truhlar, D. G., (2009). *PCCP, 11*, 10757.
29. Perdew, J. P., & Zunger, A., (1981). *Phys. Rev. B, 23*, 5048.
30. Baroni, S., & Tuncel, E., (1983). *J. Chem. Phys., 79*, 6140.
31. Jacob, C. R., & Reiher, M., (2012). *Int. J. Quantum Chem., 112*, 3661.
32. Perdew, J. P., Parr, R. G., Levy, M., & Balduz, J. L., (1982). *Jr. Phys. Rev. Lett., 49*, 1691.
33. Perdew, J. P., & Levy, M., (1997). *Phys. Rev. B, 56*, 16021.
34. Kleinman, L., (1997). *Phys. Rev. B., 56*, 16029.
35. Kleinman, L., (1997). *Phys. Rev. B., 56*, 12042.
36. Janak, J. F., (1978). *Phys. Rev. B., 18*, 7165.
37. Harbola, M. K., (1999). *Phys. Rev. B., 60*, 4545.
38. Baerends, E. J., Gritsenko, O. V., & Meer, R. V., (2013). *PCCP, 15*, 16408.
39. Ballentine, L., (1998). *Quantum Mechanics: A Modern Development*, World Scientific Publishing Co. Pte. Ltd.: London.
40. Bochicchio, R. C., & Rial, D., (2012). *J. Chem. Phys., 137*, 226101.
41. Bochicchio, R. C., Miranda-Quintana, R. A., & Rial, D., (2013). *J. Chem. Phys., 139*.
42. Miranda-Quintana, R. A., & Bochicchio, R. C., (2014). *Chem. Phys. Lett., 593*, 35.
43. Yang, W. T., Zhang, Y. K., & Ayers, P. W., (2000). *Phys. Rev. Lett., 84*, 5172.
44. Ayers, P. W., (2008). *J. Math. Chem., 43*, 285.
45. Nesbet, R. K., (1997). *Phys. Rev. A., 56*, 2665.
46. Balawender, R., & Holas, A., (2013). In *arXiv:0901.1060*.

47. Balawender, R., & Holas, A., (2013). In *arXiv:0904.3990*.
48. Balawender, R., (2013). In *arXiv:1212.1367*.
49. Malek, A., & Balawender, R., (2015). *J. Chem. Phys.*, *142*, 054104.
50. Franco-Pérez, M., Ayers, P., Gazquez, J. L., & Vela, A., (2015). *J. Chem. Phys., 143*, 244117.
51. Franco-Pérez, M., Gazquez, J. L., Ayers, P., & Vela, A., (2015). *J. Chem. Phys., 143*, 154103.
52. Mermin, N. D., (1965). *Phys. Rev.*, *137*, A1441.
53. Geerlings, P., De Proft, F., & Langenaeker, W., (2003). *Chem. Rev.*, *103*, 1793.
54. Chermette, H., (1999). *J. Comput. Chem., 20*, 129.
55. Johnson, P. A., Bartolotti, L. J., Ayers, P. W., Fievez, T., & Geerlings, P., (2012). In *Modern Charge Density Analysis*; Gatti, C., Macchi, P., Eds., Springer: New York, p. 715.
56. Ayers, P. W., Anderson, J. S. M., & Bartolotti, L. J., (2005). *Int. J. Quantum Chem.*, *101*, 520.
57. Ayers, P. W., & Parr, R. G., (2000). *J. Am. Chem. Soc., 122*, 2010.
58. Ayers, P. W., & Parr, R. G., (2001). *J. Am. Chem. Soc., 123*, 2007.
59. Ayers, P. W., Morrison, R. C., & Roy, R. K., (2002). *J. Chem. Phys., 116*, 8731.
60. Bultinck, P., Cardenas, C., Fuentealba, P., Johnson, P. A., & Ayers, P. W., (2013). *J. Chem. Theory Comp., 9*, 4779.
61. Bultinck, P., Cardenas, C., Fuentealba, P., Johnson, P. A., & Ayers, P. W., (2014). *J. Chem. Theory Comp., 10*, 202.
62. Parr, R. G., Donnelly, R. A., Levy, M., & Palke, W. F., (1978). *J. Chem. Phys., 68*, 3801.
63. Politzer, P., & Weinstein, M., (1979). *J. Chem. Phys., 71*, 4281.
64. Parr, R. G., & Pearson, R. G., (1983). *J. Am. Chem. Soc., 105*, 7512.
65. Parr, R. G., & Yang, W. T., (1984). *J. Am. Chem. Soc., 106*, 4049.
66. Yang, W. T., Parr, R. G., & Pucci, R., (1984). *J. Chem. Phys., 81*, 2862.
67. Morell, C., Grand, A., & Toro-Labbé, A., (2005). *J. Phys. Chem. A, 109*, 205.
68. Boisdenghien, Z., Van Alsenoy, C., De Proft, F., & Geerlings, P., (2013). *J. Chem. Theory Comp., 9*, 1007.
69. Geerlings, P., Fias, S., Boisdenghien, Z., & De Proft, F., (2014). *Chem. Soc. Rev., 43*, 4989.
70. Mulliken, R. S., (1934). *J. Chem. Phys., 2*, 782.
71. Ayers, P. W., Anderson, J. S. M., Rodriguez, J. I., & Jawed, Z., (2005). *PCCP, 7*, 1918.
72. Cardenas, C., Chamorro, E., Galvan, M., & Fuentealba, P., (2007). *Int. J. Quantum Chem., 107*, 807.
73. Cardenas, C., Tiznado, W., Ayers, P. W., & Fuentealba, P., (2011). *J. Phys. Chem. A., 115*, 2325.
74. Parr, R. G., Von Szentpály, L., & Liu, S. B., (1999). *J. Am. Chem. Soc., 121*, 1922.
75. Atkins, P., & Paula, J. D., (2006). *Physical chemistry*, 8th edn., W. H. Freeman and Company, New York.
76. Goldstein, H., (1980). *Classical Mechanics*, Addison-Wesley, Reading, MA.
77. Kaplan, T. A., (2006). *Journal of Statistical Physics*, *122*, 1237.
78. Perez, P., Chamorro, E., & Ayers, P. W., (2008). *J. Chem. Phys., 128*, 204108.

79. Galvan, M., Vela, A., & Gazquez, J. L., (1988). *J. Phys. Chem.*, *92*, 6470.
80. Vargas, R., Galvan, M., & Vela, A., (1998). *J. Phys. Chem. A.*, *102*, 3134.
81. Melin, J., Aparicio, F., Galvan, M., Fuentealba, P., & Contreras, R., (2003). *J. Phys. Chem. A.*, *107*, 3831.
82. Ghanty, T. K., & Ghosh, S. K., (1994). *J. Am. Chem. Soc.*, *116*, 3943.
83. Chan, G. K. L., (1999). *J. Chem. Phys.*, *110*, 4710.
84. Ólah, J., De Proft, F., Veszpremi, T., & Geerlings, P., (2004). *J. Phys. Chem. A.*, *108*, 490.
85. Garza, J., Vargas, R., Cedillo, A., Galvan, M., & Chattaraj, P. K., (2006). *Theor. Chem. Acc.*, *115*, 257.
86. Sanderson, R. T., (1951). *Science*, *114*, 670.
87. Sanderson, R. T., (1976). *Chemical Bonds and Bond Energy*, Academic, New York.
88. Morales, J., & Martinez, T. J., (2001). *J. Phys. Chem. A.*, *105*, 2842.
89. Morales, J., & Martinez, T. J., (2004). *J. Phys. Chem. A*, *108*, 3076.
90. Verstraelen, T., Ayers, P. W., Van Speybroeck, V., & Waroquier, M., (2013). *J. Chem. Phys.*, *138*.
91. Pearson, R. G., (1967). *Chem. Br.*, *3*, 103.
92. Pearson, R. G., (1968). *J. Chem. Educ.*, *45*, 581.
93. Ayers, P. W., (2005). *J. Chem. Phys.*, *122*, 141102.
94. Ayers, P. W., Parr, R. G., & Pearson, R. G., (2006). *J. Chem. Phys.*, *124*, 194107.
95. Ayers, P. W., (2007). *Faraday Discuss.*, *135*, 161.
96. Cardenas, C., & Ayers, P. W., (2013). *PCCP*, *15*, 13959.
97. Berkowitz, M., Ghosh, S. K., & Parr, R. G., (1985). *J. Am. Chem. Soc.,* *107*, 6811.
98. Shoeib, T., Gorelsky, S. I., Lever, A. B. P., Siu, K. W. M., & Hopkinson, A. C., (2001). *Inorg. Chim. Acta.*, *315*, 236.
99. Torrent-Sucarrat, M., De Proft, F., Ayers, P. W., & Geerlings, P., (2010). *PCCP*, *12*, 1072.
100. Gal, T., (2012). *Theor. Chem. Acc.*, *131*.
101. Mendez, F., & Gazquez, J. L., (1994). *J. Am. Chem. Soc.*, *116*, 9298.
102. Geerlings, P., & De Proft, F., (2000). *Int. J. Quantum Chem.*, *80*, 227.
103. Faver, J., & Merz, K. M., (2010). *J. Chem. Theory Comp.*, *6*, 548.
104. Pearson, R. G., & Palke, W. E., (1992). *J. Phys. Chem.*, *96*, 3283.
105. Chattaraj, P. K., & Nath, S., (1994). *Indian J. Chem., Sect. A: Inorg., Bio-inorg., Phys., Theor. Anal. Chem.*, *33A*, 842.
106. Chattaraj, P. K., & Ayers, P. W., (2005). *J. Chem. Phys.*, *123*, 086101.
107. Chattaraj, P. K., Ayers, P. W., & Melin, J., (2007). *PCCP*, *9*, 3853.
108. Gazquez, J. L., Vela, A., & Chattaraj, P. K., (2013). *J. Chem. Phys.*, *138*, 214103.
109. Pearson, R. G., (1999). *J. Chem. Educ.*, *76*, 267.
110. Labet, V., Morell, C., Toro-Labbe, A., & Grand, A., (2010). *PCCP*, *12*, 4142.
111. Sola, M., & Toro-Labbe, A., (1999). *J. Phys. Chem. A.*, *103*, 8847.
112. Noorizadeh, S., (2007). *J. Phys. Org. Chem.*, *20*, 514.
113. Chattaraj, P. K., (2007). *Indian Journal of Physics and Proceedings of the Indian Association for the Cultivation of Science*, *81*, 871.
114. Morell, C., Labet, V., Grand, A., & Chermette, H., (2009). *PCCP*, *11*, 3417.

CHAPTER 3

COMPUTING THE UNCONSTRAINED LOCAL HARDNESS

ROGELIO CUEVAS-SAAVEDRA,[1,2] NATALY RABI,[2] and
PAUL W. AYERS[2]

[1]*Department of Chemistry, The University of Western Ontario,
London, Ontario, N6A 5B7, Canada*

[2]*Department of Chemistry and Chemical Biology, McMaster
University, Hamilton, Ontario, L8S 4M1, Canada*

CONTENTS

ABSTRACT

The unconstrained local hardness is calculated using an exact equation in
the Kohn–Sham density functional theory. Several numerical difficulties
are encountered, mostly related to the exponential asymptotic divergence
of the unconstrained local hardness. The unconstrained local hardness

shows shell structure. Somewhat unexpectedly, it takes negative values in certain regions.

3.1 MOTIVATION

Many of the greatest successes in the density functional theory (DFT) of chemical reactivity, or conceptual DFT, are tied to the chemical hardness [1–6]. Prior to the seminal 1983 paper by Parr and Pearson, the chemical hardness was only a empirical, qualitative measure of chemical reactivity preferences [7]. By proposing a mathematical definition for the global chemical hardness,

$$\eta = \left(\frac{\partial^2 E}{\partial N^2} \right)_{v(\mathbf{r})} = \left(\frac{\partial \mu}{\partial N} \right)_{v(\mathbf{r})} \tag{1}$$

Parr and Pearson made it possible to formulate *quantitative* arguments related to the chemical hardness (In Eq. (1), N is the number of electrons, μ is the electronic chemical potential, and η is the global hardness.). By using this definition, it was possible to derive the global hard/soft acid/base (HSAB) principle [8–17]; the assumptions required in these increasingly refined mathematical derivations have, in turn, provided insights into the scope and limitations of the global HSAB principle. Although there are still robust debates about how the global hardness can be defined [7, 14, 18–32], the mere ability to define the chemical hardness was enough to generate an entirely new chemical principle: the maximum hardness principle [33–40].

The mathematical theory of the HSAB principle is, however, incomplete unless one can also consider the *local* HSAB principle [10, 41–44]. Specifically, given a molecule with two reactive sites of differing local hardness, the hard reagents will prefer to react at the site with the greater local hardness and soft reagents will prefer to react at the site with smaller local hardness. The local HSAB principle remains controversial [20, 36, 45–62]. Part of the problem is the need for a definition of the local hardness that has comparable utility to the Parr-Pearson definition for the global hardness. The most obvious way to proceed—by replacing the number of electrons in the entire molecule with the number of electrons at the point **r** (which is the electron density, $\rho(\mathbf{r})$)—gives a definition for the local hardness that is inherently ambiguous, namely [47, 48],

$$\eta(r) = \left(\frac{\delta\mu}{\delta\rho(r)} \right)_{v(r)} \tag{2}$$

The ambiguity arises because when the external potential is held constant, it is impossible to freely vary the ground-state electron density [47, 48, 50, 56, 57, 63]. The ambiguity can be removed by eliminating the constraint on the functional derivative; this defines the unconstrained local hardness [36, 50],

$$\eta_u(r) = \frac{\delta\mu}{\delta\rho(r)} \tag{3}$$

However, this definition has never been studied computationally; therefore, it is unknown whether it may assist in elucidating the local HSAB principle. The goal of this paper is to take a first step in this direction by proposing and testing a computational approach to the unconstrained local hardness.

Section 3.2 presents theoretical derivations and the computational algorithm. The method is validated in Section 3.3 by studying atomic systems, where the rather expensive calculation that are involved can be carefully tested.

3.2 COMPUTATIONAL APPROACH TO THE UNCONSTRAINED LOCAL HARDNESS

Like other reactivity indicators in the DFT reactivity theory that depend on derivatives with respect to N and/or $\rho(\mathbf{r})$, the unconstrained local hardness splits into multiple reactivity indicators [64], depending on whether the variation in the number of electrons increases the number of electrons,

$$\eta_u^+(r) = \frac{\delta\mu^+}{\delta\rho(r)} \tag{4}$$

or decreases the number of electrons,

$$\eta_u^- = \frac{\delta\mu^-}{\delta\rho(r)} \tag{5}$$

This bifurcation of reactivity indicators occurs for systems with integer electron number and is induced by the fundamental derivative discontinuity of the energy [65, 66] and its derivatives [64, 67] at integer electron number. (In actuality, there is yet another bifurcation that occurs here, because variations in the electron density that increase and decrease N give rise to different indicators in Eqs. (4) and (5). But, these indicators differ only by a constant shift [68–70], and for a Kohn–Sham DFT calculation with a generalized gradient approximation functional (GGA), there is no difference at all [71, 72]).

An equation for the unconstrained local hardness can be derived from the chain rule for functional derivatives [16, 17, 51],

$$\left(\frac{\delta\mu^{\pm}}{\delta v_s(r')}\right)_N = \int \frac{\delta\mu^{\pm}}{\delta\rho(r)}\left(\frac{\delta\rho(r)}{\delta v_s(r')}\right)_N dr \tag{6}$$

where $v_s(\mathbf{r})$ is the Kohn–Sham potential. The same equations can be written for the external potential, $v(\mathbf{r})$, but that formula is less computationally accessible [51, 63]. As exploited in some of these authors' previous works, for GGAs, the chemical potentials are identical to the frontier orbital energies [73–75]. We will restrict ourselves to GGAs here; Eq. (6) then simplifies to

$$\left|\phi^{HOMO}(r')\right|^2 = \int \eta_u^-(r)\chi_s(r,r')dr \tag{7}$$

$$\left|\phi^{LUMO}(r')\right|^2 = \int \eta_u^+(r)\chi_s(r,r')dr \tag{8}$$

in which the highest-occupied and lowest-unoccupied molecular orbital densities enter. The Kohn–Sham response function can be evaluated using the standard sum-over-states formula,

$$\chi_s\left(r,r'\right)=2\operatorname{Re}\left[\sum_{a=LUMO}^{\infty}\sum_{i=1}^{HOMO}\frac{\phi_i^*\left(r\right)\phi_a\left(r\right)\phi_a^*\left(r'\right)\phi_i\left(r'\right)}{\varepsilon_i-\varepsilon_a}\right] \quad (9)$$

It should be noted that this equation is closely related to the other approach to $\eta_u(\mathbf{r})$ that has been proposed in the literature: multiplying both sides of Eq. (9) in Ref. [76] by the Kohn–Sham response function and integrating recover Eqs. (7) and (8). This approach, however, avoids the problematic inverse-linear-response function; see the discussion in Ref. [77].

We solve Eqs. (7) and (8) on an approximate numerical integration grid,

$$\int f\left(r\right)dr\approx\sum_{i=1}^{n_{grid}}w_i f\left(r_i\right) \quad (10)$$

The equation for the unconstrained local hardness is then,

$$w_i\left|\phi^{\pm}\left(r_i\right)\right|^2=\sum_{j=1}^{n_{grid}}w_i\chi_s\left(r_i,r_j\right)w_j\eta_u^{\pm}\left(r_j\right) \quad (11)$$

The unconstrained local hardness on the grid points can then be computed by solving the linear equations,

$$\gamma^{\pm}=\Xi\eta^{\pm} \quad (12)$$

with

$$\gamma^{\pm}=\left[\gamma_i^{\pm}\right]=\left[w_i\left|\phi^{\pm}\left(r_i\right)\right|^2\right] \quad (13)$$

$$\Xi=\left[\xi_{ij}\right]=\left[w_i\chi_s\left(r_i,r_j\right)w_j\right] \quad (14)$$

$$\eta^{\pm}=\left[\eta_j^{\pm}\right]=\left[\eta_u^{\pm}\left(r_j\right)\right] \quad (15)$$

Recall that the Kohn–Sham response is a symmetric, negative-semi-definite, integral kernel with a zero eigenvalue, corresponding to the fact that a constant shift in the Kohn–Sham potential does not change the electron density,

$$\int \chi_s \left(r, r' \right) dr' = \sum_{j=1}^{n_{grid}} \chi_{ij} = 0 \tag{16}$$

Examining Eq. (9), it is clear that

$$rank\left[\Xi \right] = n_{occ} n_{vir} \tag{17}$$

where n_{occ} is the number of occupied orbitals and n_{vir} is the number of virtual orbitals. Moreover, because the spectrum of the Kohn–Sham response kernel has an accumulation point at zero (i.e., there are an arbitrary number of eigenvalues close to zero in the basis set limit) [77], the "effective rank" of the Kohn–Sham response may be even smaller than that obtained in Eq. (17). This means that Eq. (12) [equivalently, Eq. (11)] must be solved using a singular value decomposition algorithm, and the number of nonsingular values that are kept while solving the equations should be *at most* $n_{occ} n_{vir}$.

We wrote a Fortran 90 program that reads in the orbitals and orbital energies from Gaussian 09 [78], computes the frontier orbital densities [Eq. (13)] and the Kohn-Sham response [Eq. (14)] on a numerical integration grid, and then solves Eq. (12) by singular value decomposition. We used the BLYP GGA-type exchange-correlation functional [79–81].

3.3 NUMERICAL RESULTS

We generated plots for the unconstrained local hardness by solving Eq. (12) for the Beryllium atom with different basis sets. In order to show the suggested convergence of the plots, according to Eq. (17), we plotted our results by keeping various numbers of singular values of matrix Eq. (14).

Figure 3.1 shows the plot of the unconstrained local hardness with the 3-21G basis set keeping 11, 12, 13, and 14 singular values. As it can be observed, the plots are practically superimposed, suggesting convergence of the curves.

In order to visualize the changes in the curves as the number of singular values increases, we show the differences $\eta_{14}-\eta_X$ in Figure 3.2, where η_{14} represents the local hardness with $n_{occ}n_{vir}$ singular values and X=11,12,13 Note that the differences shown in Figure 3.2 are of the order of 10^{-8} and that the curves with $X = 12$ and 13 are practically the same. This confirms the convergence we inferred from Figure 3.1.

In Figure 3.3, we show how the behavior of the unconstrained local hardness becomes erratic when the number of singular values is $n_{occ}n_{vir}+1$ We show, for the sake of comparison, $n_{occ}n_{vir}$ curves for and $n_{occ}n_{vir}+1$.

Note that the local hardness with $n_{occ}n_{vir}+1$ singular values 10^{10} is orders of magnitude greater than the converged local hardness. In addition, the former displays an erratic behavior that does not even resemble the converged solution with $n_{occ}n_{vir}$ singular values. This figure illustrates the importance of the threshold on the number of singular values that is retained when solving the linear system (Eq. (12)).

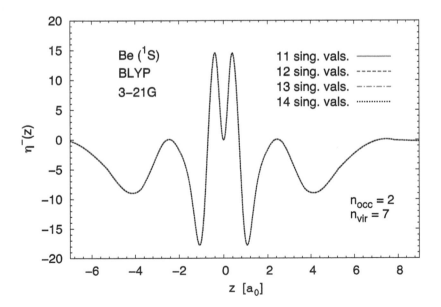

FIGURE 3.1 The unconstrained local hardness when different numbers of singular values of matrix <14> are retained when solving the linear system <12>.

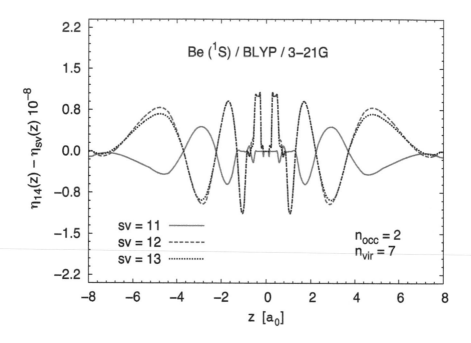

FIGURE 3.2 The convergence of the unconstrained local hardness as one retains increasing numbers of singular values is revealed by plotting the difference, $\eta_{14}-\eta_x$, which is of the order 10^{-8} a.u.

To further emphasize the robustness of the method when $n_{occ}n_{vir}$ singular values are retained, we show in Figure 3.4 the unconstrained local hardness for different basis sets with $n_{occ}n_{vir}$ singular values in all cases. We note that although the curve with the 6-311G basis set has slightly more pronounced oscillations than the other two curves, its shape is similar and it does not display a violent erratic behavior. We confirmed, however, that with $n_{occ}n_{vir}$ singular values, the erratic behavior returns. Because the unconstrained local hardness is inherently ill-conditioned, it is impossible to converge it with respect to the basis set. That is, while basis sets with similar numbers of basis functions (e.g., 3-12G and 6-31G) will give similar results, as the number of basis functions increases, $\eta_u(r)$ will (and must) diverge. Figure 3.4 reveals, however, that this divergence is slow enough for characteristic and potentially useful features in $\eta_u(r)$ to be deduced.

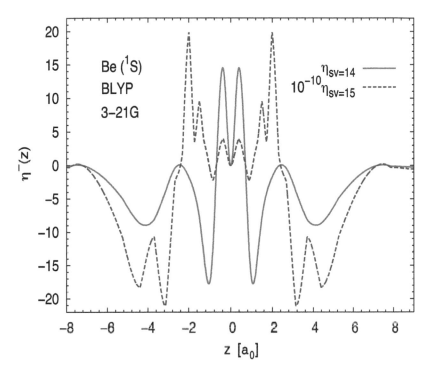

FIGURE 3.3 Unconstrained local hardness with $n_{occ} n_{vir}$ and $n_{occ} n_{vir}+1$ singular values for the Be atom. When too many singular values are retained, the local hardness diverges erratically.

One of the most surprising features of these plots is the fact that the local hardness is sometimes negative, in seeming disagreement with the tendency for increase in the electron density (ergo increase in electron number) to increase the chemical potential. Recall, however, that the external potential is not being held constant: as one increases the electron density at the point **r**, there is a compensating decrease in the external potential, which can cause the chemical potential to decrease. It seems that this effect is most pronounced in electronic shells. It is interesting that the local hardness also clearly shows shell structure and that the exponential divergence that is mandated on mathematical grounds does not appear in our calculations. This latter feature is, no doubt, due to the inadequacy of our basis set for describing the asymptotic decay of the electron density and the linear response function.

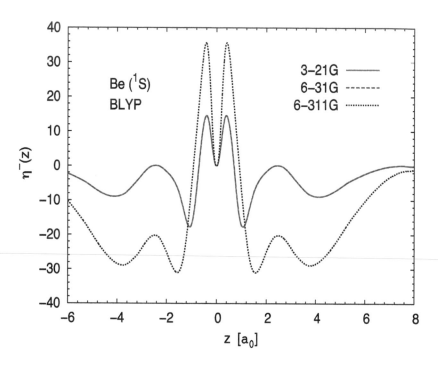

FIGURE 3.4   Unconstrained local hardness with $n_{occ} n_{vir}$ singular values in the calculations of the unconstrained local hardness of the Be atom with different basis sets.

3.4   SUMMARY

We have derived and implemented a procedure for computing the unconstrained local hardness, namely Eqs. (7) and (8). It should emphasized, once more, that this quantity—just like every other definition for the local hardness—is necessarily ill-defined. Nonetheless, by using a finite basis set and by carefully treating the singularities in the linear response kernel, we were able to obtain numerical results. We observe, perhaps unsurprisingly, that the unconstrained local hardness reveals shell structure. Surprisingly, the local hardness can be negative for certain values of $\mathbf{r}$, a result that is counterintuitive because (by generalizing the maximum hardness princi-ple) it would imply that the electron density is unstable in these regions.

Overall, we consider this to be pragmatic, proof-of-principle work. The unconstrained local hardness is too ill-behaved to be routinely considered as a chemical reactivity indicator. However, in cases where this quantity is of interest, we encourage researchers to use the method of computation developed here. Alternative definitions of the local hardness continue to be published [20, 50, 53, 82]. The most promising alternative definitions are, in our view, the frontier local hardness [50, 83],

$$\eta_f(r) = \eta \qquad (18)$$

one of the various information-theoretic approaches to resolve the inherent ambiguity in the local hardness [84], e.g.,

$$\eta_{info} \propto \frac{1}{f(r)(\rho(r))^{1/3}} \qquad (19)$$

or the definition based on the idea that interactions with hard reactive sites are usually governed by the electrostatic potential [20],

$$\eta_{ESP}(r) = \frac{\Phi(r)}{\int \Phi(r) f(r) dr} \qquad (20)$$

These latter definitions have the advantage of being comparatively easy to compute.

ACKNOWLEDGMENT

The authors acknowledge support by a Discovery Grant (PWA) and a postdoctoral fellowship (RCS) from NSERC.

KEYWORDS

- **ambiguous functional derivative**
- **conceptual**
- **linear response function**
- **unconstrained local hardness**

REFERENCES

1. Geerlings, P., De Proft, F., & Langenaeker, W., (2003). Conceptual density functional theory. *Chem. Rev., 103*, 1793–1873.
2. Liu, S. B., (2009). Conceptual Density Functional Theory and Some Recent Developments. *Acta Physico-Chimica Sinica., 25*, 590–600.
3. Gazquez, J. L., (2008). Perspectives on the density functional theory of chemical reactivity. *Journal of the Mexican Chemical Society., 52*, 3–10.
4. Ayers, P. W., Anderson, J. S. M., & Bartolotti, L. J., (2005). Perturbative perspectives on the chemical reaction prediction problem. *Int. J. Quantum Chem., 101*, 520–534.
5. Chermette, H., (1999). Chemical reactivity indexes in density functional theory. *J. Comput. Chem., 20*, 129–154.
6. Johnson, P. A., Bartolotti, L. J., Ayers, P. W., Fievez, T., & Geerlings, P., (2012). Charge density and chemical reactivity: A unified view from conceptual DFT". In: *Modern Charge Density Analysis*, Gatti, C., Macchi, P. Eds., Springer, New York, , pp. 715–764.
7. Parr, R. G., & Pearson, R. G., (1983). Absolute hardness: companion parameter to absolute electronegativity. *J. Am. Chem. Soc., 105*, 7512–7516.
8. Chattaraj, P. K., Lee, H., & Parr, R. G., (1991). HSAB principle. *J. Am. Chem. Soc., 113*, 1855–6.
9. Pearson, R.G., (1995). The HSAB Principle - more quantitative aspects. *Inorg.Chim. Acta., 240*, 93–8.
10. Gazquez, J. L., & Mendez, F., (1994). The hard and soft acids and bases principle: An atoms in molecules viewpoint. *J. Phys. Chem., 98*, 4591–4593.
11. Ayers, P. W., (2005). An elementary derivation of the hard/soft-acid/base principle. *J. Chem. Phys., 122*, 141102.
12. Chattaraj, P. K., & Ayers, P. W., (2005). The maximum hardness principle implies the hard/soft acid/base rule. *J. Chem. Phys., 123*, 086101.
13. Ayers, P. W., Parr, R. G., & Pearson, R. G., (2006). Elucidating the hard/soft acid/ base principle: A perspective based on half-reactions. *J. Chem. Phys., 124*, 194107.
14. Ayers, P. W., (2007). The physical basis of the hard/soft acid/base principle. *Faraday Discuss., 135*, 161–90.
15. Chattaraj, P. K., Ayers, P. W., & Melin, J., (2007). Further links between the maximum hardness principle and the hard/soft acid/base principle: insights from hard/soft exchange reactions. *PCCP., 9*, 3853–3856.
16. Ayers, P. W., & Cardenas, C., (2013). Communication: A case where the hard/soft acid/base principle holds regardless of acid/base strength. *J. Chem. Phys., 138*, 181106.
17. Cardenas, C., & Ayers, P. W., (2013). How reliable is the hard-soft acid-base principle? An assessment from numerical simulations of electron transfer energies. *PCCP., 15*, 13959–13968.
18. Parr, R. G., Ayers, P. W., & Nalewajski, R. F., (2005). What is an atom in a molecule? *J. Phys. Chem. A., 109*, 3957–3959.
19. Frenking, G., & Krapp, A., (2007). Unicorns in the world of chemical bonding models. *J. Comput. Chem., 28*, 15–24.

20. Torrent-Sucarrat, M., De Proft, F., Ayers, P. W., & Geerlings, P., (2010). On the applicability of local softness and hardness. *PCCP.*, *12*, 1072–1080.
21. Reed, J. L., (1997). Electronegativity: Chemical hardness .1. *J. Phys. Chem. A.*, *101*, 7396–7400.
22. Ghosh, D. C., & Islam, N., (2011). A Quest for the Algorithm for Evaluating the Molecular Hardness. *Int. J. Quantum Chem.*, *111*, 1931–1941.
23. Islam, N., & Ghosh, D. C., (2011). Spectroscopic evaluation of the global hardness of the atoms. *Mol. Phys.*, *109*, 1533–1544.
24. Islam, N., & Ghosh, D. C., (2011). A new algorithm for the evaluation of the global hardness of polyatomic molecules. *Mol. Phys.*, *109*, 917–931.
25. Pearson, R. G., (1988). Chemical hardness and bond dissociation energies. *J. Am. Chem. Soc.*, *110*, 7684–7690.
26. Cardenas, C., Tiznado, W., Ayers, P. W., & Fuentealba, P., (2011). The Fukui Potential and the Capacity of Charge and the Global Hardness of Atoms. *J. Phys. Chem. A.*, *115*, 2325–2331.
27. Cardenas, C., (2011). The Fukui potential is a measure of the chemical hardness. *Chem. Phys. Lett.*, *513*, 127–129.
28. Cardenas, C., Ayers, P. W., De Proft, F., Tozer, D. J., & Geerlings, P., (2011). Should negative electron affinities be used for evaluating the chemical hardness. *PCCP.*, *13*, 2285–2293.
29. Fuentealba, P., & Cardenas, C. (2013). On the exponential model for energy with respect to number of electrons. *J. Mol. Model.*, *19*, 2849–2853.
30. Noorizadeh, S., & Parsa, H., (2013). Evaluation of Absolute Hardness: A New Approach. *J. Phys. Chem. A.*, *117*, 939–946.
31. Kaya, S., & Kaya, C., (2015). A new method for calculation of molecular hardness: A theoretical study. *Comput. Theor. Chem.*, *1060*, 66–70.
32. Kaya, S., & Kaya, C., (2015). A new equation for calculation of chemical hardness of groups and molecules. *Mol. Phys.*, *113*, 1311–1319.
33. Pearson, R. G., (1987). Recent advances in the concept of hard and soft acids and bases. *J.Chem.Educ.*, *64*, 561–567.
34. Parr, R. G., & Chattaraj, P. K., (1991). Principle of maximum hardness. *J. Am. Chem. Soc.*, *113*, 1854–1855.
35. Pearson, R. G., & Palke, W. E., (1992). Support for a principle of maximum hardness. *J.Phys.Chem.*, *96*, 3283–3285.
36. Ayers, P. W., & Parr, R. G., (2000). Variational principles for describing chemical reactions: The Fukui function and chemical hardness revisited. *J. Am. Chem. Soc.*, *122*, 2010–2018.
37. Chattaraj, P. K., (1996). The maximum hardness principle: an overview. *Proc. Indian Natl. Sci. Acad., Part A.*, *62*, 513–532.
38. Torrent-Sucarrat, M., Luis, J. M., Duran, M., & Sola, M., (2001). On the validity of the maximum hardness and minimum polarizability principles for nontotally symmetric vibrations. *J. Am. Chem. Soc.*, *123*, 7951–7952.
39. Torrent-Sucarrat, M., Luis, J. M., Duran, M., & Sola, M., (2002). Are the maximum hardness and minimum polarizability principles always obeyed in nontotally symmetric vibrations? *J. Chem. Phys.*, *117*, 10561–10570.

40. Gazquez, J. L., Vela, A., & Chattaraj, P. K., (2013). Local hardness equalization and the principle of maximum hardness. *J. Chem. Phys., 138,* 214103.

41. Pearson, R. G., (1963). Hard and soft acids and bases. *J. Am. Chem. Soc., 85,* 3533–3539.

42. Pearson, R. G., (1966). Acids and bases. *Science., 151,* 172–177.

43. Pearson, R. G., (1997). *Chemical Hardness.* Weinheim, Germany, Wiley-VCH.

44. Mendez, F., & Gazquez, J. L., (1994). Chemical-reactivity of enolate ions - the local hard and soft acids and bases principle viewpoint. *J. Am. Chem. Soc., 116,* 9298–9301.

45. Yang, W. T., & Parr, R. G., (1985). Hardness, softness, and the fukui function in the electron theory of metals and catalysis. *Proc. Natl. Acad. Sci., 82,* 6723–6726.

46. Yang, W. T., Lee, C., & Ghosh, S. K., (1985). Molecular softness as the average of atomic softnesses: companion principle to the geometric mean principle for electronegativity equalization. *J. Phys. Chem., 89,* 5412–5414.

47. Ghosh, S. K., & Berkowitz, M., (1985). A classical fluid-like approach to the density-functional formalism of many-electron systems. *J. Chem. Phys., 83,* 2976–2983.

48. Berkowitz, M., Ghosh, S. K., & Parr, R. G., (1985). On the concept of local hardness in chemistry. *J. Am. Chem. Soc., 107,* 6811–6814.

49. Ayers, P. W., & Parr, R. G., (2008). Beyond electronegativity and local hardness: Higher-order equalization criteria for determination of a ground-state electron density. *J. Chem. Phys., 129,* 054111.

50. Ayers, P. W., & Parr, R. G., (2008). Local hardness equalization: Exploiting the ambiguity. *J. Chem. Phys., 128,* 184108.

51. Cuevas-Saavedra, R., Rabi, N., & Ayers, P. W., (2011). The unconstrained local hardness: an intriguing quantity, beset by problems. *PCCP., 13,* 19594–19600.

52. Gal, T., Geerlings, P., De Proft, F., (2011). Torrent-Sucarrat, M., A new approach to local hardness. *PCCP., 13,* 15003–15015.

53. Torrent-Sucarrat, M., De Proft, F., Geerlings, P., & Ayers, P. W., (2008). Do the Local Softness and Hardness Indicate the Softest and Hardest Regions of a Molecule? *Chemistry-a European Journal., 14,* 8652–8660.

54. Meneses, L., Araya, A., Pilaquinga, F., Contreras, R., & Fuentealba, P., (2007). Local hardness: An application to electrophilic additions. *Chem. Phys. Lett., 446,* 170–175.

55. Meneses, L., Tiznado, W., Contreras, R., & Fuentealba, P., (2004). A proposal for a new local hardness as selectivity index. *Chem. Phys. Lett., 383,* 181–187.

56. Ghosh, S. K., (1990). Energy derivatives in density functional theory. *Chem. Phys. Lett., 172,* 77–82.

57. Harbola, M. K., Chattaraj, P. K., & Parr, R. G., (1991). Aspects of the softness and hardness concepts of density-functional theory. *Isr. J. Chem., 31,* 395–402.

58. Ayers, P. W., Liu, S. B., & Li, T. L., (2011). Stability conditions for density functional reactivity theory: An interpretation of the total local hardness. *PCCP., 13,* 4427–4433.

59. Gal, T., (2012). Why the traditional concept of local hardness does not work. *Theor. Chem. Acc., 131,* 1–14.

60. Langenaeker, W., Deproft, F., & Geerlings, P., (1995). Development of Local Hardness Related Reactivity Indexes - Their Application in A Study of the Se at Monosubstituted Benzenes Within the Hsab Context. *J. Phys. Chem., 99,* 6424–6431.

61. Chattaraj, P. K., Roy, D. R., Geerlings, P., & Torrent-Sucarrat, M., (2007). Local hardness: a critical account. *Theor. Chem. Acc., 118*, 923–930.
62. Gazquez, J. L., Vela, A., & Chattaraj, P. K., (2013). Local hardness equalization and the principle of maximum hardness. *J. Chem. Phys., 138*, 214103.
63. Gal, T., (2012). Why the traditional concept of local hardness does not work. *Theor. Chem. Acc., 131*, 1–14.
64. Ayers, P. W., (2008). The continuity of the energy and other molecular properties with respect to the number of electrons. *J. Math. Chem., 43*, 285–303.
65. Perdew, J. P., Parr, R. G., Levy, M., & Balduz, J. L., (1982). Jr.; Density-functional theory for fractional particle number: derivative discontinuities of the energy. *Phys. Rev. Lett., 49*, 1691.
66. Yang, W. T., Zhang, Y. K., & Ayers, P. W., (2000). Degenerate ground states and fractional number of electrons in density and reduced density matrix functional theory. *Phys. Rev. Lett., 84*, 5172.
67. Ayers, P. W., & Levy, M., (2000). Perspective on "Density functional approach to the frontier-electron theory of chemical reactivity" by Parr R. G., & Yang, W., (1984). *Theor. Chem. Acc., 103*, 353–360.
68. Parr, R. G., & Bartolotti, L. J., (1983). Some remarks on the density functional theory of few-electron systems. *J. Phys. Chem., 87*, 2810–2815.
69. Gal, T., (2001). Differentiation of density functionals that conserves the normalization of the density. *Phys. Rev. A., 63*, 049903.
70. Gal, T., (2007). The Mathematics of Functional Differentiation Under Conservation Constraint. *J. Math. Chem., 42*, 661–676.
71. Tozer, D. J., Handy, N. C., & Green, W. H., (1997). Exchange-correlation functionals from ab initio electron densities. *Chem. Phys. Lett., 273*, 183–194.
72. Chan, G. K. L., (1999). A fresh look at ensembles: Derivative discontinuities in density functional theory. *J. Chem. Phys., 110*, 4710–4723.
73. Ayers, P. W., De Proft, F., Borgoo, A., & Geerlings, P., (2007). Computing Fukui functions without differentiating with respect to electron number. I. Fundamentals. *J. Chem. Phys., 126*, 224107.
74. Sablon, N., De Proft, F., Ayers, P. W., & Geerlings, P., (2007). Computing Fukui functions without differentiating with respect to electron number. II. Calculation of condensed molecular Fukui functions. *J. Chem. Phys., 126*, 224108.
75. Fievez, T., Sablon, N., De Proft, F., Ayers, P. W., & Geerlings, P., (2008). Calculation of Fukui functions without differentiating to the number of electrons. 3. Local Fukui function and dual descriptor. *J. Chem. Theory Comp., 4*, 1065–1072.
76. Ayers, P. W., (2001). Strategies for computing chemical reactivity indices. *Theor. Chem. Acc., 106*, 271–279.
77. Liu, S. B., Li, T. L., & Ayers, P. W., (2009). Potentialphilicity and potentialphobicity: Reactivity indicators for external potential changes from density functional reactivity theory. *J. Chem. Phys., 131*, 114106.
78. Frisch, M. J., Trucks, G. W., Schlegel, H. B., Scuseria, G. E., Robb, M. A., & Cheeseman, J. R., et al., (2009). *Gaussian 09*, Revision A.1. Gaussian Inc., Wallingford CT,.
79. Becke, A. D., (1988). Density-functional exchange-energy approximation with correct asymptotic-behavior. *Phys. Rev. A., 38*, 3098.

80. Lee, C., Yang, W., & Parr, R. G., (1988). Development of the Colle-Salvetti correlation-energy formula into a functional of the electron density. *Phys. Rev. B., 37,* 785.

81. Miehlich, B., Savin, A., Stoll, H., & Preuss, H., (1989). Results obtained with the correlation-energy density functionals of Becke and Lee, Yang and Parr. *Chem. Phys. Lett., 157,* 200–206.

82. Valone, S. M., (2011). A Concept of Fragment Hardness, Independent of Net Charge, from a Wave-Function Perspective. *Journal of Physical Chemistry Letters., 2,* 2618–2622.

83. Chattaraj, P. K., Cedillo, A., & Parr, R. G., (1995). Variational method for determining the Fukui function and chemical hardness of an electronic system. *J. Chem. Phys., 103,* 7645–7646.

84. Heidar-Zadeh, F., Fuentealba, P., Cardenas, C., & Ayers, P. W., (2014). An information-theoretic resolution of the ambiguity in the local hardness. *PCCP., 16,* 6019–6026.

CHAPTER 4

GRAND-CANONICAL INTERPOLATION MODELS

RAMÓN ALAIN MIRANDA-QUINTANA[1,2] and PAUL W. AYERS[2]

[1]*Laboratory of Computational and Theoretical Chemistry, Faculty of Chemistry, University of Havana, Havana, Cuba*

[2]*Department of Chemistry and Chemical Biology, McMaster University, Hamilton, Ontario, Canada*

CONTENTS

ABSTRACT

We analyze the physical grounds of common approaches used to model the dependency of the ground state energy on the number of electrons in molecular systems (E vs. N models). We elaborate on a recent result indicating that smooth interpolation models are inconsistent with the popular grand ensemble approach to open quantum systems, and we illustrate this discrepancy using the parabolic and exponential E vs. N models. Unlike most previous considerations of this problem, we explicitly account for the possibility of degenerate ground states and nonconvexity of the energy.

4.1 INTRODUCTION

The number of electrons in an isolated physical system is always an integer. This feature is directly reflected in the structure of the Hilbert-space wavefunction, in which the number of variables is determined by the number of electrons. However, in density functional theory (DFT) and other methods based on "reduced" descriptors, the number of variables does not depend on the number of electrons. Therefore, in density-based theories as well as in methods based on quantities like k-electron distribution functions, reduced density matrices (DMs), and electron propagators, one must allow for the possibility of noninteger electron number. For example, even the oldest and simplest density functional approximation, the Thomas–Fermi method, is defined for systems with any positive number of electrons, not just for systems with integer numbers of electrons.

A more rigorous treatment of noninteger electron number relies on the grand canonical description, where a chemical potential is used to adjust the number of electrons by placing the system in contact with an (otherwise noninteracting) reservoir of electrons with specified chemical potential and temperature. This finite-temperature open-system approach is the method by which Mermin extended the original zero-temperature Hohenberg–Kohn methodology [1, 2]. Most molecular quantum mechanics calculations are performed at zero temperature; it was Perdew, Parr, Levy, and Balduz (PPLB) who carefully took the zero-temperature limit of the grand canonical ensemble, thereby allowing the energy (and other observables) to be written as a continuous function of the number of electrons, N, in the system. Because the PPLB energy density functional is defined even for infinitesimal density variations that do not preserve electron number, their treatment solidified the common practice of using a Lagrange multiplier to impose the restriction to a specified number of electrons in the DFT variational principle [3]. More relevant to the present work, the PPLB paper revealed that noninteger electron number can be regarded as the time-average of an open system.

Aside from its formal role in the mathematical foundations of DFT, fractional electron number is also of practical importance. Probably, its greatest relevance is to the DFT of chemical reactivity, often called conceptual DFT, where noninteger electron number arises as the response of the system to partial electron transfer to/from a reagent [3–8]. It is essential to describe these process derivatives of the energy (and other state functions) with respect to the number of electrons [4, 9, 10]. These derivatives, which are commonly used as reactivity descriptors, cannot be mathematically formulated without an extension of DFT to noninteger electron number.

Despite its fundamental relevance and practical utility, there are many aspects of DFT (and quantum mechanics in general) for noninteger electron number that remain to be explored. The most fundamental problem is that the "linear mixing rule" derived by PPLB is often inconvenient or nonchemical, while the alternative models are based on ad hoc assumptions. These ad hoc models sometimes work, at other times fail badly, and are often inconsistent with known properties/features of real molecules. For these reasons, the need to differentiate with respect to the number of electrons is commonly deemed "problematic," and there have been several attempts to avoid the need to do so [11, 12]. For example, Parr proposed that the derivatives with respect to N be considered merely as "shorthand representations of finite-difference representations of the integral-N data" [13]. Substituting dy/dN with $\Delta y/\Delta N$ does indeed avoid the mathematical difficulties, but by doing so loses the rigor and insights that can be obtained by directly treating noninteger electron number.

This book chapter explores E vs. N models from first principles. It is motivated in part by our recent realization that differentiable and interpolatory E vs. N models are inconsistent with the grand canonical formalism, so that the two most prevalent approaches to noninteger-N are mutually inconsistent. Here, we present a thorough discussion of interpolation models and the grand canonical treatment, including the oft-neglected issue of degenerate ground-state energies. We will devote special scrutiny to the zero-temperature grand canonical ensemble (the PPLB model) and the popular parabolic and exponential E vs. N interpolatory models. The grand-canonical (GC) formulation is revealed to be especially appropriate for local reactivity descriptors.

4.2 INTERPOLATION AND GRAND CANONICAL ENSEMBLE MODELS: THE IRRELEVANCY OF DEGENERACY

Physically speaking, any system associated with a noninteger number of electrons must be open, implying that describing the state of the system requires the use of DMs [14, 15]. In particular, we will use rigorous mathematical framework provided by the grand canonical ensemble; thus, we are primarily interested in the grand canonical density matrices (GC-DMs) [3, 16, 17]. The GC-DM is written as a statistical average of the accessible pure states.

$$D = \sum_{M} \sum_{k} {}^{M}\omega_k \left| {}^{M}\Phi_k \right\rangle \left\langle {}^{M}\Phi_k \right| \tag{1}$$

Here, the indices k label the M-electron eigenstates, $\left| {}^{M}\Phi_k \right\rangle$. Every $\left| {}^{M}\Phi_k \right\rangle$ belongs to the M-particle antisymmetric Hilbert space, $\mathcal{F}_M$, but our interest in systems with arbitrary particle numbers forces us to work within the more general Fock space, $\mathcal{F} = \oplus_{M=0}^{\infty} \mathcal{F}_M$. ($\oplus$) indicates the direct sum of the given spaces. The coefficients, ${}^{M}\omega_k$, appearing in Eq. (1) are statistical weights, and so

$$\sum_{M} \sum_{k} {}^{M}\omega_k = 1 \tag{2}$$

$$0 \le {}^{M}\omega_k \le 1 \tag{3}$$

The expectation value of the number of particles in the system, N, is easily determined from these weights by using the usual formula for the expected value of an observable in the ensemble,

$$N = \sum_M \sum_k {}^M\omega_k M \tag{4}$$

Clearly, many different $\{{}^M\omega_k\}$ have the same N. We are interested in comparing different models for the ensemble weights for a given value of N, and we will use a sub-index X to distinguish between models for the dependence of the ensemble weights on the number of particles. For example, Eq. (1) can be rewritten as

$$D(N)_X = \sum_M \sum_k {}^M\omega_k(N)_X \left| {}^M\Phi_k \right\rangle \left\langle {}^M\Phi_k \right| \tag{5}$$

Different models of X will give the same expectation value for the number of particles, but different expressions for the mean values of other observables,

$$Q(N)_X \equiv \mathrm{Tr}\left(\hat{Q}D(N)_X\right) = \sum_M \sum_k {}^M\omega_k(N)_X \, {}^MQ_k \tag{6}$$

$$ {}^MQ_k = \left\langle {}^M\Phi_k \left| \hat{Q} \right| {}^M\Phi_k \right\rangle \tag{7}$$

Here, $\mathrm{Tr}(\hat{A})$ indicates the trace of operator $\hat{A}$.

The DMs of greatest chemical relevance are those that describe ground states (at absolute zero temperature) or equilibrium states (at nonzero temperature). Under those conditions, it can be shown that the states $\left| {}^M\Phi_k \right\rangle = \left| {}^ME_k \right\rangle$ are eigenfunctions of the system's Hamiltonian $\hat{H}$ or, more precisely, $\hat{H} - \mu\hat{N}$ where $\hat{N}$ is the particle-number operator and μ is the chemical potential [3]. Furthermore, at zero temperature, only the ground electronic states are relevant,

$$\hat{H}\left| {}^ME_k \right\rangle = {}^ME_k \left| {}^ME_k \right\rangle = {}^ME_{g.s.} \left| {}^ME_k \right\rangle \tag{8}$$

where the index k merely labels the degenerate manifold of electronic ground states. At elevated temperatures, one should consider all the eigenfunctions of the Hamiltonian, but because most electronic structure calculations are done at zero temperature and because electronic excitation energies are usually much greater than thermal energies, these "thermal mixing" effects are traditionally entirely neglected or modelled by using a different X model from the traditional ground-state treatment.

Our focus in this paper is on interpolation models: models that estimate the properties of a system with a noninteger electron number by interpolating its properties at integer values. When we are focusing our discussion on an interpolation model, as opposed to a generic energy model, we will specify $X = i$.

An interpolation model can be applied to any arbitrary observable, Q, with the defining relation that for any positive integer, the interpolated value must be equal to the observed value,

$$\forall M \in \mathbb{N}; \ Q(M)_i = \sum_k {}^M\omega_k(M)_i \, {}^M Q_k \tag{9}$$

The index k here is a sum over the (possibly degenerate) M-electron states. Even if one is only interested in ground electronic states, this is problematic as the degeneracy will ordinarily be lifted as one moves away from M electrons, and the specific choice of weights for the degenerate ground states will therefore introduce ambiguity into the entire procedure. It is therefore desirable to focus on an observable that does not present this pathological behavior: the Hamiltonian. Because all the degenerate ground states have the same energy, ${}^M E = {}^M E_1 = {}^M E_2 = \cdots$ Eq. (9) can be rewritten as:

$$\forall M \in \mathbb{N}; E(M)_i = \left(\sum_k {}^M\omega_k(N)_i \right) {}^M E \tag{10}$$

The interpolation condition can now be unambiguously written as:

$$E(M)_i = {}^M E \tag{11}$$

and therefore, the weights must be zero when evaluated at a different integer,

$$ {}^{M'}\omega_k(M)_i = 0 \qquad\qquad M \neq M' \tag{12}$$

and the sum of the weights over the degeneracy must be one,

$$\sum_k^M \omega_k (M)_i = 1 \qquad (13)$$

We will focus our treatment on interpolations of the energy, and in this context, it is permissible to sum over the degenerate states, defining a state-averaged weight,

$$^M\omega(N)_i \equiv \sum_k^M \omega_k (N)_i \qquad (14)$$

which satisfies (compare Eqs. (12) and (13)) the simple relation

$$^M\omega(N)_i \equiv \sum_k^M \omega_k (N)_i \qquad (15)$$

The interpolated energy model is now simply

$$E(N)_i = \sum_M{}^M\omega(N)_i{}^M E \qquad (16)$$

Note how these expressions are formally equivalent to those obtained in the absence of degenerate states. The presence of degeneracy does not change our considerations in any substantive way, and henceforth, we will omit any mention of degeneracy, after having established that our results hold for both degenerate and nondegenerate ground states. There is a strong formal analogy to the mathematics of DFT here: the treatment of degeneracy requires a refinement of the original Hohenberg–Kohn theorem, but in a formal sense, this is of little consequence [18–22]. In a practical sense, however, degenerate ground states can be problematic [23–28].

4.3   THE LINEAR MODEL FOR *E* VS. *N*

The simplest possible interpolation model is linear interpolation, which will be denoted by $X = li$ (For example, the linear interpolation model for the energy is denoted $E(N)_{li}$). In the linear interpolation model, one approximates the values of the energy for noninteger N by a straight line segment joining the integer below N, denoted $\lfloor N \rfloor = M$, and the integer above N, denoted $\lceil N \rceil = M+1$. This is obviously a sensible model (one is using the closest integers to N to determine the properties of the N-electron system), and many conceptual DFT practitioners will argue that in fact this is the only physically acceptable model because, as it has been discussed at length in the literature, this is the exact form of the variation of the ground state energy with respect to N at absolute zero temperature [29–35]. However, as it is also well known, this model is only exact if the energy is a nonconcave function of the number of electrons at integer numbers,

$$E(M+1)+E(M-1) \geq 2E(M) \tag{17}$$

For example, if one considers all the straight lines joining every pair of integers M_1 and M_2 satisfying $M_1 \leq N \leq M_2$, a Lemma proved by Bochicchio and Rial [36] shows that the straight line subtended by M and $M+1$ will provide the lowest energy value for all $M < N < M+1$ within the set of all allowed straight lines (see Figure 4.1 for an example). Note that these "alternative" straight lines are in fact justifiable states in an excited-state DFT context [37].

The preference for linear interpolation between adjacent integers vanishes, however, when E vs. M is concave. In general, the ground state energy of a system with N electrons can be determined using the variational principle:

$$E(N)_{gs} = \min_{\left\{ {}^{M}\omega(N) \left| \begin{array}{l} 0 \leq {}^{M}\omega(N) \\ 1 = \sum_{M} {}^{M}\omega(N) \\ N = \sum_{M} M^{M}\omega(N) \end{array} \right. \right\}} \sum_{M} {}^{M}\omega(N)^{M}E \tag{18}$$

According to our previous remark, $E(M)_{li} = E(M)_{gs}$ if and only if ${}^{M}E$ is nonconcave, but the general behavior is [38]:

$$E(M)_{li} \geq E(M)_{gs} \tag{19}$$

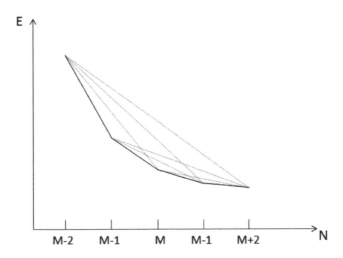

E

M-2   M-1   M   M-1   M+2   N

FIGURE 4.1   Possible linear interpolations for a system with convex E vs. M (e.g., $E(M)_{li}=E(M)_{gs}$). Linear interpolation between adjacent integers is shown by solid lines; interpolation between nonadjacent integers is shown by lightly dotted lines.

$E(N)_{gs}$ is (always) a nonconcave function, and its graph will also be formed by a set of straight segments connecting states with integer number of particles. However, as can be seen in Figure 4.2, when ^{M}E is concave, there will be states for which $^{M}E=E(M)_{li}>E(M)_{gs}$, which implies that Eqs. (11)–(15) will not be valid in this case.

For isolated Coulomb systems like molecules in the absence of an external field, it is observed, experimentally, that the ground state energy is convex. There are also theoretical justifications for the phenomena (which stop well short of a mathematical proof) [39–41]. Nonetheless, this is not the case for certain choices of inter-particle interactions [42–44]. More relevant to conceptual DFT, convexity is not true for interacting reagents; molecules embedded in a solvent; or atoms, functional groups, and other moieties within a molecule [3, 45–53]. In these situations, the line-segment interpolations $E(M)_{li}$ and $E(M)_{gs}$ are different. However, because the circumstances where the models differ are so limited, and because working with excited states is not only permitted, but necessary, when working at a finite temperature [3, 54–56], we will focus on $E(M)_{li}$.

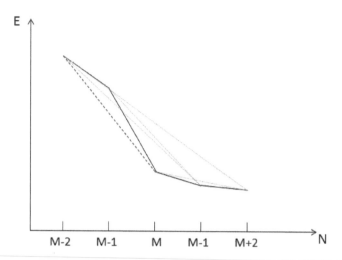

FIGURE 4.2   Possible linear interpolations for a system with nonconvex E vs. M. Linear interpolations between adjacent integers, $E(M)_{li}$, are shown as solid curves, the ground state energy is depicted as a dashed curve, $E(M)_{gs}$, and linear interpolations between nonadjacent integers are shown as lightly dotted lines.

Recalling the notation $N=\lfloor M \rfloor$ and defining x as the fractional part of the number of electrons, $x=N-\lfloor N \rfloor=N-M$, we can write the linear interpolation model as

$$E(N)_{li} = x^{\,M+1}E + (1-x)^{\,M}E \tag{20}$$

and the associated interpolation weights (cf. Eq. (5)) as

$$^{M}\omega(N)_{li} = \sum_{k}{}^{M}\omega_{k}(N)_{li} = 1 - x \tag{21}$$

$$^{M+1}\omega(N)_{li} = \sum_{k}{}^{M+1}\omega_{k}(N)_{li} = x \tag{22}$$

This is a complete description when the ground states are nondegenerate, giving the DMs as

$$D(N)_{li} = x \left| {}^{M+1}\Phi \right\rangle \left\langle {}^{M+1}\Phi \right| + (1-x) \left| {}^{M}\Phi \right\rangle \left\langle {}^{M}\Phi \right| \tag{23}$$

and therefore, the values of the observables as (cf. Eq. (6)) as

$$Q(N)_{li} = x^{M+1}Q + (1-x)^M Q \tag{24}$$

In the nondegenerate case, using other observable properties as the base for the interpolation would give the same results. However, when either the M or $M+1$ electron system (or both) has a degenerate ground state energy, we cannot infer the exact weights because there are an infinite number of weights $^{M(+1)}\omega_k(N)$ that are consistent with the interpolation requirements, and therefore, an infinite number of DMs and values for the associated properties.

The linear interpolation model is a mathematically justifiable approach to systems with noninteger electron number. Moreover, for isolated systems at zero temperature, the derivative discontinuities in the energy (and other properties) at integer values are important and are essential for describing the Kohn–Sham orbital energies and bond dissociation and for avoiding self-interaction error [57–67]. However, molecules and molecular fragments in the process of chemical reaction need not be well described by such a model, and the fact that interacting subsystems can violate the convexity postulate is already a strong indication that the linear interpolation model might not be the best choice for studying chemical reactivity.

4.4 NONLINEAR E VS. N MODELS: PARABOLIC AND EXPONENTIAL INTERPOLATION

The perturbative formulation of conceptual DFT [6] starts from the calculation of the variation of the ground state energy, ΔE, when the system experiences a change in the number of electrons, ΔN, and of the external potential that binds them, $\Delta v(r)$:

$$\Delta E = \left(\frac{\partial E}{\partial N}\right)_{v(r)} \Delta N + \int \left(\frac{\delta E}{\delta v(r)}\right)_N \Delta v(r)dr + \frac{1}{2}\left(\frac{\partial^2 E}{\partial N^2}\right)_{v(r)} \Delta N^2 +$$

$$\int \left(\frac{\delta^2 E}{\delta v(r)\partial N}\right) \Delta N \Delta v(r)dr + \int \left(\frac{\delta^2 E}{\delta v(r)\delta v(r')}\right)_N \Delta v(r')\Delta v(r)drdr' + \dots \tag{25}$$

Writing an expression like this implicitly assumes that it is possible to define derivatives with respect to the number of electrons, e.g., $\dfrac{\partial^k E}{\partial N^k}$ (We will drop the subscript explicitly indicating the constancy of the external potential to unclutter the notation.). Given an interpolation model for the energy, of course, the derivatives with respect to N are clearly defined.

The linear model from the previous section, Eq. (20), gives especially simple expressions for the derivatives. Specifically, for $0 < x < 1$,

$$\mu(N)_{li} = \left.\frac{\partial E}{\partial N}\right|_{M+x} = {}^{M+1}E - {}^{M}E = -A \tag{26}$$

where A is the electron affinity of the system with $M = \lfloor N \rfloor$ electrons. The explicit restriction to a noninteger number of electrons is inconvenient, as the common practice in conceptual DFT, and in chemistry in general, is to consider integer-charged systems (like isolated molecules or species in an integer-defined oxidation state) as the reference state. This is where the problem arises: the derivative does not exist for integer numbers of electrons, and instead, one has two one-sided derivatives,

$$^{M}\mu_{li}^{+} = \lim_{x \to 0^{+}} \frac{E(M+x)_{li} - E(M)_{li}}{x} = {}^{M+1}E - {}^{M}E = -A \tag{27}$$

$$^{M}\mu_{li}^{-} = \lim_{x \to 0^{-}} \frac{E(M+x)_{li} - E(M)_{li}}{x} = {}^{M}E - {}^{M-1}E = -I \tag{28}$$

where I is the ionization energy of the M-electron system. This means that we have different formulas for Eq. (25) for electron-increasing ($\Delta N > 0$),

$$\Delta E = \mu^{+}\Delta N + \ldots \tag{29}$$

and electron-decreasing ($\Delta N < 0$),

$$\Delta E = \mu^{-}\Delta N + \ldots \tag{30}$$

processes. These two equations can be combined into a single compact expression,

$$\Delta E = \frac{1}{2}\left[\mu^+\left(1+\text{sgn}\left(\Delta N\right)\right)+\mu^-\left(1-\text{sgn}\left(\Delta N\right)\right)+\right]\Delta N +... \qquad (31)$$

The inadequacy of the linear model is revealed when higher-order derivatives with respect to N are considered. For fractional N,

$$\left.\frac{\partial^k E}{\partial N^k}\right|_{M+x} = 0 \qquad (32)$$

Important chemical concepts like the hardness [68–72], $\eta \equiv \dfrac{\partial^2 E}{\partial N^2}$, therefore lose their ability to characterize, and distinguish between, different reagents. The behavior is even worse for an integer number of electrons, where the derivatives of the energy are no longer well defined. This motivates us to consider nonlinear models for E vs. N.

4.5 PARABOLIC MODEL

The most popular nonlinear interpolation model, by far, is the parabolic interpolation model [3, 41, 68, 73, 74]. (For quantities associated with the parabolic model, we will denote as $X = pi$.) In the parabolic model, the energy of a system with $N = M + x$ electrons is fit to a parabola,

$$E\left(N\right)_{pi} = aN^2 + bN + c \qquad (33)$$

As in the linear interpolation model, one typically restricts oneself to $|x| < 1$ and one uses the requirement that the energy be exact for $M, M+1$, and $M-1$ electrons to specify the interpolation. This leads to a system of linear equations for the interpolation coefficients,

$$^M E = aM^2 + bM + c \qquad (34)$$

$$^{M+1} E = a\left(M+1\right)^2 + b\left(M+1\right) + c \qquad (35)$$

$$^{M-1} E = a\left(M-1\right)^2 + b\left(M-1\right) + c \qquad (36)$$

which can be solved to obtain,

$$a = \left({}^{M+1}E + {}^{M-1}E - 2\,{}^{M}E \right) \Big/ 2 \qquad (37)$$

$$b = {}^{M+1}E\left(1 - \frac{2M+1}{2}\right) + {}^{M}E(2M) + {}^{M-1}E\left(-\frac{2M+1}{2}\right) \qquad (38)$$

$$c = {}^{M+1}E\left(\frac{M^2 - M}{2}\right) + {}^{M}E(1 - M^2) + {}^{M-1}E\left(\frac{M^2 + M}{2}\right) \qquad (39)$$

To be consistent with a GC treatment, we must be able to rewrite Eq. (33) in the form

$$E(N)_{pi} = {}^{M-1}\omega(N)_{pi}{}^{M-1}E + {}^{M}\omega(N)_{pi}{}^{M}E + {}^{M+1}\omega(N)_{pi}{}^{M+1}E \qquad (40)$$

Explicit expressions for the weights in this expression are obtained by substituting Eqs. (37)–(39) into Eq. (33), giving

$$E(N)_{pi} = \left[\frac{x^2 - x}{2}\right]{}^{M-1}E + \left[1 - x^2\right]{}^{M}E + \left[\frac{x^2 + x}{2}\right]{}^{M+1}E \qquad (41)$$

Differentiating this expression gives expressions for the chemical potential and chemical hardness (for $N=M$):

$$^{M}\mu_{pi} = -\frac{I + A}{2} \qquad (42)$$

$$^{M}\eta_{pi} = I - A \qquad (43)$$

Note first how, because we are working with a smooth dependence of E with respect to N, the ground state energy is differentiable at every particle number. However, in the same way that the linear model gave zero for the second- and higher-order derivatives of the energy under most circumstances (cf. Eq. (32)), the quadratic model gives zero for the third- and higher-order derivatives of the energy. This is undesirable because higher-order global descriptors like the hyperhardness, $\gamma = \frac{\partial^3 E}{\partial N^3}$ [75], can be chemically relevant in some circumstances [76, 77]. Finally, note that Eqs. (42) and (43) coincide with the formulas for the derivative that one would obtain using a finite difference approximation with $\Delta N = 1$.

Assuming that there are no degenerate ground states, the DMs for the quadratic model can be deduced directly from Eqs. (40) and (41),

$$D(N)_{pi} = \left[\frac{x^2 - x}{2}\right] \left|{}^{M-1}\Phi\right\rangle\left\langle{}^{M-1}\Phi\right| + \left[1 - x^2\right]\left|{}^{M}\Phi\right\rangle\left\langle{}^{M}\Phi\right| + \left[\frac{x^2 + x}{2}\right]\left|{}^{M+1}\Phi\right\rangle\left\langle{}^{M+1}\Phi\right| \quad (44)$$

This expression, in turn, allows one to calculate other properties of the system, of which perhaps the most interesting are those related to the molecule's regioselectivity. In the absence of degeneracies, [26, 27] one can use first order perturbation theory to identify $\left(\frac{\delta E}{\delta v(r)}\right)_N - \rho(r)$ [73], where $\rho(r)$ is the ground-state electron density of the system. Then, local descriptors like the Fukui function [78–81], $f(r) \equiv \frac{\delta^2 E}{\delta v(r)\partial N} = \frac{\partial\rho(r)}{\partial N}$, and the dual descriptor [82–84], $\Delta f(r) \equiv \frac{\delta^3 E}{\delta v(r)\partial N^2} = \frac{\partial^2\rho(r)}{\partial N^2}$, can be calculated. Specifically, differentiating Eq. (41) with respect to the external potential gives

$$^N\rho(r)_{pi} = \left[\frac{x^2 - x}{2}\right]{}^{M-1}\rho(r) + \left[1 - x^2\right]{}^{M}\rho(r) + \left[\frac{x^2 + x}{2}\right]{}^{M+1}\rho(r) \quad (45)$$

Subsequent differentiation with respect to N gives a formula for the Fukui function

$$^{M+x}f(r)_{pi} = \left[\frac{2x - 1}{2}\right]{}^{M-1}\rho(r) - 2x\,{}^{M}\rho(r) + \left[\frac{2x + 1}{2}\right]{}^{M+1}\rho(r) \quad (46)$$

which is reduced to a simple formula for the M-electron reference system $(x = 0)$,

$$^M f(r)_{pi} = \frac{{}^{M+1}\rho(r) - {}^{M-1}\rho(r)}{2} \quad (47)$$

One additional differentiation gives the dual descriptor

$$^{M}\Delta f\left(r\right)_{pi} = {}^{M+1}\rho\left(r\right) + {}^{M-1}\rho\left(r\right) - 2\,^{M}\rho\left(r\right)$$

(48)

which does not depend on the number of particles. For $k > 2$, $\dfrac{\partial^{k}\rho\left(r\right)}{\partial N^{k}} = 0$.
Once again, the finite-difference formulas with $\Delta N = 1$ for the density derivatives are exact. Further, note that there is a simple relationship between the one-sided Fukui functions that can be deduced in the linear model, $^{M}f^{+}(r)_{li} = {}^{M+1}\rho(r) - {}^{M}\rho(r)$ and $^{M}f^{-}(r)_{li} = {}^{M}\rho(r) - {}^{M+1}\rho(r)$, and the Fukui function and dual descriptor in the quadratic model,

$$^{M}f\left(r\right)_{pi} = \frac{^{M}f^{+}\left(r\right)_{li} + {}^{M}f^{-}\left(r\right)_{li}}{2}$$

(49)

$$^{M}\Delta f\left(r\right)_{pi} = {}^{M}f^{+}\left(r\right)_{li} - {}^{M}f^{-}\left(r\right)_{li}$$

(50)

4.6   EXPONENTIAL MODEL

As mentioned previously, the vanishing of third- and higher-order derivatives in the quadratic model is occasionally undesirable. Among the various models for E vs. N that have derivatives of all orders, the exponential model of Parr and Bartolotti is seemingly the most prevalent [85]. Parr and Bartolotti proposed this model in order to derive Sanderson's geometrical mean electronegativity equalization principle [86], and it has subsequently been used to justify other equalization principles also [87]. Denoting quantities from the exponential interpolation model as $X = ei$, the fundamental energy equation is

$$E\left(N\right)_{ei} = E\left(M\right) - \frac{\mu_{0}}{\alpha}\left[e^{-\alpha\left(N-M\right)} - 1\right]$$

(51)

where M is an arbitrary integer number of electrons, μ_{0} is the chemical potential at M, and α is a (universal) constant to be determined empirically. For Eq. (51) to be consistent with a GC model, it should be expressed as:

$$E(N)_{ei} = \sum_M {}^M \omega(N)_{ei} {}^M E \tag{52}$$

Substituting Eq. (51) and the interpolation condition in Eq. (52), we obtain:

$$E(M) - \frac{\mu_0}{\alpha}\left[e^{-\alpha(N-M)} - 1\right] = \sum_{M'} {}^{M'}\omega(N)_{ei}\left\{E(M) - \frac{\mu_0}{\alpha}\left[e^{-\alpha(M'-M)} - 1\right]\right\} \tag{53}$$

This expression can be greatly simplified using the normalization condition, Eq. (2)

$$e^{-\alpha N} = \sum_M {}^M \omega(N)_{ei} e^{-\alpha M} \tag{54}$$

As before, it is sensible to restrict ourselves to relatively small changes in the number of electrons and, in particular, the fewest and most nearby states that can be used to determine the unknown parameters [13]. Restricting ourselves to the states where the system has M, $M+1$, and $M-1$ electrons, Eq. (54) can be explicitly rewritten as:

$$e^{-\alpha N} = {}^{M-1}\omega(N)_{ei} e^{-\alpha(M-1)} + {}^M\omega(N)_{ei} e^{-\alpha M} + {}^{M+1}\omega(N)_{ei} e^{-\alpha(M+1)} \tag{55}$$

Combining Eq. (55) with the normalization condition:

$$^{M-1}\omega(N)_{ei} + {}^M\omega(N)_{ei} + {}^{M+1}\omega(N)_{ei} = 1 \tag{56}$$

the expression for the number of particles:

$$^{M-1}\omega(N)_{ei}(M-1) + {}^M\omega(N)_{ei} M + {}^{M+1}\omega(N)_{ei}(M+1) = M + x \tag{57}$$

it is easy to show that:

$$^{M-1}\omega(N)_{ei} = \frac{e^{-\alpha x} - \left[1 + x\left(e^{-\alpha} - 1\right)\right]}{e^{\alpha} - 2 + e^{-\alpha}} \tag{58}$$

$$^M\omega(N)_{ei} = 1 - 2\left\{\frac{e^{-\alpha x} - \left[1 + x\left(e^{-\alpha} - 1\right)\right]}{e^{\alpha} - 2 + e^{-\alpha}}\right\} - x \qquad (59)$$

$$^{M+1}\omega(N)_{ei} = \frac{e^{-\alpha x} - \left[1 + x\left(e^{-\alpha} - 1\right)\right]}{e^{\alpha} - 2 + e^{-\alpha}} + x \qquad (60)$$

The ground state energy expression is then:

$$E(N)_{ei} = \left[\frac{e^{-\alpha x} - \left[1 + x\left(e^{-\alpha} - 1\right)\right]}{e^{\alpha} - 2 + e^{-\alpha}}\right]^{M-1}E +$$

$$\left[1 - 2\left\{\frac{e^{-\alpha x} - \left[1 + x\left(e^{-\alpha} - 1\right)\right]}{e^{\alpha} - 2 + e^{-\alpha}}\right\} - x\right]^{M}E + \left[\frac{e^{-\alpha x} - \left[1 + x\left(e^{-\alpha} - 1\right)\right]}{e^{\alpha} - 2 + e^{-\alpha}} + x\right]^{M+1}E \qquad (61)$$

To obtain concise expressions for the reactivity descriptors is convenient to define the derivative of the ensemble weights with respect to the number of electrons as an auxiliary function,

$$g(N) = \frac{\partial\left(^{M-1}\omega(N)_{ei}\right)}{\partial N} = -\left(\frac{\alpha e^{-\alpha x} + e^{-\alpha} + 1}{e^{\alpha} - 2 + e^{-\alpha}}\right) \qquad (62)$$

which, for the reference system ($N = M$; $x = 0$), gives the simpler expression

$$g(M) = -\left(\frac{\alpha + e^{-\alpha} + 1}{e^{\alpha} - 2 + e^{-\alpha}}\right) \qquad (63)$$

Higher-order derivatives of the ensemble weights give the expressions,

$$\frac{\partial^k g(N)}{\partial N^k} = (-1)^{k+1}\left(\frac{\alpha^{k+1}}{e^{\alpha} - 2 + e^{-\alpha}}\right)e^{-\alpha x} \qquad (64)$$

$$\frac{\partial^k g(M)}{\partial N^k} = (-1)^{k+1} \left(\frac{\alpha^{k+1}}{e^\alpha - 2 + e^{-\alpha}} \right) \tag{65}$$

The chemical potential then has the simple expression,

$$\mu(N)_{ei} = g(N)\left[^{M-1}E + {}^{M+1}E - 2{}^M E\right]$$
$$+ \left[^{M+1}E - {}^M E\right] = g(N)\left[I - A\right] - A = g(N)^M \eta_{pi} - A \tag{66}$$

and, for the $N = M$ electron reference system,

$$^M \mu_{ei} = g(M)\left[I - A\right] - A = g(M)^M \eta_{pi} - A \tag{67}$$

The corresponding expressions for the chemical hardness are:

$$\eta(N)_{ei} = \frac{\partial g(N)}{\partial N}\left[^{M-1}E + {}^{M+1}E - 2{}^M E\right] = \frac{\partial g(N)}{\partial N}\left[I - A\right] = \frac{\partial g(N)}{\partial N}{}^M \eta_{pi} \tag{68}$$

$$^M \eta_{ei} = \frac{\partial g(M)}{\partial N}\left[I - A\right] = \frac{\partial g(M)}{\partial N}{}^M \eta_{pi} \tag{69}$$

These expressions can be further simplified if we use a particular value of α. Taking the "experimental" value $\alpha \approx 2.15$ [85, 87], $g(M) \approx -1/2$, and therefore, the parabolic model and the exponential model have nearly equivalent chemical potentials,

$$^M \mu_{ei} \approx \frac{^{M+1}E - {}^{M-1}E}{2} = {}^M \mu_{pi} \tag{70}$$

It should be emphasized that, unlike the linear interpolation model and the parabolic interpolation model, higher-order derivatives of the energy do not vanish for the exponential interpolation.

In the absence of degeneracies, there is an unambiguous expression for the DM in the exponential model,

$$D(N)_{ei} = \left[\frac{e^{-\alpha x} - \left[1 + x\left(e^{-\alpha} - 1\right)\right]}{e^{\alpha} - 2 + e^{-\alpha}} \right] \Big|^{M-1}\Phi\Big\rangle\Big\langle^{M-1}\Phi\Big|$$

$$+ \left[1 - 2 \left\{ \frac{e^{-\alpha x} - \left[1 + x\left(e^{-\alpha} - 1\right)\right]}{e^{\alpha} - 2 + e^{-\alpha}} \right\} - x \right] \Big|^{M}\Phi\Big\rangle\Big\langle^{M}\Phi\Big| \qquad (71)$$

$$+ \left[\frac{e^{-\alpha x} - \left[1 + x\left(e^{-\alpha} - 1\right)\right]}{e^{\alpha} - 2 + e^{-\alpha}} + x \right] \Big|^{M+1}\Phi\Big\rangle\Big\langle^{M+1}\Phi\Big|$$

If we treat the electron density as a property of the system, we can use Eq. (71) to obtain an expression for the electron density and its derivatives with respect to the number of electrons,

$$^{M}f(r)_{ei} = g(M)\,^{M}\Delta f(r)_{pi} + {}^{M}f^{+}(r)_{li} \qquad (72)$$

$$^{M}\Delta f(r)_{ei} = \frac{\partial g(M)}{\partial N}\,^{M}\Delta f(r)_{pi} \qquad (73)$$

Because $\dfrac{\partial g(M)}{\partial N} > 0$, the dual descriptors in the parabolic and exponential models give the same predictions for the nucleophilic and electrophilic regions of a system as well as for the relative electrophilicity/nucleophilicity of competing reactive sites.

4.7 DIFFERENTIABLE GRAND CANONICAL INTERPOLATION MODELS

At first glance, parabolic interpolation and exponential interpolation are smooth, nonlinear E vs. N models that are grounded in the grand canonical formalism through Eqs. (44) and (71). This would seem to contradict our earlier result that differentiable interpolation models are inconsistent with any type of ensemble averaging. Resolving this paradox requires closer inspection. In particular, while the DMs (and energies) can be written in the form of an ensemble average, cf. Eqs. (5) and (6), inspection of the "weighting factors" $^M\omega(N)$ reveals that the non-negativity constraint, Eq. (3) is violated. The interpolation weights cannot be interpreted as statistical weights, and therefore, the interpolation formulas cannot be considered as any sort of ensemble average.

To see this in more detail, consider the parabolic interpolation model (cf. Eq. (44)). It is clear that

$$
^M\omega\left(M+x\right)_{pi}
\begin{cases}
\geq 0 & -1 \leq x \leq 1 \\
< 0 & |x| > 1
\end{cases}
\tag{74}
$$

$$
^{M+1}\omega\left(M+x\right)_{pi}
\begin{cases}
\geq 0 & x \geq 0; x \leq -1 \\
< 0 & -1 < x < 0
\end{cases}
\tag{75}
$$

$$
^{M-1}\omega\left(M+x\right)_{pi}
\begin{cases}
\geq 0 & x \leq 0; x \geq 1 \\
< 0 & 0 < x < 1
\end{cases}
\tag{76}
$$

For any noninteger x and, in particular, in the chemically relevant interval that is bounded by the reference energies, $-1 < x < 1$, exactly one of the three interpolation weights is always negative. The parabolic interpolation model can only be considered as a statistical ensemble at the (trivial) reference points associated with the M, $M+1$, and $M-1$ electron systems. This fact, already noted by Morales and Martínez [88], implies that the parabolic model cannot be consistently formulated using a GC formalism. In fact, the only way in which this would be possible is considering complex temperatures.

Why is the parabolic model so successful in practice? Probably, the model should not be viewed as an "exact" E vs. N model, but rather a suitable Occam's razor interpretation to the energy of a subsystem, in terms of a Taylor series expansion truncated at the second order. Presuming that such a truncated Taylor series would be reliable, evaluating the coefficients of the Taylor series with finite differences is similar, and this leads to the quadratic model. Although it is less obvious, the inequalities (74)–(76) also hold for the exponential interpolation model. (This is difficult to show mathematically, but easily seen from Figure 4.3.) As such, the exponential model is also incompatible with an ensemble formulation.

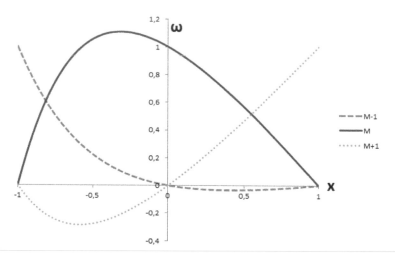

FIGURE 4.3 Variation of the weighting coefficients of the exponential model in the range $-1 < x = N - M < 1$, cf. Eq. (71). $^{M-1}\omega(N)_{ei}$ is the dotted line; $^{M}\omega(N)_{ei}$ is the solid line; $^{M+1}\omega(N)_{ei}$ is the dashed line.

The same general precept is true for any interpolation model: smooth (nonlinear) interpolation models are incompatible with the ensemble representation. In our previous work, we showed this as a theorem, with two independent proofs. Here, we present yet another approach. First, note that in the same way we could define a particle number operator, $\hat{N}$, we can define a chemical potential operator, $\hat{\mu}$. It is clear then, from the definition of μ (see Eq. (26)), that, in a particle number representation, $\hat{\mu}$, will contain a derivative with respect to N [89]. This implies that these operators do not commute, making impossible to determine at the same time the number of particles and the chemical potential. (This provides another argument in favor of the indeterminacy of the chemical potential at 0 Kelvin ($A < -\mu < I$) for integer electron number, thus pointing out once again the inherent arbitrariness of the Mulliken expression, Eq. (42), in this case.) The main conclusion in the present context is that, for the reactivity indices of conceptual DFT to be well defined, we must consider mixed states. Even if we try to calculate the derivatives at integer M, the corresponding state must have contributions from more than one pure state. Then, the fact that interpolation models have, by construction, only a contribution from a single pure state at integer M demonstrates that interpolation models are incompatible with the ensemble description.

4.8 CONCLUSIONS

The goal of this chapter is to provide a perspective on different approaches to define the energy of a system with a noninteger number of electrons. The ensemble model and the associated linear interpolation model are mathematically appealing, but chemically problematic due to the derivative discontinuity. This probably can be removed by defining a parabolic interpolation model (where still all the higher than second-order derivatives with respect to the number of electrons vanish) and the exponential model (where all derivatives with respect to N typically exist and are nonzero.) However, although the interpolation models can be written as "weighted" sums of the reference energies/densities, their "weighting factors" are sometimes non-negative (in fact, one of their weighting factors is always non-negative); therefore, these models cannot be considered as statistical ensembles. This casts some doubt on the utility of interpolation models although these interpolation models, and others like them, remain ubiquitous in the literature.

ACKNOWLEDGMENTS

RAMQ acknowledges support from Foreign Affairs, Trade and Development Canada in the form of an ELAP scholarship. PWA acknowledges support from NSERC and Compute Canada.

KEYWORDS

- **CDFT**
- **chemical reactivity**
- **density matrices**
- **DFT**
- **grand-canonical formalism**
- **interpolation models**

REFERENCES

1. Mermin, N. D., (1965). Thermal properties of the inhomogeneus electron gas. *Phys. Rev., 137,* A 1441–A 1443.
2. Kohn, W., & Hohenberg, P., (1964). Inhomogeneous electron gas. *Phys. Rev.,136,* B864.
3. Parr, R. G., & Yang, W., (1989). *Density-Functional Theory of Atoms and Molecules.* New York, Oxford University Press.
4. Geerlings, P., Prooft, F. D., & Langenaeker,W., (2003). Conceptual density functional theory. *Chem. Rev., 103,* 1793–1873.
5. Chermette, H., (1999). Chemical Reactivity Indexes in Density Functional Theory. *J. Comp. Chem., 20,* 129–154.
6. Ayers, P. W., Anderson, J. S. M., & Bartolotti, L. J., (2005). Perturbative perspectives on the chemical reaction prediction problem. *Int. J. Quantum Chem., 101,* 520–534.
7. Johnson, P. A., Bartolotti, L. J., Ayers, P. W., Fievez, T., & Geerlings, P., (2012). Charge density and chemical reactions: a unified view from conceptual DFT, in: C. Gatti, P. Macchi (Eds.) *Modern charge-density analysis,* Springer Science + Business Media.
8. Ayers, P. W., & Parr, R. G., (2000). Variational principles for describing chemical reactions: the Fukui function and chemical hardness revisited. *J. Am. Chem. Soc., 122,* 2010–2018.
9. Parr, R. G., Donnelly, R. A., Levy, M., & Palke, W. E., (1978). Electronegativity: the density functional viewpoint, *J. Chem. Phys., 68,* 3801–3807.
10. Parr, R. G., & Pearson, R. G., (1983). Absolute hardness: companion parameter to absolute electronegativity. *J. Am. Chem. Soc., 105,* 7512–7516.
11. Fievez,T., Sablon, N., Proft, F. D., Ayers, P. W., & Geerlings, P., (2008). Calculation of Fukui functions without differentiating to the number of electrons. 3. Local Fukui function and dual descriptor. *J. Chem. Theory Comput., 4,* 1065–1072.
12. Ayers, P. W., Proft, F. D., Borgoo, A., & Geerlings, P., (2007). Computing Fukui functions without differentiating with respect to the electron number. I. Fundamentals. *J. Chem. Phys., 126,* 224107.
13. Fuentealba, P., & Parr, R. G., (1991). Higher order derivatives in density functional theory, especially the hardness derivative $\partial\eta/\partial N$. *J. Chem. Phys., 94,* 5559–5564.
14. Blum, K., (1996). *Density Matrix Theory and Applications.* New York, Plenum Press.
15. Ballentine, L., (1998). *Quantum Mechanics: A Modern Development,* World Scientific Publishing Co. Pte. Ltd., London.
16. Bochicchio, R. C., Miranda-Quintana, R. A., & Rial, D., (2013). Communication: Reduced density matrices in molecular systems: Grand-canonical electron states. *J. Chem. Phys., 139,* 191101.
17. Miranda-Quintana, R. A., & Bochicchio, R. C., (2014). Energy dependence with the number of particles: Density and reduced density matrices functionals. *Chem. Phys. Lett., 593,* 35–39.
18. Hohenberg, P., & Kohn, W., (1964). Inhomogeneous electron gas. *Phys.Rev., 136,* B864–B871.

19. Levy, M., (1979). Universal Variational Functionals of Electron-Densities, 1st-Order Density-Matrices, and Natural Spin-Orbitals and Solution of the V-Representability Problem. *Proc. Natl. Acad. Sci., 76,* 6062–6065.
20. Levy, M., (1982). Electron densities in search of Hamiltonians. *Phys. Rev. A., 26,* 1200–1208.
21. Kohn, W., (1983). Upsilon-Representability and Density Functional Theory. *Phys. Rev. Lett., 51,* 1596–1598.
22. Yang, W. T., Ayers, P. W., & Wu, Q., (2004). Potential functionals: Dual to density functionals and solution to the upsilon-representability problem. *Phys. Rev. Lett., 92,* 146–404.
23. Savin, A., (1996). On degeneracy, near-degeneracy, and density functional theory, in: J. M. Seminario (Ed.) *Recent Developments and Applications of Modern Density Functional Theory,* Elsevier, New York, pp. 327.
24. Ayers, P. W., & Levy, M., (2014). Tight constraints on the exchange-correlation potentials of degenerate states. *J. Chem. Phys., 140,* 18a537.
25. Levy, M., Anderson, J. S. M., Zadeh, F. H., & Ayers, P. W., (2014). Kinetic and electron-electron energies for convex sums of ground state densities with degeneracies and fractional electron number. *J. Chem. Phys., 140,* 18a538.
26. Cardenas, C., Ayers, P. W., & Cedillo, A., (2011). Reactivity indicators for degenerate states in the density-functional theoretic chemical reactivity theory. *J. Chem. Phys., 134,* 174103.
27. Bultinck, P., Cardenas, C., Fuentealba, P., Johnson, P. A., & Ayers, P. W., (2013). Atomic Charges and the Electrostatic Potential Are Ill-Defined in Degenerate Ground States. *J. Chem. Theory Comp., 9,* 4779–4788.
28. Bultinck, P., Cardenas, C., Fuentealba, P., Johnson, P. A., & Ayers, P. W., (2014). How to Compute the Fukui Matrix and Function for Systems with (Quasi-)Degenerate States. *J. Chem. Theory Comp., 10,* 202–210.
29. Perdew, J. P., Parr, R. G., Levy, M., & Balduz, J. J. L., (1982). Density-functional theory for fractional particle number: derivative discontinuities of the energy. *Phys. Rev. Lett., 49,* 1691–1694.
30. Parr, R. G., Yang, W., (1989). *Density-Functional Theory of Atoms and Molecules,* Oxford UP, New York,.
31. Bochicchio, R. C., & Rial, D., (2012). Note: Energy convexity and density matrices in molecular systems. *J. Chem. Phys., 137,* 226101.
32. Yang, W. T., Zhang, Y. K., & Ayers, P. W., (2000). Degenerate ground states and fractional number of electrons in density and reduced density matrix functional theory. *Phys. Rev. Lett., 84,* 5172–5175.
33. Ayers, P. W., (2008). The continuity of the energy and other molecular properties with respect to the number of electrons. *J. Math. Chem., 43,* 285–303.
34. Nesbet, R. K., (1997). Fractional occupation numbers in density-functional theory. *Phys. Rev. A., 56,* 2665–2669.
35. Zhang, Y. K., & Yang, W. T., (2000). Perspective on "Density-functional theory for fractional particle number: derivative discontinuities of the energy". *Theor. Chem. Acc., 103,* 346–348.
36. Bochicchio, R. C., & Rial, D., (2012). Note: Energy convexity and density matrices in molecular systems. J. *Chem. Phys., 137,* 226101.

37. Ayers, P. W., (2001). PhD. *Dissertation*. University of North Carolina at Chapel Hill,.

38. Yang, W., Zhang,Y., & Ayers, P. W., (2000). Degenerate ground staes and a fractional number of electrons in density and reduced density matrix functional theory. *Phys. Rev. Lett., 84,* 5172–5175.

39. Sagvolden, E., Perdew, J. P., & Levy, M., (2009). Comment on "Functional derivative of the universal density functional in Fock space". *Phys. Rev. A., 79,* 026501.

40. Perdew, J. P., Parr, R. G., Levy, M., & Balduz, J. L., (1982). Jr., Density-functional theory for fractional particle number: derivative discontinuities of the energy. *Phys. Rev. Lett., 49,* 1691–1694.

41. Ayers, P. W., & Parr, R. G., (2008). Local hardness equalization: Exploiting the ambiguity. *J. Chem. Phys., 128,* 184108.

42. Phillips, P., & Davidson, E. R., (1983). Chemical potential for harmonically interacting particles in a harmonic potential. *Int. J. Quantum Chem., 23,* 185–194.

43. Lieb, E. H., (1983). Density functionals for Coulomb systems. *Int. J. Quantum Chem., 24,* 243–277.

44. Fye, R. M., Martins, M. J., & Scalettar, R. T., (1990). Binding of holes in one-dimensional Hubbard Chains. *Phys. Rev. B., 42,* 6809–6812.

45. Ayers, P. W., (2007). On the electronegativity nonlocality paradox. *Theor. Chem. Acc.,118,* 371–381.

46. Ayers, P. W., (2006). Can one oxidize an atom by reducing the molecule that contains It? *PCCP., 8,* 3387–3390.

47. Saveant, J. M., (2006). *Elements of Molecular and Biomolecular Electrochemistry: an electrochemical approach to electron transfer chemistry.* Wiley, Hoboken..

48. Evans, D. H., & Hu, K., (1996). Inverted potentials in two-electron processes in organic electrochemistry. *Journal of the Chemical Society-Faraday Transactions., 92,* 3983–3990.

49. Evans, D. H., (1998). The kinetic burden of potential inversion in two-electron electrochemical reactions. *Acta Chem. Scand., 52,* 194–197.

50. Lehmann, M. W., & Evans, D. H., (1999). Effect of comproportionation on voltammograms for two electron reactions with an irreversible second electron transfer. *Anal. Chem., 71,* 1947–1950.

51. Evans, D. H., & Lehmann, M. W., (1999). Two-electron reactions in organic and organometallic electrochemistry. *Acta Chem. Scand., 53,* 765–774.

52. Evans, D. H., (2008). One-electron and two-electron transfers in electrochemistry and homogeneous solution reactions. *Chem. Rev., 108,* 2113–2144.

53. Miller, J. S., & Min, K. S., (2009). Oxidation Leading to Reduction: Redox-Induced Electron Transfer (RIET). *Angewandte Chemie-International Edition.,48,* 262–272.

54. Malek, A., & Balawender, R., (2015). Revisiting the chemical reactivity indices as the state function derivatives. The role of classical chemical hardness. *J. Chem. Phys., 142,* 054104.

55. Franco-Pérez, M., Ayers, P. W., Gázquez, J. L., & Vela, A., (2015). Local and linear chemical reactivity response functions at finite temperature in density functional theory. *J. Chem. Phys.,143,* 244117.

56. Franco-Pérez, M., Gázquez, J. L., Ayers, P. W., & Vela, A., (2015). Revisiting the definition of the electronic chemical potential, chemical hardness, and softness at finite temperatures. *J. Chem. Phys., 143,* 154103.

57. Zhang, Y., & Yang, W., (2000). Perspective on "Density-functional theory for fractional particle number: derivative discontinuities of the energy". *Theor. Chem. Acc.*, *103*, 346–348.

58. Johnson, E. R., Salamone, M., Bietti, M., & DiLabio, G. A., (2013). Modeling noncovalent radical-molecule interactions using conventional density-functional theory: beware of erroneous charge transfer. *J. Phys. Chem. A.*, *117*, 947–952.

59. Cohen, A. J., Mori-Sánchez, P., & Yang, W., (2012). Challenges for density functional theory. *Chem. Rev.*, *112,*289–320.

60. Ruzsinszky, A., Perdew, J. P., Csonka, G. I., Vydrov, O. A., & Scuseria, G. E., (2006). Spurious fractional charge on dissociated atoms: Pervasive and resilient self-interaction error of common density functionals. *J. Chem. Phys.*, *125*, 194112.

61. Ruzsinszky, A., Perdew, J. P., Csonka, G. I., Vydrov, O. A., & Scuseria, G. E., (2007). Density Functionals that are one- and two- are not always many-electron self-interaction-free, as shown for H2+, He2+, LiH+, and Ne2+. *J. Chem. Phys.*, *126,*104102.

62. Haunschild, R., Henderson, T. M., Jimenez-Hoyos, C. A., & Scuseria, G. E., (2010). Many-electron self-interaction and spin polarization errors in local hybrid density functionals. *J. Chem. Phys.*, *133,*134116.

63. Mori-Sanchez, P., Cohen, A. J., & Yang, W. T., (2008). Localization and delocalization errors in density functional theory and implications for band-gap prediction. *Phys. Rev. Lett.*, *100*, 146401.

64. Mori-Sanchez, P., Cohen, A. J., & Yang, W. T., (2009). Discontinuous Nature of the Exchange-Correlation Functional in Strongly Correlated Systems. *Phys. Rev. Lett.*, *102,*066403.

65. Cohen, A. J., Mori-Sanchez, P., & Yang, W. T., (2008). Insights into current limitations of density functional theory. *Science.*, *321*, 792–794.

66. Cohen, A. J., Mori-Sanchez, P., & Yang, W. T., (2012). Challenges for density functional theory, *Chem. Rev.*, *112*, 289–320.

67. Cuevas-Saavedra, R., Chakraborty, D., Rabi, S., Cardenas, C., & Ayers, P. W., (2012). Symmetric Non local Weighted Density Approximations from the Exchange-Correlation Hole of the Uniform Electron Gas. *J. Chem. Theory Comp.*, *8*, 4081–4093.

68. Parr, R. G., Pearson, R. G., (1983). Absolute hardness: companion parameter to absolute electronegativity. *J. Am. Chem. Soc.,105*, 7512–7516.

69. Pearson, R. G., (2009). *The Hardness of Closed Systems*, in: P. K. Chattaraj (Ed.) Chemical reactivity theory: A density functional view, CRC Press, Boca Raton., pp. 155–162.

70. Pearson, R. G., (2005). Chemical hardness and density functional theory. *J. Chem. Sci.*, *117*, 369–377.

71. Pearson, R. G., (1997). *Chemical Hardness*, Wiley-VCH, Weinheim, Germany.

72. Ayers, P. W., (2007). The physical basis of the hard/soft acid/base principle. *Faraday Discuss*, *135*, 161–190.

73. Parr, R. G., Donnelly, R. A., Levy, M., & Palke, W. E., (1978). Electronegativity: the density functional viewpoint. *J. Chem. Phys.*, *68*, 3801–3807.

74. Parr, R. G., & Pariser, R., (2013). The parameter I - A in electronic structure theory, in: S. K. Ghosh, P. K. Chattaraj (Eds.) *Concepts and Methods in Modern Theoretical Chemistry: Electronic Structure and Reactivity.* CRC Press. Boca Raton, pp. 431–440.

75. Fuentealba, P., & Parr, R. G., (1991). Higher-order derivatives in density-functional theory, especially the hardness derivative. *J. Chem. Phys., 94,* 5559–5564.
76. Morell, C., Grand, A., Toro-Labbé, A., & Chermette, H., (2013). Is hyper-hardness more chemically relevant than expected? *J. Mol. Model., 19,* 2893–2900.
77. Geerlings, P., & De Proft, F., (2008). Conceptual DFT: the chemical relevance of higher response functions. *PCCP., 10,* 3028–3042.
78. Parr, R. G., & Yang, W. T., (1984). Density functional approach to the frontier-electron theory of chemical reactivity. *J. Am. Chem. Soc., 106,* 4049–4050.
79. Yang, W. T., Parr, R. G., & Pucci, R., (1984). Electron density, Kohn-Sham frontier orbitals, and Fukui functions. *J. Chem. Phys., 81,* 2862–2863.
80. Ayers, P. W., abd Levy, M., (2000). Perspective on "density functional approach to the frontier-electron theory of chemical reactivity" by Parr, R. G., & Yang, W. *Theor. Chem. Acc., 103,* 353–360.
81. Ayers, P. W., Yang, W. T., Bartolotti, L. J., (2009). Fukui Function, in: P. K. Chattaraj (Ed.) *Chemical Reactivity Theory: A Density Functional View,* CRC Press, Boca Raton, pp. 255–267.
82. Morell, C., Grand, A., & Toro-Labbé, A., (2005). New dual descriptor for chemical reactivity. *J. Phys. Chem. A., 109,* 205–212.
83. Ayers, P. W., Morell, C., De Proft, F., & Geerlings, P., (2007). Understanding the Woodward-Hoffmann rules using changes in the electron density. *Chem, Eur. J.,13,* 8240–8247.
84. Morell, C., Grand, A., & Toro-Labbé, A., (2006). Theoretical support for using the $\Delta f(r)$ descriptor. *Chem. Phys. Lett., 425,* 342–346.
85. Parr, R. G., & Bartolotti, L. J., (1982). On the geometric mean principle of electronegativity equalization. *J. Am. Chem. Soc., 104,* 3801–3803.
86. Sanderson, R. T., (1951). An interpretation of bond lengths and a classification of bonds. *Science., 114,* 670–672.
87. Fuentealba, P., Cárdenas, C., (2013). On the exponential model for energy with respect to number of electrons. *J. Mol. Model., 19,* 2849–2853.
88. Morales, J., & Martinez, T. J., (2001). Classical fluctuating charge theories: The maximum entropy valence bond formalism and relationships to previous models. *J. Phys. Chem. A.,105,* 2842–2850.
89. Dirac, P. A. M., (1947). *The Principles of Quantum Mechanics.* Oxford University Press. New York.

CHAPTER 5

CHEMICAL EQUALIZATION PRINCIPLES AND THEIR NEW APPLICATIONS

SAVAŞ KAYA,[1] CEMAL KAYA,[1] and IME BASSEY OBOT[2]

[1]Cumhuriyet University, Faculty of Science, Department of Chemistry, Sivas, 58140, Turkey

[2]Centre of Research Excellence in Corrosion, Research Institute, King Fahd University of Petroleum and Minerals, Dhahran 31261, Kingdom of Saudi Arabia

CONTENTS

ABSTRACT

In this chapter, some new applications of chemical equalization principles related to hardness, electronegativity, electrophilicity, and nucleophilicity are presented. Molecular hardness, molecular electronegativity, lattice energy, and bond force constant equations introduced in the chapter are very useful in chemical world.

5.1   INTRODUCTION

Chemical reactivity can be simply defined as the tendency of a chemical matter to undergo chemical reaction with another chemical matter. It is well known that the understanding of the nature of chemical interactions and the prediction of chemical reactivity of atoms, ions, or molecules are some of the challenging issues in chemistry. The principal aim of theoretical chemistry is to develop rules to explain chemical reactivity and chemical reactions. Within the framework of this aim, theoretical chemists have produced many theories, electronic structure principles, and semi-empirical equations. Especially, conceptual density functional theory (CDFT) [1, 2] that provides great contributions in the advancement of quantum chemistry has been quite successful to achieve the above-mentioned aim.

Chemical concepts such as chemical potential, chemical hardness, softness, electronegativity, electrophilicity, and nucleophilicity have important applications in the topics like prediction of reaction mechanisms and analysis of chemical reactions. The concept of electronegativity was introduced by Pauling [3] as "the power of an atom in a molecule to attract electrons to itself." Pauling determined the electronegativity values of elements using thermochemical data, and after her this study, the electronegativity concept has become one of the most widely used concepts in chemistry. It is apparent that the properties of constituent atoms can be used to determine the characteristics of the molecules formed. From the light of the idea that molecular properties can be estimated using atomic data, Sanderson put forward the electronegativity equalization principle (EEP) [4] and proposed the geometric mean principle for the calculation of electronegativities of molecules. Since then, this principle has been used in many topics such as calculation of atomic charges and determination of reactive sites of molecules, calculation of electronegativities of molecules, drug design, corrosion, material design, conductivity, and analysis of chemical reactions.

In 1963, there was a novel development in acid-base chemistry. R. G. Pearson [5, 6] put forward the chemical hardness concept that is currently an important parameter in the chemical reactivity theory in connection with generalized acid-base reactions of Lewis, and he introduced the hard and soft acid-base theory (HSAB). Then, the maximum hardness principle (MHP) [7, 8] was proposed by Pearson, and Parr and Chattaraj presented important proofs about the relationship between chemical hardness and stability. It is apparent that there are thousands of publications based on chemical hardness in the literature. One of them has been written by Dipankar Datta [9] who used the geometric mean principle to calculate the molecular hardness assuming that the chemical hardness of constituent atoms becomes equalized during molecule formation. In recent times, we derived two new formulas [10, 11] to calculate the chemical hardness of molecules with the help of the global hardness equalization principle and published some important applications based on this electronic structure principle in solid-state chemistry and physical inorganic chemistry. These new applications will be explained in detail in the following parts of this chapter.

In 2010, P. K. Chattaraj and his students proposed the electrophilicity equalization principle [12] specifying that electrophilicity also becomes equalized like electronegativity and hardness, and later, we [13] proposed the nucleophilicity equalization principle in accordance with the results of Chattaraj. Although many discussions about the validity of the aforementioned chemical equalization principles are available, we showed in our several articles that these principles are very useful especially for small molecules.

In this chapter, we introduce some new applications proposed to calculate the chemical hardness, electronegativity, electrophilicity and nucleophilicities of molecules; lattice energies of inorganic ionic solids; bond force constants of diatomic molecules; and electronic structure principles such as EEP, global hardness equalization principle, electrophilicity equalization principle, nucleophilicity equalization principle, and MHP. We strongly believe that the new developments described in this chapter will be useful for our colleagues and students.

5.2   CHEMICAL REACTIVITY INDICES IN CONCEPTUAL DENSITY FUNCTIONAL THEORY

The main objective of a general theory of chemistry should be to explain the relative stabilities and tendencies to undergo chemical change under certain conditions. Based on Hohenberg–Kohn theorems [14], density functional theory (DFT) focuses on electron density as a basic marker for chemical reactivity analysis in all atomic or molecular systems. Parr and Pearson conducted important studies to quantitatively describe the chemical reactivity descriptors such as chemical hardness (η), electro-negativity (χ), and chemical potential (μ) with the help of DFT. First, Parr put forward the concept of the slope of the energy (E) versus the number of electrons (N) at a constant external potential, $v(r)$. Then, Parr defined the electronegativity as the negative of chemical potential [1,15].

$$\mu = \left(\frac{\partial E}{\partial N} \right)_{v(r)} \tag{1}$$

$$\chi = -\mu \tag{2}$$

At this stage, it is important to note that this relation given by Eq. (2) has been criticized by Pearson [16] and Allen [17]. Both Pearson and Allen proposed that Pauling's electronegativity and the chemical potential should be regarded as two distinct properties. Later on, similar criticisms related to this topic were made by Politzer [18] and his co-workers, and they stated that the validity of $\chi=-\mu$ is not the case.

At first, chemical hardness introduced by Pearson in the 1960s in con-nection with a study related to Lewis acid-bases was used to provide quali-tative justifications to exchange reactions, to predict the products that will form in a Lewis acid-base reaction, and to explain the geochemistry of elements (for instance, why hard acids such as Fe^{3+} and Mg^{2+} tend to occur as oxides, carbonates, and fluorides, while soft acids such as Hg^{2+} and Pb^{2+} tend to occur as sulfides, selenides, and tellurides). In the following years, by using DFT, Pearson and Parr defined the chemical hardness as the sec-ond derivative of electronic energy (E) with respect to number of electron (N) at a constant external potential $v(r)$. Furthermore, Pearson presented the softness (σ) as multiplicative inverse of the chemical hardness [19].

$$\eta = \left(\frac{\partial^2 E}{\partial N^2} \right)_{\upsilon(r)} \tag{3}$$

$$\sigma = 1/\eta \tag{4}$$

Applying the finite difference approximation to Eq. (1) and Eq. (3) and assuming that there is a quadratic relationship between E and N, Pearson and Parr presented the following mathematical formulations that are associated with the first vertical ionization energy (I) and the first vertical electron affinity (A) of chemical species for the calculation of electronegativity and chemical hardness, respectively [20, 21].

$$\chi = \frac{I + A}{2} \tag{5}$$

$$\eta = I - A \tag{6}$$

Ionization energies and electron affinities of chemical species can be predicted by Koopman's theorem [22]. This theorem states that the negative values of highest occupied and lowest unoccupied molecular orbital energies correspond to ionization energy and electron affinity, respectively. Namely, this theory is quite important in terms of the prediction through molecular orbital (MO) theory of chemical reactivity descriptors based on DFT. Within the framework of Koopman's theorem, in 1986, Pearson gave the electronegativity and hardness as follows and noted that a hard species has a large HOMO-LUMO gap and a soft species has a small HOMO-LUMO gap.

$$\chi = \frac{-(E_{HOMO} + E_{LUMO})}{2} \tag{7}$$

$$\eta = E_{LUMO} - E_{HOMO} \tag{8}$$

The terms electrophile and nucleophile introduced by Ingold [23] are taken into account in many studies related to reaction mechanisms by organic chemists. In organic chemistry, an electrophile is a chemical species that is bonded to a nucleophile taking an electron pair in a chemical reaction. Electrophilicity [24] may be stated as the measure of reactivity toward attracting electrons from a nucleophile of a chemical species. Parr and co-workers proposed an electrophilicity index [25] considering the electron transfer process between an electrophile that is immersed into a sea of electrons and a sea of free electrons at zero temperature and zero chemical potential. According to Parr's study, the electrophilicity index (ω) is given by the following equation in accordance with the suggestion of Maynard and co-workers [26].

$$\omega = \frac{\mu^2}{2\eta} = \frac{\chi^2}{2\eta} \tag{9}$$

The prediction of electron donating or accepting abilities of chemical species is a key part of reactivity analysis because fundamental particles having the most major role in chemical reactions are electrons. One of the useful studies in this topic was published by Gazquez [27]. He defined the electron accepting, ω^+, and electron donating, ω^-, capabilities as:

$$\omega^+ = \frac{A^2}{2(I - A)} \tag{10}$$

$$\omega^- = \frac{I^2}{2(I - A)} \tag{11}$$

where ω^+ is the measure of the propensity of a given system to accept charge and ω^- represents the propensity to donate charge. It is important to note here that a greater ω^+ value corresponds to a better capability of accepting charge, whereas a smaller value of ω^+ of a system makes it a better electron donor.

In recent times, considering Parr's electrophilicity index and examining several molecular systems, Kiyooka [28] presented a parabolic correlation between electrophilicity and nucleophilicity and presented the intrinsic reactivity index (IRI) as IRI=2ε to identify the relationship between chemical hardness and chemical potential. Here, ε is the nucleophilicity. With the help of simplest approach proposed for the evaluation of chemical potential and chemical hardness in terms of the frontier orbital energies, namely considering Koopman's theorem, Kiyooka proposed that the nucleophilicity (ε) and IRI values of molecules can be calculated by the following equations, respectively.

$$\varepsilon = \frac{\mu}{\eta} - \frac{1}{2}\frac{E_{LUMO} + E_{HOMO}}{E_{LUMO} - E_{HOMO}} \tag{12}$$

$$IRI = \frac{E_{LUMO} + E_{HOMO}}{E_{LUMO} - E_{HOMO}} \tag{13}$$

Shortly after, this proposal of Kiyooka about electrophilic and nucleophilic reactivity was criticized by Chamorro [29]. By analyzing with caution the results obtained using the IRI index, Chamorro noted the limitations of the IRI model and showed that the results obtained using the IRI index may not always be compatible with experimental electrophilicity and nucleophilicity scales.

5.3 ELECTRONEGATIVITY EQUALIZATION PRINCIPLE

Electronegativity is one of the chemical reactivity descriptors that is most widely used to understand the nature of chemical interactions. This concept introduced first by Pauling is known as a measure of an atom in a molecule to attract electronic charge. It is apparent that many physical and chemical properties of atoms and molecules are closely related to this concept.

The paradigm that molecular characteristics can be determined with the help of the properties of constituent atoms is exceptionally useful in chemistry. In this sense, a significant development in the electronegativity concept was provided by Sanderson's EEP. According to this principle, when two or more atoms with initially different electronegativities combine chemically, the electronegativities of the said atoms become equalized as a consequence of electron transfer among them. The equalized value gives the electronegativity of the formed molecule (χ_M) and is equivalent to the geometric mean of the electronegativity values of constituent atoms, as given in Eq. (14) [4].

$$\chi_M = (\prod_{i=1}^{N} \chi_i)^{1/N} \tag{14}$$

where χ_M is the electronegativity of a molecule, χ_i ($i=$ *1, 2, 3,....N*) represents the isolated atom electronegativity, and N is the total number of atoms in any molecule.

As is known, the group electronegativity concept is an important tool in structure and reactivity studies. Especially, in organic chemistry, the electronegativity of functional groups is considered significantly to predict the reaction mechanisms and inductive effects of functional groups. In the literature, many methods proposed to calculate the electronegativities of functional groups and molecules based on Sanderson's electronegativity equalization principle are available. Group electronegativity prediction methods widely used are given below in detail.

Iczkowski and Margrave [30] defined the electronegativity as the derivative of the energy with respect to charge, and then, Hinze et al. [31] introduced the orbital electronegativity concept. It is important to note that the DFT definition of electronegativity (vide Eq. (1) and Eq. (2)) show similarity with these earlier electronegativity studies. In the light of studies of scientists mentioned above, electronegativity is given as:

$$\chi = a + 2b\delta \tag{15}$$

where $a=(I+A)/2$ and $b=(I-A)/2$, I and A are valence-state ionization energy and electron affinity values of atoms, respectively, and δ is partial charge.

For the estimation of group electronegativities and partial charges of atoms, Huheey [32] suggested a simple scheme based on Iczkowski and Margrave study and Sanderson's EEP. He used the following expressions to estimate the partial charges of atoms that form these substituents and the electronegativities of AB_n type substituents:

$$\chi_{AB_n} = \chi_A = a_A + 2b_A\delta_A \tag{16}$$

$$\chi_{AB_n} = \chi_B = a_B + 2b_B\delta_B \tag{17}$$

$$\delta_A + n\delta_B = q \tag{18}$$

Bratsch [33] first developed a simple method for partial charge estimation of atoms through the application of Sanderson's EEP considering Huheey's observation, and subsequently, he proposed the following equation to calculate the group electronegativities in Pauling units.

$$\chi_G = \frac{N+q}{\sum\left(\dfrac{\upsilon}{\chi}\right)} \tag{19}$$

In this equation, χ_G represents the group electronegativity, $N=\Sigma(\upsilon)$ is the number of atoms in the group formula, q is the charge of species, and χ is the Pauling electronegativity of pre-bonded or isolated atoms in the species.

Orbital electronegativity definition states that the electronegativities of all orbitals on an atom are different from each other. A comprehensive study to calculate the orbital electronegativity was conducted by John Mullay [34]. Mullay showed that the electronegativity of an orbital i on atom A can be given as:

$$\chi_{A,i}(\delta_{A,i}) = \chi_{A,i}{}^0\left(1+0.5\sum_{j\neq i}\delta_{A,j} +1.5\delta_{A,i}\right) \tag{20}$$

Here, $\chi_{A,i}{}^0$ represents the neutral atom electronegativity of orbital i on atom A, $\delta_{A,i}$ is the charge on orbital i, and $\delta_{A,j}$ is the charges of other orbitals on atom A. Total charge (δ_A) on atom A is equal to sum of the charges in each orbital.

$$\delta_A = \sum_k \delta_{A,k} \tag{21}$$

In any molecule, the electronegativities of both atomic orbitals involved in the bond are equalized. For every bond in this molecule between A and B atoms, we can write the following equations with the help of Eqs. (20)

and (21). Using these equations given below, Mullay predicted the group electronegativities of functional groups and molecules.

$$\chi_A(\delta_{A,i}) = \chi_B(\delta_{B,i}) \tag{22}$$

$$\delta_{A,i} + \delta_{B,i} = 0 \tag{23}$$

5.3.1   A NEW EQUATION FOR THE CALCULATION OF GROUP ELECTRONEGATIVITY

In recent times, considering the relationship with charge of the electronic energy for atoms, Sanderson's EEP and DFT definition of the electronegativity, Kaya et al. [35] derived a new equation by which group electronegativity can be calculated from ionization energies (I) and electron affinities (A) of atoms that constitute the group. The important points about the derivation and features of this equation are given in detail below. Deriving the above-mentioned equation, actually, Kaya et al. combined the characteristics of previous group electronegativity equations.

Assuming that the energy of neutral atom is zero, if the relationship with charge of atomic energy is examined, a curve such as that in Figure 5.1 is obtained. This parabolic correlation between atomic energy and charge can be given as:

$$E(A) = a\delta + b\delta^2 + c\delta^3 + d\delta^4 + ... \tag{24}$$

where E is the total energy of A atom, δ is the charge of the atom, and a, b, c, and d are coefficients. It is important to note that a, b, c, and d coefficients can be determined considering successive ionization energy and electron affinity reactions and assuming that the energy of neutral atom is zero. As mentioned above, according to Iczkowski and Margrave, the electronegativity is defined as the first derivative with respect to charge of total energy.

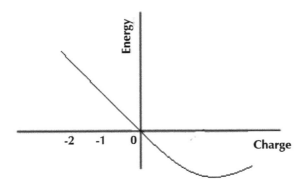

FIGURE 5.1   Atomic energy change with charge for any atom.

$$(\partial E \,/\, dq) = a + 2b\delta + 3c\delta^2 + 4d\delta^3 + \ldots\ldots\ldots \tag{25}$$

To define the electronegativity, Iczkowski and Margrave used only the first two terms of Eq. (25). Thus, they presented the electronegativity of atoms as:

$$(\partial E \,/\, dq) - a + 2b\delta \tag{26}$$

where E is the total energy of the atom, which is obtained from ΣI or ΣA, and I and A are the ionization energy and electron affinity of the atom, respectively. In Eq. (9), if δ is taken (+1), E will be the energy of the (+1) cation energy or first ionization energy. Likewise, for $\delta = -1$, the energy will be negative of the first electron affinity (note that the definition of electron affinity does not follow the usual thermodynamic convention in that a positive electron affinity is exothermic). This argument leads to the following relations:

$$I = a + b \tag{27}$$

$$-A = -a + b \tag{28}$$

From these equations, the following relations are obtained:

$$a = (I + A)/2 \qquad (29)$$

$$b = (I - A)/2 \qquad (30)$$

It can be clearly seen from Eq. (25) that a is equal to the electronegativity of neutral atom, and this parameter is similar to Mulliken electronegativity formula. In this stage, it is important to note that Mulliken used the valence-state ionization energy and electron affinity values of atoms to calculate the electronegativity. However, the parameter is calculated from ground-state ionization energy and electron affinities of atoms. Actually, this parameter represents the absolute electronegativity of any atom.

With the help of Eq. (26), for each atom in a chemical species (molecule or group) that contains N atoms, we can write

$$\chi_1 = a_1 + 2b_2\delta_1$$

$$\chi^2 = a_2 + 2b_2\delta_2$$

$$\chi^3 = a_3 + 2b_3\delta_3$$

$$\vdots \qquad \vdots \qquad \vdots$$

$$\chi_N = a_N + 2b_N\delta_N$$

On the basis of Sanderson's principle, one can write the following relation:

$$\chi_1 = \chi_2 = \chi_3 = \dots\dots\dots\dots = \chi_N = \chi_G \qquad (31)$$

where χ_G is the electronegativity of the species. It is known that the sum of partial charges of atoms in a molecule is equal to charge of this molecule.

$$\delta_1 + \delta_2 + \delta_3 + \dots\dots\dots\dots + \delta_N = q \qquad (32)$$

The partial charge of any atom in the species can be given as follows:

$$\delta_i = \frac{\chi_G - a_i}{2b_i} = \frac{\chi_G}{2b_i} - \frac{a_i}{2b_i} \tag{33}$$

From Eqs. (32) and (33), one can obtain the following equation:

$$\chi_G = \frac{\sum_{i=1}^{N}\left(\dfrac{a_i}{b_i}\right) + 2q}{\sum_{i=1}^{N}\left(\dfrac{1}{b_i}\right)} \tag{34}$$

Furthermore, by inserting Eqs. (29) and (30) into Eq. (34), we obtained the following equation that allows direct calculation of electronegativity from the ground-state ionization energies and electron affinities of the constituent atoms.

$$\chi_G - \frac{\sum_{i=1}^{N}\left(\dfrac{I_i + A_i}{I_i - A_i}\right) + 2q}{\sum_{i=1}^{N}\left(\dfrac{2}{I_i - A_i}\right)} \tag{35}$$

where I_i and A_i are the ionization energy and electron affinity of i-th atom, respectively, and q is the charge of chemical species considered.

By using these equations, both absolute electronegativity and Mulliken and Pauling electronegativities of functional groups and molecules can be calculated. The absolute electronegativity concept was first used by Pearson, and he used the ground-state ionization energy and electron affinities to predict the absolute electronegativities of chemical species through Eq. (5). If the ground-state ionization energy and electron affinities of atoms are used in Eq. (35), we obtain the absolute electronegativity of group. For the Mulliken electronegativity, it is necessary to use valence-state ionization energy and electron affinity in Eq. (35). By using the substitutions, $a_i = \chi_i$ and $b_i = \chi_i/2$, we can find Pauling electronegativity by Eq. (34). The values calculated in this way are equal to those obtained from Bratsch equation. The results of group electronegativity equation derived by Kaya et al. are compatible with the results obtained using other methods existing in the literature. In the Table 5.1, we presented the results obtained using Sanderson's geometric mean equation and Eq. (35) for some selected molecules and functional groups.

In Table 5.2, the electronegativity values calculated using Eq. (5) and Eq. (35) are compared. As is known, the electronegativities of charged molecules cannot be calculated using Sanderson's geometric mean equation. The advantage of the new equation proposed by Kaya et al. with respect to Sanderson's equation is that it can be used to calculate the electronegativities of charged molecules. The values calculated using Eq. (35) for some charged functional groups are given in Table 5.3.

TABLE 5.1   Comparison of Electronegativities Calculated via Some Methods (eV)

Molecule	Kaya	Sanderson	Molecule	Kaya	Sanderson
CH_3	6.90	6.94	PBr_3	7.15	7.04
CH_3CH_2	6.87	6.90	PCl_3	7.67	7.52
CH_3NO_2	7.14	7.16	$POCl_3$	7.63	7.53
CH_2F	7.63	7.62	CH_3I	6.87	6.90
CHFCl	7.91	7.90	HNO_3	7.43	7.42
CHClBr	7.37	7.30	H_2O	7.30	7.30
SiH_3	6.24	6.48	H_2S	6.76	6.84
SiF_3	8.10	8.56	NH_3	7.20	7.20
NF_2	9.39	9.25	CO_2	7.06	7.09
NCl_2	8.06	7.95	C_5H_5N	6.71	6.76
NH_2	7.22	7.22	C_6H_5SH	6.62	6.67
NHOH	7.19	7.30	$HCONH_2$	7.07	7.09
PH_2	6.56	6.62	CH_4	6.96	6.99
PCl_2	7.47	7.29	CS_2	6.23	6.23
PF_2	8.40	8.47	COS	6.59	6.65
BF_2	7.56	7.74	SO_2	6.98	7.07
BCl_2	6.82	6.66	SO_3	7.11	7.18

TABLE 5.2   Electronegativities of Some Molecules

Molecules	I (eV)	A (eV)	Electronegativity	
			Eq. (5)	Eq. (35)
IBr	9.79	2.55	6.17	7.15
S_2	9.40	1.66	5.53	6.22
Br_2	10.56	2.60	6.58	7.59
Cl_2	11.48	2.40	6.94	8.30
O_2	12.06	0.44	6.25	7.54
CS_2	10.08	1.00	5.54	6.23
COS	11.18	0.46	5.82	6.59
SO_2	12.34	1.05	6.70	6.98
O_3	12.67	1.82	7.25	7.54
N_2O	12.89	1.47	7.18	7.41
PBr_3	9.85	1.60	5.73	7.15
PCl_3	9.91	0.80	5.34	7.67
$POCl_3$	11.40	1.40	6.40	7.63
CIl_3I	9.54	0.20	4.87	6.87
SO_3	11.00	1.70	6.35	7.11
C_2H_2	11.41	0.43	5.92	6.67
HNO_3	11.03	0.57	5.80	7.43
CS	11.71	0.20	5.90	6.24
CO	14.00	-1.80	6.10	6.84
H_2	15.40	-2.00	6.70	7.18

5.3.2   CALCULATION OF PARTIAL CHARGES OF ATOMS IN MOLECULES BASED ON ELECTRONEGATIVITY EQUALIZATION PRINCIPLE

Difference in electronegativities of atoms in a molecule causes to transfer electrons between them. As a result of electron transfer between atoms, electronegative atoms have negative partial charges and electropositive atoms have positive partial charges. Sanderson who put forward the EEP proposed a formalism to predict the atomic charges in a molecule based on the change in electronegativity of an atom from isolated atom value to its equalized value after the molecule formation.

TABLE 5.3   Electronegativities of Some Common Ionic Functional Groups (eV)

Group	χ_G	Group	χ_G
OH	1.12	ClO_3^-	4.94
CO_3^{2-}	1.40	CH_3CO^-	4.94
SH^-	1.56	ClO_4^-	5.43
AsH_2^-	2.65	CF_3^-	5.90
ClO^-	2.68	BF_4^-	6.28
PH_2^-	2.68	$C_6H_5^+$	7.64
SO_4^{2-}	2.74	CH_3CO^+	8.82
NH_2^-	2.77	SiH_3^+	8.87
NCS^-	3.04	$NH_2NH_3^+$	9.11
NO_2^-	3.19	CH_3^+	9.91
CH_3^-	3.91	H_3O^+	10.44
ClO_2^-	4.15	CF_3^+	12.27

The charge distribution is one of the fundamental tools in the explanation of chemical structures and properties of molecules. In addition, by analyzing the partial charges of atoms in a molecule, its reactive sites that play a major role for electrophilic and nucleophilic attack can be determined.

Electronegativity equalization has important quantitative applications. It is seen from the past publications in the literature that different formalisms which use the EEP are successful in the determination of partial charges of atoms in molecules. The first attempts in this regard were made by Huheey, Parr, and Pearson. In the determination of partial charges of atoms in a chemical species, Kaya et al. used Eq. (33). The values calculated by this equation are given in Table 5.4 together with the results obtained by the method of Huheey. It should be noted that there is a remarkable agreement between the results of both methods.

In 1980, Gasteiger and Marsili [36] presented PEOE (partial equalization of orbital electronegativity) method for the rapid calculation of atomic charges in molecules. This method that was initially used to calculate the partial charges of atoms in σ-bonded and nonconjugated π systems was extended later to conjugated systems by Gasteiger, Marsili, and Saller. Mortier [37] developed an electronegativity equalization method for par-

TABLE 5.4   Comparison of Partial Charges Calculated Using Some Methods

Molecule	Atom	Kaya	Huheey	Molecule	Atom	Kaya	Huheey
NaF	Na	+0.41	+0.43	KCl	K	+0.45	+0.46
	F	−0.41	−0.43		Cl	−0.45	−0.46
CaCl$_2$	Ca	+0.48	+0.58	SrBr$_2$	Sr	+0.48	+0.58
	Cl	−0.24	−0.29		Br	−0.24	−0.29
CF$_4$	C	+0.30	+0.24	SiI$_4$	Si	+0.23	+0.07
	F	−0.07	−0.06		I	−0.06	−0.02
NF$_3$	N	0.16	+0.09	H$_2$S	H	−0.03	+0.02
	F	−0.05	−0.03		S	+0,06	−0.04
HF	H	+0.12	+0.17	IF	I	+0.17	+0.15
	F	−0.12	−0.17		F	−0.17	−0.15
BrCl	Br	+0.04	+0.05	BF$_4^-$	B	⁺0.23	⁺0.14
	Cl	−0.04	−0.05		F	−0.31	−0.29
OH⁻	H	−0.48	−0.46	CH$_3$O⁻	C	−0.16	−0.22
	O	−0.52	−0.54		H	−0.20	−0.16
					O	−0.24	−0.30

tial charge calculations. This method derives a useful equation to calculate the effective electronegativity of an atom in a molecule, and considering DFT definitions of electronegativity and chemical potential, it is quite successful in terms of the accurate prediction of atomic charges. Another successful method for the calculation of the charge distribution in large molecules was proposed by Yang and Wang [38]. Within the framework of Sanderson's EEP and DFT, they derived useful equations to predict the effective electronegativities of an atom and a bond in a molecule. The charge distributions calculated by Yang's method are compatible with the results of *ab initio* quantum chemical calculations. Although group electronegativity methods introduced above (Kaya, Huheey, Mullay, Bratsch, and Sanderson) are useful for the determination of partial charges of atoms in small molecules, the methods presented by Mortier and Yang can be successfully applied to large molecular systems. In addition, it should be noted that Ionescu and co-workers [39] used the electronegativity equalization principle for the rapid calculation of accurate atomic charges in proteins in recent years.

In 1998, Smith [40] presented a simple scheme for the calculation of fractional atomic charges and group electronegativities and showed that the charge distributions obtained within the framework of electronegativity equalization can be used to rationalize the relative stabilities of primary, secondary, and tertiary carbocations. To calculate the electronegativity of a group (χ_G) $AB_1B_2...B_i...B_n$, Smith derived the following equation based on the prebonded electronegativities (χ^0) in Pauling units. (It is important to note that the correlation between χ and ξ is given as $\xi = \chi^1$.)

$$\xi_G = (n+2)^{-1}[2\xi_A^0 + \sum \xi_{B_i}^0] \qquad (36)$$

For instance, the Pauling electronegativities of H, C, and Br atoms are 2.1, 2.5, and 2.8, respectively. If so, ξ values for these atoms are 0.476, 0.4, and 0.357, respectively. The electronegativity of CH_2Br is calculated as follows in the light of Eq. (36).

$$\zeta_{CH_2Br} = (2+3)^{-1}\left[2\times0.4+(2\times0.476+0.357)\right] = 0.422$$

$$f_k = \left(\frac{\partial \rho(\vec{r})}{\partial N}\right)_{\upsilon(\vec{r})}$$

According to Bratsch, fractional charge on an atom A in a molecule can be determined using the following equation. In this equation, χ_A and χ_A^0 represent the equalized electronegativity value and isolated atom electronegativity value for atom A, respectively.

$$q_A = (\chi_A - \chi_A^0)/\chi_A^0 \qquad (37)$$

Smith [40] considered the equation given above presented by Bratsch to determine the partial charges of atoms in a molecule. He derived a new formula to calculate the electronegativity of an atom in a molecule in terms of the prebonded electronegativities and determined the atomic fractional charges considering the results obtained using this new formula and Eq. (37). The equation derived by Smith to calculate in Pauling units the electronegativity of an atom (χ_a) in a molecule is given as:

$$\xi_i = (m_i + 1)^{-1}(\xi_i^0 + \sum_{j \neq i} b_j \xi_j^0) \tag{38}$$

where m_i is the coordination number of the atom considered and the coefficients b_j are determined with the help of the following expressions.

$$b_j^{-1} = (m_1 + 1)(m_2 + 1)......(m_j + 1) \tag{39}$$

$$\sum b_j = m_i \tag{40}$$

where m_j represents the coordination number of j-th atom in a chain leading from i to j. In this chain, the atoms are numbered from 1 to j and atom 1 is directly bonded to i.

5.4 CHEMICAL HARDNESS EQUALIZATION PRINCIPLE

Chemical hardness is an important property of chemical species and is defined as the resistance toward electron cloud polarization or deformation of chemical species. This concept has important applications in topics such as chemical reactivity analysis, solubility of matters, prediction of complex stabilities, and estimation of formed products in a chemical reaction. In the literature, popular electronic structure principles regarding quantum chemical descriptors such as electronegativity, hardness, and electrophilicity are available. One of them is the hardness equalization principle. As mentioned above, Sanderson put forward the EEP considering the relationship between electronegativity and charge. It is important to note that all such equalization processes about chemical reactivity descriptors are explained by means of the correlation between the said reactivity descriptor and charge. In 1986, Datta [9] proposed that hardness of all atoms in a molecule during molecule formation is equalized like their electronegativities, and he used the following equation based on the assumption that the chemical hardness is equalized globally to calculate the molecular hardness. In this equation, N is the total number of atoms in a molecule and η_i ($i=1,2, 3...N$) is isolated atom hardness value.

$$\eta_M = (\prod_{i=1}^{N}\eta_i)^{1/N} \qquad (41)$$

In addition to Datta's geometric mean equation, S. K. Ghosh and coworkers [41] first used Eq. (42) to calculate the softness of molecules (σ_M). In this study, molecular softness was described as the average of the atomic softness. Then, Ghosh [42] obtained and used Eq. (43) for the calculation of the molecular hardness considering the electron transfer energy formula derived by Pearson and Parr.

$$\sigma_M = \frac{1}{N}\sum_{i=1}^{N}\sigma_i \qquad (42)$$

where N is the total number of atoms in a molecule, σ_M is the softness of molecules, and σ_i ($i=1, 2, 3...N$) represents the isolated atom softness value.

$$\eta_{AB} = \frac{(\eta_A^2 + \eta_B^2 + 3\eta_A\eta_B)}{\eta_A + \eta_B} \qquad (43)$$

In this equation, η_A, η_B, and η_{AB} are the chemical hardness value of atoms A and B and the molecule AB, respectively.

Some authors mentioned that the chemical hardness is equilibrated locally and criticized the global hardness equalization principle. However, Ayers and Parr [43] confirmed that the local hardness is equal to the global hardness at every point in space by introducing the higher-order global softness and higher-order global hardness concepts. In the recent past, Ghosh and Islam [44-46] conducted important studies on whether there is hardness equalization principle for molecules similar to electronegativity equalization, and they derived useful algorithms from the light of the idea that the chemical hardness is equilibrated globally during molecule formation. The results obtained in mentioned studies belong to Ghosh and Islam [45] supported that the hardness equalization principle is a valid law in nature like the EEP. Ghosh and Islam [45] first defined the atomic hardness as a function of atomic radius ($\eta_a = f(r)$) and calculated the hardness of many elements in the periodic table. In the sequel of this study, they derived the following equation for the calculation of hardness of molecules relying upon the global hardness equalization principle and with the help of the idea that the molecular hardness is a function of the internuclear distance of the atoms ($\eta_M = f(R_{AB})$).

$$\eta_{AB} = C\left(\frac{14.4}{R_{AB}}\right) \tag{44}$$

where η_{AB} and R_{AB} are the molecular hardness (in eV units) and internuclear distance of atoms, respectively. It is important to note that R_{AB} is expressed in Angstrom units. In the equation, C is the constant depending on the fundamental nature of the hardness, e.g., bond type and steric factors, and its numerical value is approximately 0.75.

Kaya et al. derived two useful equations [10, 11] to calculate the molecular hardness based on the hardness equalization principle. The details about derivation of these equations are given below.

Kaya Molecular Hardness Equation 1: In the first stage of the study, we proposed the following equation, which is based on charges, ionization energies, and electron affinities of atoms for the calculation of atomic hardness (η_a).

$$\eta_a = 2b_a + a_a q_a \tag{45}$$

In this equation, a_a and b_a are parameters depending on ground-state ionization energy (I) and electron affinity (A) values of atom, respectively, and q_a represents the charge of atom. a_a and b_a parameters known as Mulliken-Jaffe parameters are described by the following equations. For some selected atoms, ionization energy, electron affinity, and a_a and b_a values are given in Table 5.5.

$$a_a = \frac{I + A}{2} \tag{46}$$

$$b_a = \frac{I - A}{2} \tag{47}$$

TABLE 5.5   Experimental Values for Some Selected Atoms

Atom	I	A	a	b
H	13.598	0.754	7.176	6.422
Li	5.392	0.618	3.005	2.387
Be	9.323	0.295	4.809	4.514
B	8.298	0.280	4.289	4.009
C	11.260	1.262	6.261	4.999
N	14.534	0.07	7.302	7.232
O	13.618	1.461	7.539	6.078
F	17.423	3.401	10.412	7.011
Na	5.139	0.548	2.843	2.295
Mg	7.646	0.541	4.093	3.552
Al	5.986	0.433	3.209	2.776
Si	8.152	1.390	4.771	3.381
P	10.487	0.747	5.617	4.870
S	10.360	2.077	6.218	4.141
Cl	12.968	3.613	8.290	4.677
K	4.341	0.501	2.421	1.920
Ca	6.113	0.024	3.068	3.044
Sc	6.561	0.188	3.374	3.186
Cr	6.767	0.666	3.716	3.050
Mn	7.434	-0.498	3.468	3.96
Fe	7.902	0.151	4.026	3.875
Ni	7.640	1.156	4.398	3.242
Cu	7.726	1.235	4.480	3.245
Zn	9.390	-0.490	4.450	4.940
As	9.789	0.814	5.301	4.487
Se	9.752	2.020	5.886	3.866
Cs	3.894	0.471	2.180	1.710
Br	11.814	3.363	7.588	4.225
I	10.451	3.059	6.755	3.696

Chemical hardness of an atom can be calculated by Pearson-Parr equation given as Eq. (6). It should be pointed out that Pearson-Parr equation is recovered when q_a=0 in our atomic hardness equation. Already, we noted above that equalization processes regarding chemical reactivity descriptors such as electronegativity, electrophilicity, and hardness are explained considering the relationship between the said reactivity descriptor and charge. Ghosh and Islam suggested that equalization processes do not work in the formation of homonuclear molecules. Within the framework of their suggestion, the hardness equalization principle can be expressed as "when a molecule or group is formed by atoms initially different in chemical hardness, the hardness of atoms in the said molecule or group will be equal as a consequence of electron transfer between them because chemical hardness is a charge-dependent property."

The following expressions for each atom in a chemical species (molecule or group) that contains N atoms can be written considering atomic hardness equation proposed by us.

$$\eta_1 = 2b_1 + a_1 q_1$$

$$\eta_2 = 2b_2 + a_2 q_2$$

$$\eta_3 = 2b_3 + a_3 q_3$$

$$\vdots \qquad \vdots \qquad \vdots$$

$$\eta_N = 2b_N + a_N q_N$$

According to the chemical hardness equalization principle mentioned above, the equalized chemical hardness value of atoms corresponds to the chemical hardness value of molecule (η_M). Furthermore, the sum of the partial charges of atoms in a molecule is equal to the charge of molecule (q_M). In that case, we can write the following mathematical correlations.

$$\eta_1 = \eta_2 = \eta_3 = \cdots\cdots\cdots = \eta_N = \eta_M \qquad (48)$$

$$q_1 + q_2 + q_3 + \cdots\cdots + q_N = q_M \qquad (49)$$

It is known that the electronic structure principles such as EEP and hardness equalization principle have been widely considered to calculate the partial charges of atoms in a molecule. In our study, the partial charge of any atom (q_i) in molecules can be calculated using the following equation.

$$q_i = \frac{\eta_M - 2b_i}{a_i} = \frac{\eta_M}{a_i} - \frac{2b_i}{a_i} \tag{50}$$

Combining Eqs. (49) and (50), we obtain the following equation.

$$\left(\frac{\eta_M}{a_1} - \frac{2b_1}{a_1}\right) + \left(\frac{\eta_M}{a_2} - \frac{2b_2}{a_2}\right) + \left(\frac{\eta_M}{a_3} - \frac{2b_3}{a_3}\right) + \ldots\ldots + \left(\frac{\eta_M}{a_N} - \frac{2b_N}{a_N}\right) = q_M \tag{51}$$

If we make necessary arrangements on Eq. (51), the molecular hardness equation given below is obtained.

$$\eta_M = \frac{(2\sum_{i=1}^{N} \frac{b_i}{a_i}) + q_M}{\sum_{i=1}^{N} \frac{1}{a_i}} \tag{52}$$

Kaya Molecular Hardness Equation 2: We already mentioned in the section that explains the EEP and its applications that there is a parabolic correlation as given in Eq. (25) between total electronic energy and charge of an atom. Initially, we obtained a new atomic hardness equation using only first three terms of Eq. (25) and DFT definition of chemical hardness (namely Eq. (3)).

$$E = aq + bq^2 + cq^3 \tag{53}$$

$$\eta_a = \left(\frac{\partial^2 E}{\partial q^2}\right) = 2b + 6cq \tag{54}$$

For any X atom, a, b, and c coefficients can be determined by considering successive ionization energy and electron affinity reactions of the said atom and assuming that the energy of neutral atom is zero. First ionization energy, second ionization energy, and first electron affinity reactions that will be considered to determine the coefficients are given below.

$$X \rightarrow X^+ + e^- \quad I_1$$

$$X^+ \rightarrow X^{2+} + e^- \quad I_2$$

$$X^- \rightarrow X + e^- \quad A_1$$

Considering Eq. (25), ionization energy, and electron affinity reactions given above for any X atom and the assumption that the energy of neutral atom is zero, we can now write the following expressions.

$$E\left(X^+\right) - E\left(X\right) = \left[a+b+c\right] - \left[0\right] = a+b+c = I_1$$

$$E\left(X^{2+}\right) - E\left(X^+\right) = \left[2a+4b+8c\right] - \left[a+b+c\right] = a+3b+7c = I_2$$

$$E\left(X\right) - E\left(X^-\right) = \left[0\right] - \left[a+b+c\right] = a+b+c = A_1$$

Performing necessary mathematical operations, we find the $2b$ and $6c$ as:

$$2b = I_1 - A_1$$

$$6c = I_2 - 2I_2 + A_1$$

Now, we can apply the atomic hardness equation in which its 2b and 6c parameters were obtained based on ionization energy and electron affinity values of atoms to a molecule containing N atoms as follows.

$$\eta(1) = 2b_1 + 6c_1 q_1 = \left[I(1)_1 - A(1)_1 \right] + \left[I(2)_1 - 2I(1)_1 + A(1)_1 \right] q_1$$

$$\eta(2) = 2b_2 + 6c_2 q_2 = \left[I(1)_2 - A(1)_2 \right] + \left[I(2)_2 - 2I(1)_2 + A(1)_2 \right] q_2$$

$$\eta(3) = 2b_3 + 6c_3 q_3 = \left[I(1)_3 - A(1)_3 \right] + \left[I(2)_3 - 2I(1)_3 + A(1)_3 \right] q_3$$

$$\vdots \qquad\qquad \vdots \qquad\qquad \vdots$$

$$\eta(N) = 2b_N + 6c_N q_N = \left[I(1)_N - A(1)_N \right] + \left[I(2)_N - 2I(1)_N + A(1)_N \right] q_N$$

With a similar reasoning to the derivation of Kaya molecular hardness equation 1 (please see Eqs. (48–51)), the molecular hardness equation given below based on first ionization energy, second ionization energy, and first electron affinity of the constituent atoms for the prediction of chemical hardness of a molecule is obtained.

$$\eta_M = \frac{\displaystyle\sum_{i=1}^{N} \left(\frac{I_{1_0} - E_{1_0}}{I_{2_0} - 2I_{1_0} + A_{1_0}} \right) + q_M}{\displaystyle\sum_{i=1}^{N} \left(\frac{1}{I_{2_0} - 2I_{1_0} + A_{1_0}} \right)} \tag{55}$$

It is apparent that Eq. (55) cannot be applied to molecules containing hydrogen atoms because second ionization energy for hydrogen is out of question. Making a simple approach, we extended Eq. (55) to calculate the molecular hardness of molecules containing hydrogen atoms. To obtain a specific 6c parameter for a hydrogen atom, first ionization energy (I_1), first electron affinity (A_1), and second electron affinity (A_2) reactions given below for the aforementioned atom are considered.

$$H \rightarrow H^+ + e^- \quad I_1(H)$$

$$H^- \rightarrow H + e^- \quad A_1(H)$$

$$H^{2-} \rightarrow H^- + e^- \quad A_2(H)$$

Assuming again that the energy of neutral atom is zero and considering Eq. (53), we can write the following statements.

$$I(H) = E(H^+) - E(H) = [a+b+c] - [0] = a+b+c$$

$$A_1(H) = E(H) - E(H^-) = [0] - [-a+b-c] = a-b+c$$

$$A_2(H) = E(H^-) - E(H^{2-}) = [-a+b-c] - [-2a+4b-8c] = a-3b+7c$$

If necessary mathematical operations are made, 2b and 6c parameters for the hydrogen atom are obtained as follows:

$$2b(H) = I_1(H) - A_1(H)$$

$$6c(H) = I_1(H_1) - 2A_1(H) - A_2(H)$$

At this stage, a simple approach can be applied for 6c values of hydrogen. It is known that noble gases have the most stable electron configuration because their sub-shells are filled. For this reason, electron affinities of noble gases are considered as zero in many articles. The second electron affinity of hydrogen A_2 (H) is the first electron affinity A (H⁻) of hydride ion (H⁻) at the same time. The hydride ion has noble gas configuration, and its electron affinity values can be accepted as zero like electron affinities of noble gases. By using this approach, 6c parameter for the hydrogen atom can be presented as:

$$6c(H) = I_1(H_1) - 2A_1(H)$$

With the help of these parameters obtained for hydrogen, Eq. (55) can be rewritten as:

$$\eta_M = \frac{\left[\sum\limits_{i=1}^{P}\left(\dfrac{I_{1_0} - A_{1_0}}{I_{2_0} - 2I_{1_0} + A_{1_0}}\right) + (N-P)\left(\dfrac{I_1(H) - A_1(H)}{I_1(H) - 2A_1(H)}\right)\right] + q_M}{\sum\limits_{i=1}^{P}\left(\dfrac{1}{I_{2_0} - 2I_{1_0} + A_{1_0}}\right) + \left(\dfrac{N-P}{I_1(H) - 2A_1(H)}\right)} \tag{56}$$

In both Eqs. (55) and (56), N and q_M are the total number of atoms in a molecule and charge of molecule, respectively. In Eq. (56), P is the number of atoms different from hydrogen and $(N$-$P)$ is the number of hydrogen atoms in the molecule.

Geometric mean equations proposed to calculate the global reactivity descriptors such as electronegativity and chemical hardness can be used for only neutral species. For this reason, the chemical hardness values of charged molecules cannot be predicted by Datta's geometric mean equation, but it is apparent that our new molecular hardness equations given above can be used for charged molecules. The calculated chemical hardness values for some selected molecules by using Datta's equation and new equations derived by us are given in Table 5.6. The results obtained show that there is a remarkable agreement between the results of all equations. Table 5.7 shows the calculated chemical hardness values using Eq. (52) for some charged chemical species. The concept of chemical hardness as one of the fundamental descriptors used in theoretical chemistry and condensed matter physics is not physically observable. In the literature, the chemical hardness values calculated using the experimental ionization energy and electron affinities of chemical species and Pearson-Parr formula given by Eq. (8) are accepted as experimental data. In Table 5.8, the experimental results and the results obtained using Eqs. (52) and (56) are compared. In recent years, ab initio methods have been widely used in the prediction of molecular properties and calculation of quantum chemical descriptors. To test the reliability and validity of molecular hardness equation derived by us (Eq. (52)), we also compared our results with the ab initio data calculated using Gaussian 09 Revision-A.02 at the Hartree-Fock

TABLE 5.6   Comparison of Molecular Hardness Values (eV) Calculated from Datta Equation and Present New Equations for Some Selected Molecules

Molecule	Kaya 1 (Eq. 52)	Kaya 2 (Eq. 56)	Datta (Eq. 41)
C_5H_5N	11.59	12.67	11.59
C_6H_5SH	11.08	10.42	11.07
CH_3CN	12.06	13.50	12.06
MgO	8.88	7.24	9.29
CaO	7.84	5.88	8.60
BeS	8.70	9.05	8.63
NH	13.64	14.53	13.64
CS_2	8.84	9.03	8.82
COS	10.94	9.81	10.02
NH_3	13.24	14.43	13.24
H_2O	12.62	12.60	12.62
SO_3	11.04	10.64	11.04
N_2O	13.70	14.53	13.65
SiH_3	10.82	8.32	10.95
SiF_3	10.96	10.68	11.68
CHFCl	11.38	10.68	11.39
CH_3CH_2	11.96	11.12	11.96
$N(CH_3)_2$	12.32	13.44	12.32
$Be(CH_3)_2$	11.60	8.57	11.68

TABLE 5.7   Calculated Chemical Hardness Values (eV) for Some Charged Functional Groups

Functional Group	Hardness	Functional Group	Hardness
CF_3^+	14.82	NH_2^-	10.98
SiH_3^+	12.42	PH_2^-	9.46
CH_3CO^+	12.85	NCS^-	8.54
CH_3^+	13.80	CH_3O^-	10.68
NO^+	17.03	NO_2^-	10.46
OH^+	16.20	$C_2H_3^-$	10.26
OH^-	8.84	CF_3^-	10.36

TABLE 5.8   Experimental and Calculated Chemical Hardnesses for Some Molecules

Molecule	I	A	η (experimental)	η (Eq. 52)	η (Eq. 56)
I_2	9.40	2.42	6.98	7.38	7.39
IBr	9.79	2.55	7.24	7.89	7.90
S_2	9.40	1.66	7.74	8.28	8.29
F_2	15.70	3.08	12.62	14.02	14.03
Cl_2	11.48	2.40	9.08	9.36	9.35
SO	10.00	1.13	8.87	9.84	9.60
CN	14.50	3.82	10.68	12.06	14.01
NH	13.10	0.38	12.72	13.64	14.53
CS_2	10.08	1.00	9.08	8.84	9.03
COS	11.18	0.46	10.72	10.94	9.81
SO_2	12.34	1.05	11.29	10.69	10.24
NH_2	12.80	0.78	12.02	13.38	14.47
N_2O	12.89	1.47	11.42	13.70	14.53
CH_3I	9.54	0.20	9.34	11.10	9.00
SO_3	11.00	1.70	9.30	11.04	7.38
C_2H_2	11.41	0.43	10.98	11.34	10.60
CF_3Br	11.82	0.91	10.91	11.64	11.16
HNO_3	11.03	0.57	10.46	12.76	14.24

(HF) method/3-21 basis set. *Ab*-initio molecular hardness (η_{ab}) values given in Table 5.9 are calculated considering Koopman's Theorem. The table clearly shows an agreement between *ab* initio results and the results of Eq. (52).

5.4.1   MAXIMUM HARDNESS AND MINIMUM POLARIZABILITY PRINCIPLES

One of the aims of theoretical chemists is to also present general principles or laws that provide compatible results with experiments. It is known that the rationalization of chemical reactions is a major task of chemists. The

TABLE 5.9   The Comparison of Calculated Molecular Hardness Values of Some Selected Molecules

Molecule	E_{HOMO}	E_{LUMO}	I	A	η_{ab}	η(Eq.52)
IBr	−0.3826	−0.0385	10.4130	1.0484	9.3646	7.89
S_2	−0.3593	−0.0793	9.7793	2.1592	7.6200	8.28
Br_2	−0.4081	−0.0270	11.1066	0.7366	10.3700	8.44
Cl_2	−0.4737	−0.0444	12.8920	1.2106	11.6814	9.36
SO	−0.3746	−0.0203	10.1937	0.5537	9.6399	9.84
O_2	−0.4771	0.0233	12.9851	−0.6353	13.6204	12.16
NH	−0.4198	0.0795	11.4250	−2.1641	13.5892	13.64
CS_2	−0.3766	−0.0452	10.2481	1.2299	9.0181	8.84
OCS	−0.4097	0.1180	11.1499	−3.2120	14.3620	10.94
SO_2	−0.4490	−0.0227	12.2201	0.6179	11.6022	10.69
O_3	−0.4723	−0.0538	12.8534	1.4656	11.3877	12.16
PBr_3	−0.3933	−0.0100	10.7039	0.2726	10.4312	8.84
PCl_3	−0.4433	−0.0068	12.0650	0.1858	11.8792	9.48
$POCl_3$	−0.4944	−0.0616	13.4550	1.6775	11.7774	10.00
CH_3I	−0.3553	0.1121	9.6690	−3.0509	12.7200	11.10
SO_3	−0.5215	−0.0397	14.1927	1.0827	13.1100	11.04

ultimate aim of the reactivity theory is to provide an answer to the fundamental questions regarding stability and chemical reactivity of atoms, molecules, and ions. Chemists have the aim to rationalize experimental facts with the help of simple chemical theories. HSAB [47, 48] principle, MHP [7, 8], and Minimum Polarizability principle (MPP) [49] are among the most widely accepted electronic structure principles of chemical reactivity. The MHP introduced by Pearson affirms that, at a given temperature, a chemical system tends to arrange itself so as to achieve the maximum hardness and the chemical hardness can be considered as a measure of stability. Then, in the light of inverse relationship between hardness and polarizability (α), Chattaraj and Sengupta [50] proposed the MPP which states that the natural evolution of any system is toward a state of minimum polarizability. Both principles have been applied successfully to the study of molecular vibrations, internal rotation excited states, aromaticity, and different types of chemical reactions. It has been found in most of these cases that the conditions of maximum hardness and minimum polarizability complement the minimum energy criterion for molecular stability. A formal proof of the MHP based on statistical mechanics and the fluctuation–dissipation theorem was given by Parr and Chattaraj [7]. Then, Pearson and Palke [51] presented a support for the validity of MHP. Theoretical chemists published many papers on the validity of MHP and MPP. Although some of them proposed that the aforementioned principles are useful and valid for chemical reactions, others opined that the applications of these principles are limited. For example, investigating the validity of MHP and MPP for nontotally symmetric vibrations, Torrent-Sucarrat et al. [52, 53] noted that the generalized maximum hardness and minimum polarizability principles may not be obeyed in not only chemical reactions but also in the favorable case of nontotally symmetric vibrations. As is known, aromaticity can be considered as a measure of the stability of molecules and "an aromatic molecule is one in which electrons are free to cycle around circular arrangements of atoms connected via identical bonds which are resonance hybrids of single and double bonds. It displays enhanced chemical stability compared to similar nonaromatic molecules and possesses significant local magnetic field, a planar structure, and (4n + 2, $n \geq 0$) π-electrons in a single ring. On the other hand, an antiaromatic molecule contains $4n$ ($n{\neq}0$) π-electrons in a cyclic planar, or nearly planar, system of alternating single and double bonds." Chattaraj et al. [54] studied some metal-aromatic and antiaromatic systems and showed that the most stable isomer with minimum energy among their various possible isomers studied aromatic compounds is the hardest and the least polarizable. In addition to the abovementioned studies, Ghanty and Ghosh [55] considered some exchange and dissociation reactions whose enthalpies are known to be discussed on the validity of the aforementioned electronic structure principles, namely MHP and MPP. In this study, they analyzed how the energy changes are related to changes in hardness or polarizability in these reactions and observed that the condition of minimum polarizability can largely be associated with maximum hardness or energetically more stable situations.

In the recent past, we presented a simple method to calculate the lattice energies of inorganic ionic compounds with the help of maximum hardness and minimum polarizability principles [56]. In the subsequent section, we will first mention past lattice energy calculation methods that are well known and then give detailed information about new methods based on MHP and MPP.

5.4.2 LATTICE ENERGY OF IONIC CRYSTALS

Lattice energy (U) is a key parameter in the thermodynamic analysis of the existence and stability of ionic crystals. This energy is also considered in the matter of whether new inorganic materials can be synthesized or in the estimation of various synthetic routes toward the preparation of new ionic salts. For a crystal with $A_m B_n$ as the general formula, the lattice energy can be expressed as change of energy for the following process.

$$A_m B_{n(c)} \rightarrow m A b^{n+}{}_{(g)} + n B^m{}_{(g)}$$

Direct experimental determination of lattice energies of ionic crystals is not possible. The lattice energy values evaluated by means of Born–Haber–Fajans (BHF) thermochemical cycle for ionic crystals are regarded as experimental values. This thermodynamic cycle can be considered as a useful technique in which Hess Law is applied to standard enthalpy changes observed when an ionic compound is formed.

Several theoretical methods have been proposed to calculate the lattice energies of ionic compounds. One of the most frequent methods mentioned in inorganic chemistry books was proposed by Kapustinskii. He developed a useful equation based on Born-Mayer [57] and Born-Lande [58] equations but more comprehensive than their equations for lattice energy calculations. The advantage of Kapustinskii [59] equation in comparison with the said other two equations is that it can be used for lattice-type uncertain solids. Kapustinskii equation and the meanings of the parameters appearing in the equation are given below.

$$U = \frac{B z^+ z^- n}{(r_+ + r_-)} \left(1 - \frac{0.345}{(r_+ + r_-)} \right) \tag{57}$$

where $B = 2427.8/2$, n is the total number of ions in formula of ionic crystal, and $r^+ + r^-$ is the sum of radii of anion and cation

An interesting study in this topic was conducted by Kudriavtsev [60]. Kudriavtsev developed a new theory to calculate the lattice energies of ionic crystals. According to this theory, the lattice energy of an ionic crystal has been associated with the mean sound velocity within the crystal. Kudriavtsev's theory has been applied successfully for the prediction of many ionic crystals based on sound velocity data. In 1993, Reddy and co-workers [61] computed the lattice energy values of alkali halide crystals taking advantage from interrelations between interionic separation, plasma energy, and lattice energy and obtained good agreement with literature values. Recently, Jenkins et al [62]. developed the Volume Based Thermodynamics (VBT) approach to predict their thermodynamic properties based on molar volumes (V_m) of ionic compounds. Jenkins who considered the study of Barlett [63] as the starting point proposed the following equation to calculate the lattice energies of ionic compounds. It is important to note that this equation that can be applied to the majority of ionic salts gives quite compatible results with the BHF thermodynamic cycle.

$$U / kJ \ mol^{-1} = 2I[\frac{\alpha}{V_m^{1/3}} + \beta] \tag{58}$$

Here, α and β are coefficients that depend on stoichiometry of salt. V_m represents the molecular volume and I is ionic strength of lattice, and it can be calculated by means of the equation below. In the following equation, n_i is the number of ions of type i in the formula unit with charge z_i.

$$I = \frac{1}{2}\sum n_i z_i^2 \tag{59}$$

In addition to all popular studies mentioned above, recently, we derived an equation that supports the validity of both MHP and MPP to calculate the lattice energies of inorganic ionic crystals. As mentioned above, chemical hardness defined as the resistance toward electron cloud polarization or deformation of chemical species is considered as a stability criterion. Electronic polarizabilities of compounds are considered as important data for investigating and predicting their dielectric, thermal, and lattice dynamical behaviors. It is important to note that Clausius-Mossotti equation indicates that polarizabilities of compounds are directly proportional to their volumes and dielectric constants. Namely, molar volumes of compounds can be considered as a measure of their polarizabilities. It can be said with the help of this information that a stable compound should have high chemical hardness and low polarizability. Lattice energies of ionic compounds give important hints about their stabilities or reactivities. If so, there should be a remarkable correlation between their lattice energies and chemical hardness values of ionic compounds. Actually, the first clues that such correlation might be related to the chemical hardness of some thermodynamic quantities were presented by Pearson. Pearson pointed out that there is a significant correlation between the chemical hardness and cohesive energy that determines the stability, reactivity, and structure of a solid. In addition to this correlation between cohesive energy and chemical hardness, Ghanty [64] demonstrated that there is a good linear correlation between softness and $\alpha^{1/3}$ (α=polarizability) for metal clusters and carbon clusters. In the light of these developments mentioned, we thought that a remarkable correlation can exist between its lattice energy with chemical hardness and molar volume of an ionic compound. On the basis of the studies performed to verify this idea, we presented the following equation for lattice energy calculations.

$$U \ / \ kjmol^{-1} = 2I[a \frac{\eta_M}{V_m^{1/3}} + b] \tag{60}$$

where U represents the lattice energy of ionic crystal in kJ/mol unit, I is ionic strength of lattice and is calculated by Eq. (59), η_M is molecular hardness value of ionic crystal calculated by Eq. (52), and the list of a and b constants for various stoichiometries is given in Table 5.10. Eq. (60) can be considered as a new application of this electronic structure

TABLE 5.10 Constants for Use in Eq. (60)

Ionic Crystal (Charge ratio)	Ionic Strength (I)	a kJ nm/mol eV	b (kJ mol⁻¹)
MX (1:1)	1	12.245±0.0025	195.15±0.40
M₂X (1:2)	3	14.193±0.035	195.5±0.49
MX₂ (2:1)	3	11.598±0.021	182.16±0.31
MX (2:2)	4	7.186±0.028	239.37±0.56

principle because Eq. (52) derived by us is based on the hardness equalization principle.

The new chemical hardness and molar volume-based method proposed for the lattice energy calculation is quite reliable. In the Table 5.11, lattice energy values calculated using Eq. (60) and theoretical methodologies published in the past years and experimental lattice energies for alkali halides are given. The values given in Table 5.11 show that Eq. (60) is more accurate than the other existing methods for simple inorganic crystals. In Table 5.12, chemical hardness calculated using Eq. (52); density,

TABLE 5.11 Comparison of Lattice Energies Determined with Various Methods for Alkali Halides (kJ/mol)

Alkali halide crystal	Born-Fajans-Haber	Born-Lande	Born-Mayer	Kapustinskii	Reddy	Kudri-avtsev	Jenkins	Eq. 60
LiF	1036	1005	1000	952.9	968.0	1085.2	1029	1046
LiCl	853	810	818	803.9	851.7	880.5	827	839
LiBr	807	765	772	792.6	813.9	844.5	780	800
LiI	757	713	710	713.1	755.8	-	721	746
NaF	923	899	894	885.1	905.8	1011.9	905	924
NaCl	787	753	756	752.9	799.9	879.3	764	787
NaBr	747	717	719	713.5	765.0	844.1	727	753
NaI	704	671	670	673.4	712.8	768.8	678	713
KF	821	795	792	788.9	831.2	863.8	796	803
KCl	715	686	687	680.9	732.0	799.3	695	701
KBr	682	658	659	675	699.8	772.1	667	679
KI	649	622	620	613.9	651.1	699.3	630	650
RbF	785	758	756	760.4	799.7	853.7	723	761
RbCl	689	659	661	662.1	701.5	740.7	668	679
RbBr	660	634	635	626.5	670.4	685.9	644	659
RbI	630	601	600	589.7	622.5	680.1	610	623
CsF	740	724	714	713.1	760.0	693.9	723	714
CsCl	659	621	621	625.2	644.0	672.1	672	664
CsBr	631	598	598	602.2	612.1	669.2	648	646
CsI	604	568	565	563.7	562.6	659.1	616	623

TABLE 5.12   Chemical Hardness, Density, Formula Mass, Molar Volume and Experimental Lattice Energy and Calculated Lattice Energy Values for Some Selected Inorganic Ionic Compound

Ionic Compound	η_M(eV)	Density (g/cm^3)	Molar mass (g/mol)	Molar volume (nm^3)	U_{cal} (Eq.60) (kJ/mol)	U_{BFH} (kJ/mol)
BeO	10.25	3.01	25.01	0.01380	4371	4443
MgO	8.88	3.58	40.30	0.01869	3838	3791
CaO	7.85	3.34	54.07	0.02688	3421	3401
SrO	7.45	4.70	103.62	0.03661	3204	3223
BaO	6.92	5.72	153.32	0.04450	3037	3054
MgS	7.57	2.84	56.38	0.03296	3272	3238
CaS	6.81	2.59	72.43	0.04644	3004	2966
PbSe	7.32	8.10	286.16	0.05866	2998	3144
TiO	8.45	4.95	63.86	0.02142	3664	3811
MnO	9.25	5.43	70.93	0.02168	3822	3745
FeO	9.28	5.74	71.84	0.02077	3855	3865
ZnO	10.72	5.60	81.38	0.02413	4047	3971
HgO	10.92	11.14	216.59	0.03228	3886	3806
SnO	8.36	5.75	66.94	0.01931	3706	3652
Li$_2$O	6.00	2.013	29.88	0.02458	2930	2814
Na$_2$O	5.79	2.27	61.98	0.04533	2555	2478
K$_2$O	4.99	2.32	94.20	0.06742	2217	2232
Rb$_2$O	4.82	4.00	186.94	0.07760	2135	2161
Li$_2$S	5.45	1.66	45.95	0.04596	2468	2472
Na$_2$S	5.27	1.86	78.04	0.06981	2262	2203
K$_2$S	4.56	1.74	110.26	0.10521	1995	1979
Rb$_2$S	4.41	2.91	203.00	0.11571	1943	1949
Tl$_2$O	6.99	10.45	424.77	0.06748	2635	2575
Tl$_2$Se	6.29	9.05	487.72	0.08949	2370	2326
Cu$_2$O	7.75	6.00	143.09	0.03960	3109	3189
Cu$_2$S	6.93	5.60	159.16	0.04719	2806	2865
Cu$_2$Te	6.65	7.27	254.69	0.05817	2634	2683
Ag$_2$O	7.62	7.14	231.74	0.05390	2890	2910
Ag$_2$S	6.81	7.23	247.8	0.05687	2681	2677

TABLE 5.12   (Continued)

Ionic Compound	η_M(eV)	Density (g/cm³)	Molar mass (g/mol)	Molar volume (nm³)	U_{cal} (Eq.60) (kJ/mol)	U_{BFH} (kJ/mol)
Ag_2Te	6.50	8.32	341.33	0.06813	2528	2600
Cs_2O	4.52	4.65	281.81	0.10063	2000	2063
BeF_2	11.05	1.986	47.01	0.03930	3354	3464
$MgCl_2$	8.23	2.32	95.21	0.06813	2495	2540
$CaCl_2$	7.47	2.15	110.98	0.08570	2271	2271
$CaBr_2$	7.14	3.353	199.89	0.09898	2167	2134
CaI_2	6.70	3.956	293.89	0.12336	2029	2087
$BaCl_2$	6.74	3.856	208.23	0.08967	2140	2069
$BaBr_2$	6.47	4.78	297.14	0.10322	2052	1995
BaI_2	6.09	5.15	391.14	0.12612	1938	1890
$MgBr_2$	7.80	3.72	184.113	0.08218	2341	2451
$SrBr_2$	6.85	4.21	247.43	0.09744	2128	2040
CrF_2	9.4	3.79	89.99	0.03942	3014	2939
$MnBr_2$	8.17	4.385	214.75	0.08132	2405	2482
NiF_2	9.93	4.72	96.69	0.03402	3225	3089
CuF_2	9.97	4.23	101.54	0.03985	3124	3102
$FeCl_2$	8.54	3.16	126.75	0.06660	2559	2641
$PbBr_2$	7.76	6.66	367.01	0.09149	2291	2230
PbI_2	7.24	6.16	461.01	0.12428	2102	2177
CuF	8.75	7.10	82.54	0.01929	1189	1088
$CuCl$	7.49	4.15	98.99	0.03960	928	996
AgF	8.59	5.85	126.87	0.03600	1027	974
$AgCl$	7.35	5.56	143.32	0.04279	905	918

molar volume, formula mass, and lattice energy calculated using Eq. (60); and experimental lattice energy values for some ionic compounds with various stoichiometries are presented.

Actually, Eq. (60) can be considered as a proof of the validity of maximum hardness and minimum polarizability principles in ionic compounds. It is apparent from this equation that lattice energy, as a measure of the stability of ionic compounds, increases as hardness increases. On the other hand, there is an inverse correlation between lattice energy and $V_m^{1/3}$ for ionic compounds as can be seen from Jenkins's equation.

5.4.3 PREDICTION OF THE INTERIONIC DISTANCES IN IONIC SOLIDS WITH THE HELP OF $\eta_M / V_M^{1/3}$ RATIO

Interionic distance is defined as distance between oppositely charged two ions that form any crystal. According to Coulomb's law, electrostatic attraction force (F) between oppositely charged two ions is given as follows and is directly proportional to multiplication of charges of ions ($q_1 . q_2$) and inversely proportional to the square of interionic distance (R).

$$F = \frac{q_1 . q_2}{R^2} \tag{61}$$

With the help of Coulomb's law, the energy (E_{att}) of electrostatic attraction between oppositely charged two ions can be given as:

$$E_{att} = \int_{\infty}^{R} F . dR = -\frac{q_1 . q_2}{R} = \frac{(Z^+ Z^-)}{R} \tag{62}$$

In this equation, Z^+ and Z^- are oxidation states of cation and anion, respectively, and are defined as $Z^+ = q_1/e$ and $Z^- = q_2/e$. Here, e represents the charge of electron.

It is important to note that electrostatic repulsion also occurs between their electron clouds at a particular distance as oppositely charged two ions come close to each other. To predict the electrostatic energy (E_{rep}) originating such repulsion, two mathematical equations are proposed. One of them is equation proposed by Born. The second equation proposed for this purpose was obtained from radial wave functions. These equations mentioned are, respectively, given as follows.

$$E_{rep} = \frac{B}{R^n} \tag{63}$$

$$E_{rep} = be^{-\delta/R} \tag{64}$$

In the first equation, B and n are constants. n is known as the repulsion exponent and is based on electron configuration of ions that form the crystal. In the second equation, b and δ are constants. The numerical value of δ for ionic compounds is approximately 34.5 pm. In both equations, R represents the interionic distance.

Total electrostatic energy (E_{tot}) is the sum of the electrostatic repulsion energy (E_{rep}) and electrostatic attraction energy (E_{att}).

$$E_{tot} = E_{att} + E_{rep} \tag{65}$$

It can be understood from the equations given above that E_{tot}, E_{att}, and E_{rep} are a function of interionic distance (R). Born-Lande and Born-Mayer equations proposed to calculate the lattice energies of ionic compounds have been derived considering total electrostatic energy in any ionic crystal. For this reason, one of the parameters of the aforementioned equations is interionic distance. Namely, there is a remarkable correlation between lattice energy and interionic distance. Assuming that there is a linear relationship between lattice energy and interionic distances of the molecular group halides, Reddy et al. [65] obtained the following relationships between lattice energy, U (in kcal/mol) and interionic distances, R in A^0 within molecular group halides

$$U = -49.51(R) + 324.72, (\text{for alkali halides})$$

$$U = -144.30(R) + 901.00, (\text{for divalent halides})$$

$$U = -47.75(R) + 286.54, (\text{for Ga, In, T1 halides})$$

The $\eta_M/V_m^{1/3}$ ratio was used by us not only for the estimation of lattice energies of ionic compounds but also for the calculation of interionic distances in ionic solids. Recently, we again investigated the correlation between lattice energy and interionic distance for molecular group halides mentioned above considering the lattice energy equation published by us

(Eq. (60)) and showed that interionic distances for these halide molecules can be calculated using the following equations.

$$r_0 = -1,744 \ln\left(\eta_M / V_m^{1/3}\right) + 7,6649, \text{(for alkali halides)}$$

$$r_0 = -1,769 \ln\left(\eta_M / V^{1/3}\right) + 7,5144, \text{(for divalent halides)}$$

$$r_0 = -1,740 \ln\left(\eta_M / V^{1/3}\right) + 7,342, \text{(for Ga, In T1 halides)}$$

In Table 5.13, based on the calculated interionic distances by using equations derived by us, spectroscopic interionic distances for some selected molecules and the standard deviations between two sets of interionic distances are given. The calculated standard deviations between the two sets of data prove unequivocally that the chemical hardness and molar volume-based equations proposed to predict the interionic distances of ionic compounds are valid and reliable.

5.4.4 CHEMICAL HARDNESS AND BOND FORCE CONSTANTS OF DIATOMIC MOLECULES

A primary goal of theoretical chemistry is to present useful approaches for the prediction of chemical properties whose determination by means of experimental techniques is difficult or time-consuming. Bond force constant can be considered as a measure of the stiffness of a bond. Vibrational spectroscopy does not provide direct information about bond force constant, but bond force constants of diatomic molecules can be determined by means of the data of IR or Raman spectroscopy. The bond force constant values obtained from the mentioned spectroscopic methods are considered experimental values. The correlation between the vibrational frequency and force constant of a bond is presented using the following equation [66].

$$\bar{v} = \frac{1}{2\pi c}\sqrt{\frac{k}{\mu}} \tag{66}$$

where c, k, and μ are speed of light, bond force constant, and reduced mass, respectively. In addition to this equation, several theoretical meth-

TABLE 5.13 Calculated and Experimental Interionic Distances for Some Selected
Molecules

| Molecule | Calculated | Interionic Distance (R) | |
		Experimental	SD%
LiF	1.900	1.996	4.80
LiCl	2.557	2.539	0.70
LiBr	2.716	2.713	0.11
LiI	2.961	2.951	0.33
NaF	2.254	2.295	1.78
NaCl	2.771	2.789	0.64
NaBr	2.925	2.954	0.98
NaI	3.129	3.194	2.03
KF	2.703	2.648	2.07
KCl	3.197	3.116	2.59
KBr	3.326	3.262	1.96
KI	3.509	3.489	0.57
RbF	2.887	2.789	3.51
RbCl	3.327	3.259	2.08
RbBr	3.447	3.410	1.08
RbI	3.699	3.628	1.95
CsF	3.126	2.975	5.07
CsCl	3.415	3.523	3.06
CsBr	3.535	3.668	3.62
CsI	3.700	3.900	5.12
BeF_2	1.356	1.400	3.14
$MgCl_2$	2.201	2.176	1.14
$MgBr_2$	2.406	2.335	3.04
MgI_2	2.674	2.515	6.32
$CaCl_2$	2.508	2.508	0.00
$CaBr_2$	2.673	2.668	0.18
$BaCl_2$	2.716	2.820	3.68
$BaBr_2$	2.872	2.986	3.81
InBr	2.544	2.545	0.03
TlCl	2.323	2.486	6.55
TlBr	2.436	2.617	6.91

ods have been proposed to predict the bond force constants of molecules. Below, we present detailed information of these theoretical methods.

In 1934, Badger [67] found that the relationship between bond force constant (k) and internuclear distance (r_e) for diatomic molecules can be quite accurately given as:

$$k = \frac{1.86 \times 10^5}{(r_e - d_{ij})^3} \tag{67}$$

where k is bond stretching force constant in dyn cm^{-1} units, r_e represents the internuclear distance in Angstrom (A^0) units, and d_{ij} is a function of the position of bonded atoms in the periodic table.

A useful equation indicating the relation between bond force constants, bond orders, bond lengths, and the electronegativities of bonded atoms was obtained by Gordy [68] as follows:

$$k = aN(\chi_A \chi_B / d^2)^{3/4} + b \tag{68}$$

In this equation, k is the bond force constant, d is the bond length, N is the bond order, and, χ_A and χ_B are electronegativities of bonded atoms. a and b have different numerical values for different molecular groups.

Pearson [69] who introduced the chemical hardness concept showed that there is a linear correlation between bond dissociation energy and $k^{1/2}$ (k is force constant) in diatomic hydride molecules and derived the following equation for the calculation of bond force constants of diatomic hydrides.

$$kr_e = 77\chi_A^2 + 117 \tag{69}$$

Considering diatomic molecule, Arulmozhiraja and Kolandaivel [70] showed that with increasing chemical hardness, either the force constants or binding energies of molecules increase. They also pointed out that the chemical hardness is a better indicator of stability of the molecules than the chemical potential. In recent years, taking inspiration from this study of Arulmozhiraja et al. [71], we thought that bond force constants of diatomic molecules would be proportional to their η_M/r_e^2 ratio. In the subsequent careful analysis, Arulmozhiraja et al. [71] derived the following equations to calculate the bond force constants of diatomic molecules.

$$k = \frac{a\eta_M}{r_e^2} + b \qquad (70)$$

$$k = \frac{aV_e\eta_M}{r_e^2} + b \qquad (71)$$

In both equations, k is bond force constant in N/m unit, η_M is the molecular hardness (using Eq. (52)) in eV units, r_e represents the bond length of diatomic molecules, and a and b are the correlation parameters. Eq. (71) is used for only binary compounds of hydrogen, and in this equation, V_e represents the number of electrons in the valence shell of an atom that is different from hydrogen for any binary compound of hydrogen. In Table 5.14, numerical values of a and b parameters for different molecule groups are given. In order to prove the reliability of the chemical hardness-based equations proposed to calculate the bond force constants of diatomic molecules, we calculated the force constants of some diatomic molecules and compared our results with bond force constant values observed experimentally for these molecules. The calculated and observed bond force constants for some diatomic molecules and the percentage deviations between two sets of bond force constants are given in Table 5.15.

TABLE 5.14   Numerical Values of a and b Parameters for Different Molecule Groups

Molecules	a	b
Alkalihalides	85.72	18.63
Binary compounds of hydrogen	8.6	78.77
A_2 type molecules	199.2	-4.119

TABLE 5.15   Observed and Calculated Bond Stretching Bond Force Constants for Some Selected Molecules (N m^{-1})

Molecule	k_{obs}	k_{cal}	Deviation (%)
LiF	249.5	259.91	4.17
LiCl	150.8	144.46	4.20
LiBr	126.4	125.12	1.01
NaF	175.4	170.97	2.52
NaCl	110.12	108.04	1.88
NaBr	95.82	95.84	0.02
RbCl	76.57	73.03	4.62
RbBr	67.09	64.23	4.26
RbI	49.31	55.45	12.45
CsCl	71.94	62.42	13.23
CsBr	56.85	57.35	0.87
LiH	102.6	102.97	0.360
HBr	415.04	403.07	2.88
HI	314.93	311.94	0.94
NH	596.7	621.43	4.14
PH	322.4	315.47	2.14
GeH	197.53	202.18	2.35
SnH	170.56	173.01	1.43
AsH	269.26	273.84	1.700
OH	783.36	764.83	2.36
SH	425.69	377.63	11.28
SeH	338.31	315.05	6.87
SiH	240.0	215.60	10.16
C_2	1220	1286.97	5.48
N_2	2300	2385.08	3.69
P_2	558	537.31	3.70
As_2	409	400.08	2.18
Sb_2	262	270.44	3.22
S_2	499	458.22	8.17

5.5 ELECTROPHILICITY EQUALIZATION PRINCIPLE AND NUCLEOPHILICITY EQUALIZATION PRINCIPLE

Electrophilicity is an important property of atoms and molecules. This quantity has important role in the prediction of chemical reactivity of chemical species and in the understanding of organic reaction mechanisms in particular. The electrophiles that act as Lewis acids are electron lover or electron-deficient chemical species. For this reason, they tend to accept electrons from a nucleophile (electron rich). It can be easily understood from this information that electrophilicity and electron affinity are quantum chemical parameters that are closely related to each other. Parr et al. [25] defined an electrophilicity index (ω) considering a sea of free electrons at zero temperature and zero chemical potential. It is not difficult to guess that the chemical species will start to accept the electrons in the free electron sea if a chemical species is immersed in such a chemical medium. The electron transfer process continues until the chemical potential of the said chemical species becomes equal to that of the electron sea. Regarding this electron transfer process, Parr proposed the electrophilicity index as given in Eq. (9) based on electronegativity and chemical hardness.

In the recent past, Chattaraj and his students [12] proposed the electrophilicity equalization principle considering Parr's electrophilicity index and specifying that the equalization of electronegativity and hardness implies the equalization of electrophilicity. According to them, the electrophilicity of the constituent atoms becomes equalized during molecule formation, and molecular electrophilicity (ω_M) may be expressed as the geometric mean of the electrophilicities of the isolated atoms.

$$\omega_M = \frac{\chi_M^{\,2}}{2\eta_M} = (\prod_{i=1}^{N} \omega_i)^{1/N} \tag{72}$$

Here, N is the total number of atoms in the molecule and ω_i ($i = 1, 2, 3, \ldots N$) represents the isolated atom electrophilicity value.

After the publication of the electrophilicity equalization principle, Szentpaly [72] severely criticized it considering the results obtained for large metal clusters and fullerenes. In his paper, Szentpaly indicated that the electrophilicity equalization principle is not useful and reasonable for molecule formation and showed that electrophilicity increases with cluster size. This statement is sufficient to understand the effect of molecular size on equalizations of chemical reactivity indexes such as chemical hardness, electronegativity, and electrophilicity. It is apparent that equalization assumptions for electrophilicity, electronegativity, and chemical hardness in large molecules are not reasonable. Islam and Ghosh have [73] critically analyzed the comment of Szentpaly that there is no support for "hardness equalization principle" and "electrophilicity equalization principle" and pointed out that Szentpaly erred in conceiving proper domain of the equalization phenomenon. The process of charge equalization occurs only during the chemical event of heteronuclear molecule formation. This charge equalization phenomenon cannot be used in case of the formation of homonuclear molecules because there is no question of charge transfer in the formation of homonuclear molecules. In addition, Islam and Ghosh [74] proposed the following equation based on the electrophilicity equalization principle to predict the electrophilicity index of molecules in eV units.

$$\omega_M = \frac{K.7,2(n+1)}{\sum_i r_i} \tag{73}$$

In this equation, n is the number of atoms connected to the central atom in the molecule, r_i is atomic radius in angstrom (A^o) unit, and the standardized value of K is 0.382516 for diatomic molecules and 0.712 for polyatomic molecules.

As is known, chemical equalization principles about chemical reactivity descriptors such as electronegativity and hardness have been put forward with the help of the relation with the charge of the said descriptors. Nucleophilicity is a measure of the strength of nucleophile, and hence, it is a charge-dependent chemical property like chemical hardness, electronegativity, and electrophilicity. Electrophiles and nucleophiles act as Lewis acids and bases, respectively. For that reason, it is not difficult to guess that there is an inverse correlation between electrophilicity and nucleophilicity. Namely, in general, low ω value corresponds to high nucleophilic power. Chattaraj and Maiti [75] defined the nucleophilicity as the multiplicative inverse of electrophilicity as follows in the context of the correlation between softness (σ) and hardness ($\sigma = 1/\eta$).

$$\varepsilon = 1/\omega \tag{74}$$

In the recent times, we proposed a new electronic structure principle, namely nucleophilicity equalization principle, and presented new algorithms for the prediction of molecular nucleophilicity. To compute the molecular nucleophilicity (ε_M), we derived two new equations based on assumptions that nucleophilicity, which is a charge-dependent property, becomes equalized like electrophilicity, electronegativity, and hardness. To prove the validity of the nucleophilicity equalization principle, we considered the nucleophilicity-electrophilicity relationship presented by Chattaraj and Maiti (Eq. (74)) and molecular hardness and group electronegativity equations derived by us in recent times (Eqs. (35) and (52)). The detailed information about this is given below.

By means of molecular hardness and group electronegativity equations published by us, the charge effect on the molecular hardness and group electronegativity can be easily investigated. In order to propose a general method for the evaluation of nucleophilicity of ions, groups, and molecules, we first analyzed the nature of variation of molecular hardness and molecular electronegativity with the charges of some selected groups with the help of Eqs. (52) and (35). Figure 5.2 shows that chemical hardness and electronegativity of chemical species increase linearly as its charge increases.

By performing a careful analysis of the linear graphs given in Figure 5.2 (a), we determined that the slope of the graph drawn to investigate

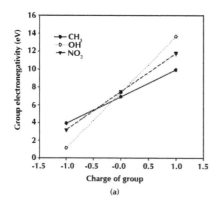
(a)

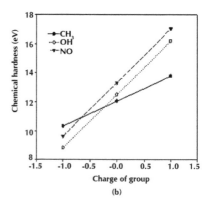
(b)

FIGURE 5.2 Graphs of variation with charge of group chemical hardness and group electronegativity for some selected functional groups.

the relationship between group electronegativity and charge for any functional group is equal to the β_M/N value (here, N is the number of atoms that form the functional group) of the said group. The cut-off point of the graph (while q=0) is equal to the α_M value of the group. Likewise, if the linear graphs given in Figure 5.2 (b) for some selected functional groups are analyzed, it is seen that the slope of the graph drawn to investigate the relationship between charge and chemical hardness for any functional group corresponds to the α_M/N value of the said group. The cut-off point value of this graph that shows the correlation between charge and chemical hardness of any group is equal to the β_M value of the said group. In this way, we obtain the following new equations to calculate the electronegativity and chemical hardness of functional groups.

$$\chi_M = \alpha_M + \frac{\beta_M}{N}q \tag{75}$$

$$\eta_M = \beta_M + \frac{\alpha_M}{N}q \tag{76}$$

In both equations, α_M and β_M parameters correspond to group electronegativity and molecular hardness values obtained by geometric mean equations, respectively. If so, with the help of this information, we can present the molecular nucleophilicity by the following equation. Here, it should be again noted that we considered Eq. (74) to explain the electrophilicity and nucleophilicity.

$$\varepsilon_M = \frac{2\eta_M}{\chi_M{}^2} = \frac{2[\beta_M + \frac{\alpha_M}{N}q]}{[\alpha_M + \frac{\beta_M}{N}q]^2} = 2\frac{\left([\prod_{i=1}^{N}(\eta_i)^{1/N}] + \frac{[\prod_{i=1}^{N}(\chi_i)^{1/N}]q}{N}\right)}{\left([\prod_{i=1}^{N}(\chi_i)^{1/N}] + \frac{[\prod_{i=1}^{N}(\eta_i)^{1/N}]q}{N}\right)^2} \tag{77}$$

where ε_M is molecular nucleophilicity, N is the total number of atoms in the molecule, and q stands for the charge of molecule. As is known, geometric mean equations are used for equalized chemical reactivity descriptors. This means that molecular electrophilicity can also be expressed as the geometric mean of nucleophilicity values of isolated atoms as follows.

$$\varepsilon_M = \frac{2\eta_M}{\chi_M^2} = (\prod_{i=1}^{N} \varepsilon_i)^{1/N} \tag{78}$$

where ε_M is the nucleophilicity of molecules, ε_i $\{i = 1,2,3....N\}$ represents isolated atom nucleophilicity values, and N stands for the number of atoms in

TABLE 5.16 Calculated and Experimental Nucleophilicity Values for Some Selected Molecules

Molecule	I(eV)	A(eV)	η_{ex}	β_M	χ_{ex}	α_M	ε_{ex}	$\varepsilon(77)$	$\varepsilon(78)$
I_2	9.400	2.420	6.980	7.392	5.910	6.755	0.39	0.32	0.32
BrI	9.790	2.550	7.240	7.903	6.170	7.159	0.38	0.31	0.30
S_2	9.400	1.660	7.740	8.282	5.530	6.218	0.50	0.43	0.42
Br_2	10.560	2.600	7.960	8.450	6.580	7.588	0.36	0.30	0.29
Cl_2	11.480	2.400	9.080	9.354	6.940	8.290	0.37	0.27	0.27
P_2	9.600	0.650	8.950	9.740	5.125	5.617	0.68	0.62	0.61
SO	10.00	1.130	8.870	10.030	5.565	6.846	0.57	0.43	0.42
CH	10.640	1.240	9.400	11.332	5.940	6.703	0.53	0.50	0.50
O_2	12.060	0.440	11.620	12.156	6.250	7.539	0.59	0.43	0.42
OH	13.180	1.830	11.350	12.495	7.505	7.355	0.40	0.46	0.45
NH	13.100	0.380	12.720	13.629	6.740	7.238	0.56	0.52	0.51
F_2	15.700	3.080	12.620	14.022	9.390	10.412	0.28	0.26	0.25
CS_2	10.080	1.000	9.080	8.818	5.540	6.232	0.59	0.45	0.45
COS	11.180	0.460	10.720	10.022	5.820	6.646	0.63	0.46	0.45
SO_2	12.340	1.050	11.290	10.696	6.695	7.070	0.50	0.43	0.42
O_3	12.670	1.820	10.850	12.156	7.745	7.539	0.41	0.42	0.42
NH_2	12.800	0.780	12.020	13.362	6.790	7.217	0.52	0.52	0.51
N_2O	12.890	1.470	11.420	13.649	7.180	7.380	0.44	0.50	0.49
$POCl_3$	11.400	1.400	10.00	9.937	6.400	7.524	0.48	0.35	0.35
SO_3	11.000	1.700	9.300	11.318	6.350	7.184	0.46	0.44	0.42
C_2H_2	11.410	0.430	10.980	11.33	5.920	6.702	0.62	0.51	0.50
CH_3	9.840	0.080	9.760	12.064	4.940	6.935	0.79	0.50	0.49
HNO_3	11.030	0.570	10.460	12.725	5.800	7.417	0.62	0.46	0.46
$C_6H_4O_2$	9.670	1.890	7.780	11.228	5.780	6.758	0.46	0.49	0.48

the molecule. Taking into consideration the correlation between electrophilicity and nucleophilicity as $\omega = 1/\varepsilon$, we can say that all equations proposed to calculate the electrophilicity of molecules can also be used for nucleophilicity calculations. In Table 5.16, the comparison with experimental data (ε_{ex}) of nucleophilicity values calculated using Eqs. (77) and (78), $\varepsilon(77)$ and $\varepsilon(78)$, respectively, for some selected molecules are given. Table 5.17 provides the nucleophilicity and electrophilicity values obtained from various theoretical methods for some alkali halides. In Table 5.17, KAYA, C, P, and ND labels represent the results obtained considering Eqs. (78), (72), (9), and (73), respectively. The data given in related tables show that the nucleophilicity equalization principle is a law of nature like the hardness equalization principle, EEP, and electrophilicity equalization principle.

5.6   CONCLUSION

As is known, the prediction of reactivity and stability of chemical species is one of the most important issues in chemistry. The aim of the proposed theoretical approaches in this area is to facilitate the estimation in accordance with their experimental trends of numerical values of quantum chemical parameters that provide remarkable hints about the reactivity or stability of chemical species, such as hardness, electronegativity, electrophilicity, and nucleophilicity. In this chapter, some new applications of chemical equal-

TABLE 5.17   The Comparison of Nucleophilicity and Electrophilicity Values That Are Calculated Using Various Methods for Some Alkali Halides

Molecule	ω_{ND}	ω_P	ω_C	ω_{KAYA}	ε_{ND}	ε_P	ε_C	ε_{KAYA}
LiF	2.796	2.411	2.014	1.905	0.357	0.415	0.496	0.525
LiCl	2.374	2.083	1.905	1.857	0.422	0.480	0.525	0.538
LiBr	2.230	1.776	1.878	1.788	0.448	0.563	0.534	0.559
NaF	2.515	2.263	2.014	1.843	0.397	0.442	0.496	0.543
NaCl	2.168	2.782	1.905	1.797	0.462	0.359	0.525	0.556
NaBr	2.047	1.955	1.850	1.729	0.488	0.512	0.541	0.578
KF	2.118	2.063	1.850	1.712	0.472	0.485	0.541	0.584
KCl	1.867	2.285	1.769	1.670	0.534	0.437	0.565	0.598
KBr	1.776	2.002	1.714	1.607	0.563	0.499	0.584	0.623

ization principles related to hardness, electronegativity, electrophilicity, and nucleophilicity are presented. Molecular hardness, molecular electronegativity, lattice energy, and bond force constant equations introduced in the chapter are very useful. We strongly believe that the new developments described in this chapter will be useful to our colleagues and students.

ACKNOWLEDGMENT

We thank to Prof. Cemal Kaya for his qualified contributions in this chapter. The chapter is dedicated to Professor Dr. Cemal Kaya on the occasion of his 40 years in science.

KEYWORDS

- bond force constants of diatomic molecules
- electronegativity equalization principle
- electrophilicity equalization principle
- global reactivity descriptors
- hardness equalization principle
- lattice energies of ionic solids
- nucleophilicity equalization principle
- partial charges of atoms in molecules

REFERENCES

1. Parr, R. G., & Yang, W., (1989). *Density Functional Theory of Atoms and Molecules*, Oxford University Press, New York.
2. Dreizler, R. M., & Gross, E. K. U., (1990). *Density Functional Theory*, Springer, Berlin, Germany.
3. Pauling, L., (1960). *The Nature of the Chemical Bond*, vol. 3. Ithaca, NY: Cornell University Press.
4. Sanderson, R. T., (1971). *Chemical Bonds and Bond Energy*, New York, London, Academic Press.
5. Pearson, R. G., (1963). Hard and Soft Acids and Bases. *J. Am. Chem. Soc., 85*, 3533–3539.

6. Pearson, R. G., (1997). *Chemical Hardness*, Wiley, New York.
7. Parr, R. G., & Chattaraj, P. K., (1991). Principle of Maximum Hardness. *J. Am. Chem. Soc., 113,* 1854–1855.
8. Pearson, R. G., (1993). The Principle of Maximum Hardness. *Acc. Chem. Res., 26,* 250–255.
9. Datta, D., (1986). Geometric Mean Principle for Hardness Eualization: a corollary of Sanderson's geometric mean principle of electronegativity equalization. *J. Phys. Chem., 90,* 4216–4217.
10. Kaya, S., & Kaya, C., (2015). A new equation for calculation of chemical hardness of groups and molecules. *Mol. Phys., 113,* 1311–1319.
11. Kaya, S., & Kaya, C., (2015). A new method for calculation of molecular hardness: A theoretical study, *Computational and Theoretical Chemistry, 1060,* 66–70.
12. Chattaraj, P. K., Giri, S., & Duley, S., (2010). Electrophilicity equalization principle. *J. Phys. Chem. Lett., 1,* 1064–1067.
13. Kaya, S., Kaya, C., & Islam, N., (2016). The nucleophilicity equalization principle and new algorithms for the evaluation of molecular nucleophilicity, *Computational and Theoretical Chemistry, 1080,* 72–78.
14. Hohenberg, P., & Kohn, W., (1964). Inhomogeneous electron gas, *Phys. Rev. B., 136,* 864.
15. Parr, R. G., Donnelly, R. A., Levy, M., & Palke, W. E., (1978). Electronegativity: the density functional viewpoint, *J. Chem. Phys., 68,* 3801–3807.
16. Allen, L. C., (1990). Electronegativity Scales. *Acc. Chem. Res., 23,* 175–176.
17. Pearson, R. G., (1990). Electronegativity Scales. *Acc. Chem. Res., 23,* 1 2.
18. Politzer, P., Grice, M. E., & Murray, J. S., (2001). Electronegativities, electrostatic potentials and covalent radii. *J. Mol. Struct. (THEOCHEM), 549,* 69–76.19.
19. Pearson, R. G., (1999). Maximum Chemical and Physical Hardness. *J. Chem. Edu., 76,* 267.
20. Pearson, R. G., (1987). Recent Advances in the Concept of Hard and Soft Acids and Bases. *J. Chem. Educ., 64,* 561−567.
21. Pearson, R. G., (1988). Absolute electronegativity and hardness: application to inorganic chemistry. *Inorg. Chem., 27,* 734−740.
22. Koopmans, T. A., (1933). About the assignment of wave functions and Eigenvalues for the individual electrons of an atom, *Physica., 1,* 104–113.
23. Ingold, C. K., (1934). Principles of an Electronic Theory of Organic Reactions. *Chem. Rev., 15,* 225–274.
24. Chattaraj, P. K., Sarkar, U., & Roy, D. R., (2006). Electrophilicity Index. *Chem. Rev., 106,* 2065–2091.
25. Parr, R. G., Szentplay, L. V., & Liu, S., (1999). Electrophilicity Index. *J. Am. Chem. Soc., 121,* 1922–1924.
26. Maynard, A. T., Huang, M., Rice, W. G., & Covell, D. G., (1998). Reactivity of the HIV-1 nucleocapsid protein p7 zinc finger domains from the perspective of density-functional theory, *Proc. Natl. Acad. Sci. USA., 95,* 11578–11583.
27. Gazquez, J. L., Cedillo, A., & Vela, A., (2007). Electrodonating and electroaccepting powers. *J. Phys. Chem. A.,111,* 1966–1970.

28. Kiyooka, S. I., Kaneno, D., & Fujiyama, (2013). Intrinsic reactivity index as a single scale directed toward both electrophilicity and nucleophilicity using frontier molecular orbitals. *Tetrahedron., 69,* 4247–4258.

29. Chamorro, E., & Melin, J., (2015). On the intrinsic reactivity index for electrophilicity/nucleophilicity responses. *J. Mol. Model., 21,* 1–3.

30. Iczkowski, R. P., & Margrave, J. L., (1961). Electronegativity. *J. Am. Chem. Soc., 83,* 3547–3551.

31. Hinze, J., & Jaffe, H. H., (1962). Electronegativity. I. Orbital electronegativity of neutral atoms. *J. Am. Chem. Soc., 84,* 540–546.

32. Huheey, J. E., (1965). The electronegativity of groups. *J. Phys. Chem., 69,* 3284–3291.

33. Bratsch, S. G., (1985). A group electronegativity method with Pauling units. *J. Chem. Edu., 62,* 101–103.

34. Mullay, J., (1985). Calculation of group electronegativity. *J. Am. Chem. Soc., 107,* 7271–7275.

35. Kaya, S., & Kaya, C., (2015). A new equation based on ionization energies and electron affinities of atoms for calculating of group electronegativity. *Computational and Theoretical Chemistry., 1052,* 42–46.

36. Gasteiger, J., & Marsili, M., (1980). Iterative partial equalization of orbital electronegativity-a rapid access to atomic charges. *Tetrahedron., 36,* 3219–3228.

37. Mortier, W. J., Ghosh, S. K., & Shankar, S., (1986). Electronegativity-equalization method for the calculation of atomic charges in molecules. *J. Am. Chem. Soc., 108,* 4315–4320.

38. Yang, Z. Z., & Wang, C. S., (1997). Atom-bond electronegativity equalization method. 1. Calculation of the charge distribution in large molecules. *J. Phys. Chem. A., 101,* 6315–6321.

39. Ionescu, C. M., Geidl, S., Svobodová Vareková, R., & Koca, J., (2013). Rapid calculation of accurate atomic charges for proteins via the electronegativity equalization method. *J. Chem. Inf. Model., 53,* 2548–2558.

40. Smith, D. W., (1998). Group electronegativities from electronegativity equilibration applications to organic thermochemistry. *Journal of the Chemical Society,* Faraday Transactions., *94,* 201–205.

41. Yang, W., Lee, C., & Ghosh, S. K., (1985). Molecular Softness as the Average of Atomic Softnesses: Companion principle to the geometric mean principle for electronegativity equalization. *J. Phys. Chem., 89,* 5412–5414.

42. Ghosh, S. K., (1994). Electronegativity, hardness, and a semiempirical density functional theory of chemical binding. *Int. J. Quant. Chem., 49,* 239–251.

43. Ayers, P. W., & Parr, R. G., (2008). Beyond electronegativity and local hardness: Higher-order equalization criteria for determination of a ground-state electron density. *J. Chem. Phys., 129,* 054111.

44. Ghosh, D. C., & Islam, N., (2011). Whether electronegativity and hardness are manifest two different descriptors of the one and the same fundamental property of atoms—A quest. *Int. J. Quant. Chem., 111,* 40–51.

45. Ghosh, D. C., & Islam, N., (2011). Whether there is a hardness equalization principle analogous to the electronegativity equalization principle—A quest. *Int. J. Quant. Chem.,111,* 1961–1969.

46. Ghosh, D. C., & Islam, N., (2011). A quest for the algorithm for evaluating the molecular hardness. *Int. J. Quant. Chem., 111*, 1931–1941.
47. Ayers, P. W., Parr, R. G., & Pearson, R. G., (2006). Elucidating the hard/soft acid/base principle: a perspective based on half-reactions. *J. Chem. Phys., 124*, 194107.
48. Ayers, P. W., (2007). The physical basis of the hard/soft acid/base principle, *Faraday Discussions, 135*, 161–190.
49. Chattaraj, P. K., Roy, D. R., Elango, M., & Subramanian, V., (2005). Stability and reactivity of all-metal aromatic and antiaromatic systems in light of the principles of maximum hardness and minimum polarizability. *J. Phys. Chem. A., 109*, 9590–9597.
50. Chattaraj, P. K., & Sengupta, S., (1996). Popular electronic structure principles in a dynamical context. *J. Phys. Chem., 100*, 16126–16130.
51. Pearson, R. G., & Palke, W. E., (1992). Support for a principle of maximum hardness. *J. Phys. Chem., 96*, 3283–3285.
52. Torrent-Sucarrat, M., Luis, J. M., Duran, M., & Sola, M., (2002). Are the maximum hardness and minimum polarizability principles always obeyed in nontotally symmetric vibrations?. *J. Chem. Phys., 117*, 10561–10570.
53. Torrent-Sucarrat, M., Luis, J. M., Duran, M., & Solà, M., (2001). On the validity of the maximum hardness and minimum polarizability principles for nontotally symmetric vibrations. *J. Am. Chem. Soc., 123*, 7951–7952.
54. Chattaraj, P. K., Roy, D. R., Elango, M., & Subramanian, V., (2005). Stability and reactivity of all-metal aromatic and antiaromatic systems in light of the principles of maximum hardness and minimum polarizability. *J. Phys. Chem. A., 109*, 9590 9597.
55. Ghanty, T. K., & Ghosh, S. K., (1996). A density functional approach to hardness, polarizability, and valency of molecules in chemical reactions. *J. Phys. Chem., 100*, 12295–12298.
56. Kaya, S., & Kaya, C., (2015). A simple method for the calculation of lattice energies of inorganic ionic crystals based on the chemical hardness. *Inorg. Chem., 54*, 8207–8213.
57. Born, M., & Mayer, J. E., (1932). The lattice theory of ionic crystals, *Journal for Physics, 75*, 1–18.
58. Born, M., & Landé, A., (1988). About calculation of compressibility of regular crystals from the lattice theory, *Fundamental Theories of Physics*. In: Selected Scientific Papers of Alfred Landé, Edition 1988, Springer Netherlands, pp. 58–64.
59. Kapustinskii, A. F., (1956). Lattice energy of ionic crystals. *Quarterly Reviews. Chemical Society., 10*, 283–294.
60. Kudriavtsev, B. B., (1956). A relation connecting between the ultrasonic velocity in an electrolytic solution and the lattice energy. *Sov. Phys. Acoust., 2*, 36–45.
61. Reddy, R. R., Kumar, M. R., & Rao, T. V. R., (1993). Interrelations between interionic separation, lattice energy, and plasma energy for alkali halide crystals. *Crystal Research and Technology., 28*, 973–977.
62. Jenkins, H. D. B., Roobottom, H. K., Passmore, J., & Glasser, L., (1999). Relationships among ionic lattice energies, molecular (formula unit) volumes, and thermochemical radii. *Inorg. Chem., 38*, 3609–3620.
63. Mallouk, T. E., Rosenthal, G. L., Mueller, G., Brusasco, R., & Bartlett, N., (1984). Fluoride ion affinities of germanium tetrafluoride and boron trifluoride from ther-

modynamic and structural data for (SF3)2GeF6, ClO2GeF5, and ClO2BF4. *Inorg. Chem., 23,* 3167–3173.

64. Ghanty, T. K., & Ghosh, S. K., (1993). Correlation between hardness, polarizability, and size of atoms, molecules, and clusters. *J. Phys. Chem., 97,* 4951–4953.

65. Reddy, R. R., Rao, T. V. R., & Viswanath, R., (1989). Lattice energies and optoelectronic properties of some ionic crystals. *Proc. Indian. Natl. Sci. Acad., 55,* 901–905.

66. Kaya, C., (2011). *Inorganic Chemistry 1.* Palme Publishing. Ankara.

67. Badger, R. M., (1934). A relation between internuclear distances and bond force constants. *J. Chem. Phys., 2,* 128–131.

68. Gordy, W., (1946). A relation between bond force constants, bond orders, bond lengths, and the electronegativities of the bonded atoms. *J. Chem. Phys., 14,* 305–320.

69. Pearson, R. G., (1993). Bond Energies, force constants and electronegativities. *J. Mol. Struct., 300,* 519–525.

70. Arulmozhiraja, B. S., & Kolandaivel, P., (1997). Force constants and chemical hardnesses. *Mol. Phys., 92,* 353–358.

71. Kaya, S., Kaya, C., Obot, I. B., & Islam, N., (2016). A novel method for the calculation of bond stretching force constants of diatomic molecules. *Spectrochimica Acta Part A: Molecular and Biomolecular Spectroscopy., 154,* 103–107.

72. Von Szentpály, L., (2011). Ruling out any electrophilicity equalization principle. *J. Phys. Chem. A., 115,* 8528–8531.

73. Islam, N., & Ghosh, D. C., (2015). Comment on "Ruling out Any Electrophilicity Equalization Principle" and Hardness Equalization Principle. *J. Phy. Chem. Biophys., 5,* 2161–0398.

74. Islam, N., & Ghosh, D. C., (2012). On the electrophilic character of molecules through its relation with electronegativity and chemical hardness. *Int. J. Mol. Sci., 13,* 2160–2175.

75. Chattaraj, P. K., & Maiti, B., (2001). Reactivity dynamics in atom-field interactions: a quantum fluid density functional study. *J. Phys. Chem. A., 105,* 169–183.

CHAPTER 6

INHIBITION OF METALLIC CORROSION BY N, O, S DONOR SCHIFF BASE MOLECULES

SOURAV KR. SAHA[1,2] and PRIYABRATA BANERJEE[1,2]

[1]*Surface Engineering and Tribology Group, CSIR-Central Mechanical Engineering Research Institute, Mahatma Gandhi Avenue, Durgapur 713209, West Bengal, India*

[2]*Academy of Scientific and Innovative Research (AcSIR), CSIR-CMERI Campus, West Bengal, Durgapur 713209, India*

CONTENTS

ABSTRACT

In today's world, the prevention of metallic corrosion is a very challenging job and would therefore be a mammoth task considering the enormous role of metals and their alloys in several industrial applications. To prevent solution state metallic corrosion suitable amount of organic molecules as inhibitor is widely used as additives to the corrosive solution. Among the organic inhibitors, different N, O, S donor Schiff base molecules are considerably used due to their low-cost starting precursor material, easy to follow synthetic route, and environmental friendly nature. Traditionally, scientists have identified new and prospective corrosion inhibitors either by the modifying the structure of the existing inhibitors or by following up the hardcore wet chemical synthetic routes, which in many instances have become laborious, expensive, time consuming and unable to reveal their microcosmic inhibition process. To overcome these shortfalls, in modern age, newly explored avenues like hardware and software advancements have opened up the door for powerful and worthy use of theoretical chemistry in corrosion inhibitory research domain following low-cost and time-saving processes. The use of computational chemistry in designing and development of organic corrosion inhibitors has been greatly enhanced due to the development of density functional theory (DFT). DFT is a modern tool that is capable enough to mostly predict accurately the relative order of inhibition efficiencies of concerned corrosion inhibitors, based on their molecular and electronic structure and reactivity indices. Furthermore, nucleophilic and electrophilic sites of the inhibitor molecules have also been determined from the analysis values of the DFT-based Fukui indices. Along with DFT, molecular dynamic (MD) simulation has also emerged as an important tool for investigating complex systems that actually need to be faced and experienced during corrosion prevention studies. MD simulation reasonably predicts the actual interfacial configuration of the surface adsorbed inhibitor molecules. Several parameters related to inhibition measurements *like* interaction energy and binding energy between the inhibitor molecule and metallic surfaces can also be calculated from MD simulation. From the obtained interaction energy and binding energy values it is possible to tell that which inhibitor molecule have superior adsorption ability on the metallic surfaces.

6.1   INTRODUCTION

A preventive step toward metallic corrosion inhibition is an important and highly concerning issue due to enormous role of metals and their alloys in several industrial applications. In fact, in all mechanical industries, iron and its alloys play a dominating role, and most unfortunately, their corrosion is the most common and alarming issue. In general, in several industrial processes, the use of acid solution (e.g., HCl, H_2SO_4, and HNO_3) in numerous techniques such as acid cleaning, acid pickling, and acid descaling is very common [1–3]. These acid solutions create serious metallic corrosion on the metallic surfaces. Due to such an adverse impact on metallic bodies, currently, almost all the industries are suffering from a huge and incredible amount of economic losses. Various attempts have been made to prevent or to minimize the metallic corrosion. Among the numerous techniques, the use of organic inhibitors is one of the most common and efficient methods to minimize the solution state metallic corrosion and to maximize the benefit. Generally, organic molecules containing polar groups have been used as effective corrosion inhibitors. An exhaustive survey of literature to date reveals that organic molecules having (i) polar functional group including N, O, and S donor sites, (ii) conjugated double bond, and (iii) planar aromatic rings are considered as an effective corrosion inhibitor [4–7]. In this context, several organic molecules such as imidazoline, triazole, pyridine, quinoline, thiourea, ionic liquid, benzimidazole, pyrimidine, pyrazine, and Schiff bases were used as a corrosion inhibitor material in many industrial applications [8–30]. Among all the inhibitors, Schiff base type inhibitors are widely used due to their low cost, easy to synthesize route, and environmental friendly nature [31–33]. Moreover, the most important advantages of using Schiff base are that one can easily modify their backbone according to the situation need. Owing to these characteristic properties, Schiff base molecules are potential materials to be used as corrosion inhibitors. In this chapter, the authors have mainly discussed different Schiff bases that were widely used as corrosion inhibitors in the aggressive acidic solution.

Organic inhibitor molecules in general adsorb on the metallic surfaces by the physical or chemical adsorption route. In physical adsorption, inhibitor molecules adsorb on the metallic surfaces by the electrostatic interaction between the inhibitor molecule and metallic surfaces. On the other hand, in chemical adsorption, inhibitor molecules donate the electron to the vacant d-orbital of iron and form a coordinate bond, and subsequently, it can also accept electrons from the metallic surfaces by backbonding. These two simultaneous adsorption processes facilitate a uniform film on the metallic surface, which in turn prevent the metallic surface from the aggressive acidic attacks.

In general, the evaluation of corrosion inhibition performance of the inhibitors is conducted experimentally such as weight loss determination, potentiodynamic polarization, and electrochemical impedance spectroscopy (EIS). By these techniques, it is possible to predict about the corrosion inhibition performance of the inhibitor molecules. However, these experimental methodologies are a bit costly, time consuming, and in some cases, they are unable to predict the actual mechanistic pathway acting on the metal solution interface [34, 35]. In view of above, in recent time, computer simulations are used to obtain a preformed idea about the molecules. Computer simulation has emerged as a powerful tool to investigate complex systems such as those experienced in corrosion. By using computer simulation, structures of the inhibitors, electron distribution, and the adsorption sites of the molecules responsible for the adsorption on the metallic surfaces have now been deeply explored. Among the computer simulation technique, density functional theory (DFT) and molecular dynamic (MD) simulation are widely used to investigate the complex systems due to their accuracy, low cost, and less time consumption [36, 37]. DFT and MD simulation techniques can exactly predict inhibition efficiency of the inhibitor molecules without performing any experimental works. A correlation of experimental and theoretical corrosion inhibition from where a clear idea regarding the molecular behavior of a molecule can be predicted is hereby discussed.

6.2 MATERIALS USED IN INDUSTRY

In the last few decades, a large number of metals (mild steel, aluminum, copper, etc.) have been utilized for different purposes in industry. Steel is the most important metal, used in oil and gas industry, from production and processing to the distribution of refined products [38]. After steel, aluminum is the most widely used metal on the planet. It is one of the key ingredients in the rapid expansion of network building and infrastructural expansion worldwide. Apart from steel and aluminum, copper and its alloys such as bronze have excellent electrical and thermal conductivity and cryogenic or cold-resistant properties. These metals are used in valves, stems, seals, and heat transfer applications. A bronze alloy with traces of nickel and aluminum can be used in wellheads and blowout prevention valves. In view of above, metals have gained a considerable attention due to their high societal impact.

In steel industries, steel plays an important role for pipeline construction. These pipelines are used for carrying liquids, gases, and aggressive solutions. The steel used for these pipes are API N80 (API-American Petroleum Institute), L80, J55, and Cr-containing alloys such as austenitic-ferritic steel [39–41]. Moreover, during the transportation of aggressive solutions, the surfaces of steel pipelines face serious threats of corrosion. In this perspective, in the last few decades, there have been many developments in manufacturing new corrosion-resistant alloys. Carbon steels are widely used in all industrial fields due to their cost-effectiveness [42, 43]. The cost of corrosion-resistant austenitic steels is several times higher than that of carbon steels [44]. It is therefore recommended for considering carbon steel along with chemical treatment rather than costly corrosion-resistant alloys as the most cost-effective methods.

6.3   WHY DO WE NEED INHIBITORS?

It is quite impossible to prevent corrosion completely; however, it is possible to control or restrict it in a limited range [45]. To prevent the unwanted corrosion in the inner wall of pipelines, tanks, coiled tubings, and in many other inner wall metallic surfaces, the floated acids need to be inhibited by using an effective corrosion inhibitor solution [46]. Corrosion inhibitor is the chemical substance that effectively reduces or prevents metallic corrosion in the aggressive medium. When corrosion inhibitors are added to water, acid, and steam in small amounts, they produce a surface film and thereby reduce the corrosion rate of the exposed metallic surface.

6.3.1 CLASSIFICATION OF INHIBITORS

Inhibitors can be classified according to their inhibition mechanism (cathodic, anodic, or both), application (pickling, acid cleaning, descaling, etc.), or based on their chemical nature. According to the chemical nature, inhibitor molecules are of two types: (i) inorganic inhibitor and (ii) organic inhibitor. In general, inorganic inhibitors have cathodic or anodic actions. The organic inhibitors have both the actions, i.e., cathodic and anodic (*vide* Figure 6.1).

6.3.1.1 Inorganic Inhibitors

In the nineteenth century, inorganic compounds such as nitrite, nitrate, chromate, arsenic, bismuth, and dichromate were extensively used as corrosion inhibitors. These inhibitors have shown better corrosion protection capability, although their uses were limited due to their toxicity and environmental hazards [47]. Inorganic inhibitors are of two types: anodic inhibitor and cathodic inhibitor.

(a) **Anodic inhibitor:** Anodic inhibitor increases anode polarization to the critical passivation potential of the metal or alloy. Anodic inhibitor mainly reduces anodic reaction by blocking the anodic reaction.

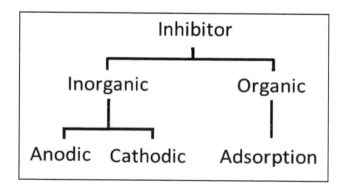

FIGURE 6.1   Classification of inhibitors.

These inhibitors are strong oxidizing agents and shift the corrosion potential of the metal in the noble direction with the formation of a passive film. Anodic inhibitors react with metallic ions (M^{n+}) and produce an insoluble hydroxide layer, which in turn deposits as an insoluble film on the metallic surfaces. The examples of anodic inhibitors are chromate, nitrite, nitrates, molybdates, etc.

(b) **Cathodic inhibitor:** Cathodic inhibitor decreases the corrosion rate by increasing the cathodic depolarization overvoltage. It is suggested that cathodic inhibitor affects the hydrogen evolution reaction or oxygen reduction reaction. These inhibitors generally have metal ions, and they can form an insoluble material that is precipitated selectively on the cathodic sites. The precipitated materials deposits over the metal surface and form a compact and adherent film. This barrier film restricts the diffusion of reducible species in these areas. The examples of cathodic inhibitors are arsenic, bismuth, antimony, etc.

6.3.1.2   Organic Inhibitor

Organic molecules containing sulphur, nitrogen, or oxygen atoms and organic heterocycles containing polar groups have been used as corrosion inhibitors. Organic inhibitors adsorb on the metallic surfaces and generally form a strong covalent bond. A thick film consisting of several monolayers are formed that cover total surface area of the corroded metal surface and change the structure of the double layer at the metal-solution interface. This layer may also act as a barrier film by blocking anodic and cathodic active sites or decreasing electroactive species transport rate to or from the metal surface. In this way, film-forming organic compounds may exhibit anodic, cathodic, or mixed properties.

6.3.2   *SELECTION OF CORROSION INHIBITORS*

It is very unfortunate to tell that inhibitors are effective only for a particular metallic material. Minor changes in the composition of the alloys or inhibitors change the inhibition effectiveness of the inhibitors used in the acid solutions. Earlier results reflected that many inhibitors that worked well in 15% HCl solution failed to work well in 28–30% HCl, and it is also valid for other acid solutions (e.g., H_2SO_4, HNO_3, and H3PO$_4$, etc.) [48]. In 1978, Smith and his co-workers reported that inhibitors for concentrated acid solution do not work at all in dilute acid solution [49]. Inhibitors also have sustained period up to which they can work. Another important point is that temperature effects also play a key role in the selection of corrosion inhibitors as inhibitors can work actively up to a certain range of temperature. Thus, the selection of inhibitors mainly depend on the type of acid and its strength, the steel type, the desired protection period, and the expected temperature.

6.3.3 DIFFERENT TYPES OF ORGANIC CORROSION INHIBITORS

Organic molecules are adsorbed on the metal surface due to existing heteroatoms and presence of unsaturated bonds. These favor electronic binding with the metal substrate. On this basis, a large number of organic molecules are used as corrosion inhibitors for metals in acidic solution. Among these inhibitors, the most common organic inhibitors are imidazoline [8,9], triazoles [10–12], pyridine and their derivatives [13, 14], quinoline derivatives [15, 16], thioureas [17, 18], ionic liquids [19], benzimidazole [20, 21], primidine [22–24], pyrazine [25, 26], and several condensation products of carbonyls and amines (Schiff base) [27–30].

Now, from the aforementioned structures of the concerned molecules (*vide* Table 6.1), it can be firmly said that all the molecules contain a good many number of heteroatoms, pi-electrons, and aromatic rings in the backbone of the molecule. In view of above, inhibitors containing a good many numbers of heteroatoms, pi-electrons, and aromatic rings behave as good corrosion inhibitors. However, before the selection of organic inhibitors, one has to consider that organic inhibitors should be low cost, easy to synthesize, and environmental friendly in nature. Among all the inhibitors presented above, Schiff base inhibitors are widely acclaimed due to their low-cost starting materials, relatively easy synthetic route, high purity, low toxicity, and eco-friendly nature. Several Schiff bases have already been used as corrosion inhibitors to protect various metals from the corrosive

TABLE 6.1   Chemical Structures of Different Types of Inhibitors

Imidazoline-based	2-(2-trifluoromethyl-4,5-dihydro-imidazol-1-yl)-ethylamine	2-(2-trichloromethyl-4,5-dihydro-imidazol-1-yl)-ethylamine	4,5-dihydro-2-tetradecylimidazole-1-carboxamide
Triazole based	3-Amino-1,2,4-triazole	3,5-bis(n-pyridyl)-4-amino-1,2,4-triazoles	3,5-bis(2-thienylmethyl)-4-amino-1,2,4-triazole
Pyridine based	Pyridine-2-thiol	2-Pyridyl disulfide	4-phenylpyrimidine
Quinoline based	8-hydroxyquinoline	8-aminoquinoline	3((phenylimino)methyl)quinoline-2-thiol
Thiourea based	Thiourea	1-p-tolylthiourea	1,3-diphenyl thiourea

TABLE 6.1   (Continued)

Ionic liquids	1-ethyl-3-methylimidazolium tetrafluoroborate	1-butyl-2,3-dimethylimidazolium tetrafluoroborate
Benzimidazole based	2-aminomethyl benzimidazole	bis (2-benzimidazolylmethyl) amine

Pyrimidine based	7-methoxypyrido [2,3-d]pyrimidin-4-amine	4,6-Diamino-2-pyrimidinethiol	2,4-diamino-6-hydroxy-pyrimidine

Pyrazine based	acenaphtho[1,2-b]quinoxaline	2-aminopyrazine	2-amino-5-bromopyrazine

Schiff base	N,N′-1,3-propylen-bis(3-methoxysalicylidenimine)	2-(2-hydroxy-5-nitrobenzylideneamino)phenol

environments. It has been targeted to unveil the corrosion inhibition properties of such Schiff bases. Before we cover all the results of inhibitors, it is quite essential to know how inhibitors work and how to materialize and finally conclude the corresponding inhibition efficiency of such materials.

6.4   EFFECTIVENESS OF CORROSION INHIBITORS

The inhibition efficiency of the materials is obtained from two different approaches. One is the experimental approach and the other is the theoretical approach. In the experimental approach, inhibition efficiency of the inhibitors is obtained using systematic laboratory methods such as weight loss measurement, potentiodynamic polarization, and electrochemical impedance spectroscopy, and in the theoretical approach, inhibition effectiveness is obtained from computational simulation approaches such as the DFT and MD simulation.

6.4.1   EXPERIMENTAL APPROACH

6.4.1.1   Weight Loss Measurement

Weight loss measurement is very useful due to its good reliability. In this process, metal coupons are immersed in an acidic solution with and without inhibitors for a certain period of time. In the first step before immersion in the acidic solution, weights of the polished, cleaned, and dried specimens are measured, and in the next step, mild steel coupons are placed in the aggressive acidic medium with and without inhibitors. After a fixed period of experimental investigation, all the specimens are taken out, washed thoroughly with distilled water to remove the corrosion product, dried with a hot air stream, and weighed to calculate the weight loss.

Percentage of inhibition efficiency ($\eta_{\%w}$) is calculated by the following formula:

$$\eta_{\%w} = \frac{w_0 - w}{w_0} \times 100 \tag{1}$$

where W_0 and W are the weight loss of metal specimens in the acid solution with and without inhibitor for the same immersion time.

6.4.1.2 Electrochemical Measurements

Electrochemical techniques of corrosion measurement are currently gaining increasing popularity due to the rapidity and quick predictability with which these measurements can be made. Long-term corrosion studies such as weight loss determinations may take several days or weeks, while an electrochemical experiment will require likely an hour.

In electrochemical measurement, potentiodynamic polarization and electrochemical impedance measurements are mainly carried out to determine inhibition efficiency of the inhibitors. These two measurements are carried out using a conventional three-electrode cell system. Here, metal specimens are used as a working electrode (WE), platinum sheet is used as a counter electrode, and saturated calomel electrode (SCE) or Ag/AgCl electrode is used as a reference electrode.

(a) **Potentiodynamic polarization:** Potentiodynamic polarization is used to know the corrosion current density in the presence and absence of inhibitors. When metal specimens are immersed in the aggressive medium, both the reactions (oxidation and reduction) take place on its surface. In the oxidation process, the metal surface is oxidized, and in the reduction process, H^+ ions are reduced. But when the metal specimen is in contact with an aggressive acidic medium and the whole setup is connected with the instrument, the metal specimen starts to function as both anode and cathode and therefore both the anodic and cathodic reactions occur on its surface.

In this study, before the experiment, the WE is kept in contact with the test solution for at least 45 minutes to achieve a steady state. When the steady state is achieved, the metal specimen is polarized both anodically and cathodically from the resting potential of the system as shown in Figure 6.2. Corrosion current density (i_{corr}) is determined from the extrapolation of cathodic

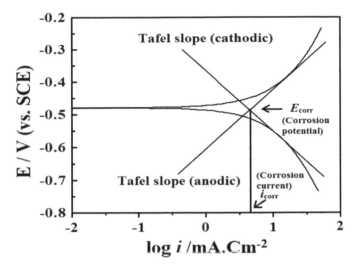

FIGURE 6.2   Potentiodynamic polarization curve.

and anodic tafel lines at the corrosion potential E_{corr}. To obtain
the inhibition efficiency of the inhibitors, this experiment should
be carried out in both presence and absence of the inhibitors. The
obtained corrosion current density in the presence and absence
of inhibitors is used to obtain the inhibition efficiency of the
inhibitor molecule.

By using the corrosion current density in the absence and presence of inhibitors, the degree of surface coverage (θ) and percentage of inhibition efficiencies ($\eta_{\%P}$) can be calculated as follows:

$$\theta = \frac{i_{corr} - i_{corr(inh)}}{i_{corr}} \tag{2}$$

$$\eta_{\%P} = \frac{i_{corr} - i_{corr(inh)}}{i_{corr}} \times 100 \tag{3}$$

where i_{corr} and $i_{corr(inh)}$ are the values of the corrosion current densities of uninhibited and inhibited specimens, respectively.

(b) Electrochemical impedance spectroscopy (EIS): In electrochemical impedance spectroscopy, the metal specimens are polarized by the application of alternating potential. For monitoring corrosion in this technique, the frequency range applied is typically in the range of 0.1 Hz to 100 kHz with a.c. amplitude of ± 10 mV (r.m.s.) at the open circuit potential (OCP). The application of small sinusoidal potential ($\Delta E \sin wt$) on the corroding metallic specimen results in a signal along with the harmonics 2ω, 3ω, and so on. Then, the impedance $\Delta I \sin (\omega t + \phi)$ is the relation between $\Delta E/\Delta I$ and phase ϕ [50]. In order to analyze the corrosion results, a randless circuit is fitted, which consists of solution resistance (R_s), polarization resistance (R_p), and double layer capacitance (C_{dl}). The measured impedance plots appear in the form of a semicircle, and it is commonly known as Nyquist plot. The Nyquist plot along with the equivalent circuit is shown in Figure 6.3. The obtained polarization resistant data are used to obtain the inhibition efficiency of the inhibitor molecule.

The following equation is used to obtain the inhibition efficiency of the inhibitor:

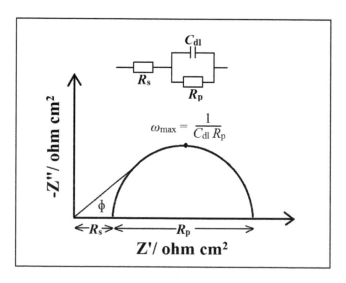

FIGURE 6.3   Nyquist plot with equivalent circuit.

$$\eta_{\%z} = \frac{R_p - R_p^0}{R_p} \times 100 \qquad (4)$$

where R_p^0 and R_p are the polarization resistance in absence and presence of an inhibitor molecule, respectively.

All the above-mentioned experimental techniques are generally used for the determination of corrosion inhibition performance of the inhibitor molecules.

6.4.1.3 Inhibition Mechanism

Inhibitor molecules adsorb on the metallic surfaces and form a layer on the surfaces that acts as a barrier against the corrosive agent in the aggressive solution. Inhibitor molecules adsorb on the metallic surface mainly by two adsorption phenomena. One is physical adsorption and the second is chemical adsorption. In physical adsorption, electrostatic interaction occurs between the charged surface of metals and the charged inhibitor molecule. In chemical adsorption, charge sharing or charge transfer occurs between the inhibitor molecule and metallic surfaces [51]. This section attempts to explain these two adsorption phenomena elaborately.

(a) **Physical adsorption:** Generally, all the inhibitor molecules contain heteroatoms in their backbone, and when these molecules are placed in an acidic solution, they are protonated in the acid medium.

$$\text{Inhibitor} + x\text{H}^+ \rightleftharpoons \left[\text{InhibitorH}_x\right]^{x+}$$

Thus, the inhibitor molecule becomes cationic in nature and exists in equilibrium with the corresponding molecular form. The charged inhibitor molecule may adsorb on the metallic surfaces by two processes.

(i) If the metal surface is negatively charged, the charged inhibitor molecule can directly adsorb on the metal surface by simple electrostatic interaction.

(ii) If the metal surface is positively charged, then Cl⁻ or SO_4^- ions (coming from the acids) first adsorb on the metal/solution interface through the electrostatic attraction force. The charge of the solution side interface will be changed from positive to negative. Thus, the cationic form of inhibitor molecule is capable to adsorb electrostatically on the metallic surfaces over the primarily adsorbed negatively charged ions.

The charge of the metal surfaces can be determined by the $(E_{corr} - E_q)$ value, where E_{corr} is the corrosion potential and E_q is the zero charge potential [52]. The E_q of iron is −530 mV vs. SCE in HCl medium [52]. In H_2SO_4 medium, the E_q of iron is −550 mV vs. SCE [53].

Metal surface is positively charged when $(E_{corr} - E_q)$ is greater than zero and negatively charged when $(E_{corr} - E_q)$ is less than zero. Thus, from this equation, it is possible to analyze the charge of the metallic surfaces. The charge determination helps to predict how the electrostatic interactions facilitate physical adsorption of the inhibitor molecules with the metal surface atoms.

(b) **Chemical adsorption:** Adsorption of the inhibitor molecules on the metallic surface occurs by the donation or acceptance of electrons between the organic inhibitor and vacant d-orbitals of the metallic atom surface. In this adsorption process, a coordinate type bond is formed. Chemical adsorption is mainly related to the density of the frontier molecular orbitals (FMOs) and their energies, that is, highest occupied molecular orbital (HOMO) and lowest unoccupied molecular orbital (LUMO) [54–57]. To calculate the frontier molecular orbital energies and their electronic distribution, theoretical calculation such as DFT is used. The theoretical section covers how chemical adsorption occurs and which factors play the crucial role for the adsorption of inhibitor molecules on the metallic surfaces.

6.4.2 THEORETICAL APPROACH

All the experimental techniques for determining corrosion inhibition capability are laborious, time consuming, expensive, and unable to reveal microcosmic inhibition process. To overcome all these limitations, several scientists are trying to determine an alternative route that can predict which molecule can be treated as a good corrosion inhibitor and which one not. In recent times, several scientists have used computational simulations to overcome these difficulties. Among the computer simulation techniques, DFT and MD simulation are the most authentic techniques due to their accuracy and capability to analyze the complex nature of these sorts of corrosion systems. These two techniques have enormous advantages in evaluating microcosmic inhibition performance as well as their mechanistic exploration.

6.4.2.1 Density Functional Theory (DFT)

The basis of DFT has been described in many books and articles [58–66] in the past 25 years. Therefore, it is not very wise to cumbersome this chapter with each and every detail of DFT. Instead of providing the details of the basis of DFT, it will be better to summarize the output obtained from DFT. It will be helpful to gather knowledge regarding the chemical reactivity of the molecules.

6.4.2.1.1 DFT-Based Chemical Reactivity Concepts and Their Application to Corrosion Inhibitor Chemistry

(a) **Frontier molecular orbitals:** The frontier orbitals of the inhibitor molecules are very useful in determining chemical reactivity. In 1982, Fukui first recognized the importance of frontier orbitals [67], and Parr and Yang demonstrated that most frontier theories can be rationalized from DFT [68]. They suggested that a good correlation is found between the frontier molecular orbital energies and corrosion inhibition efficiency of the molecule. The adsorptions of the inhibitor molecule on the metallic surfaces occur on the basis of donor-acceptor type interaction between the pi-electrons of the inhibitor molecule and vacant d-orbitals of the concerned metals. Therefore, higher the value of E_{HOMO}, higher will be the tendency of the molecule to donate their electrons to the vacant d-orbital of the metal. On the other hand, lower the energy of E_{LUMO}, higher is the tendency of the inhibitors to accept electrons from the metallic surfaces. Thus, lower the value of ΔE, higher the chemical reactivity of the inhibitor molecule [69, 70]. From Figure 6.4, it can be observed that the energy (ΔE) of the inhibitors plays the most crucial role for the adsorption of the inhibitor molecule.

Apart from E_{HOMO}, E_{LUMO}, and ΔE, other quantum chemical parameters such as Mulliken atomic charges, electron affinity (A), ionization potential (I), electronegativity (χ), hardness (η), softness (S), and fraction of electron transfered from the inhibitor molecule to the metallic surfaces (ΔN) also play the vital role. Electron affinity (A), ionization potential (I), elctronegativity (χ), hardness (η), softness (S), and fraction of electron transfer from the inhibitor molecule to the metallic surfaces (ΔN) can be calculated by the application of Koopmans' theorem [71]. According to the Koopmans' theorem, ionization potential is related to E_{HOMO},

$$I = -E_{HOMO} \tag{5}$$

The electron affinity is similarly related to E_{LUMO},

$$A = -E_{LUMO} \tag{6}$$

(b) Electronegativity: Electronegativity (χ) signifies electron attracting capability of the molecule. Higher the electronegativity of

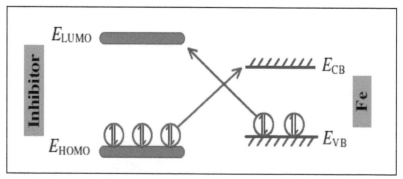

FIGURE 6.4   Chemical adsorption phenomenon. (Reprinted with permission from Guoa, L., Zhub, S., Zhanga, S., Hea, Q., & Li, W., (2014). Theoretical studies of three triazole derivatives as corrosion inhibitors for mild steel in acidic medium. *Corros. Sci., 87*, 366–375. © 2014 Elsevier.)

the inhibitor, stronger is the attracting power to accept electrons from the metallic surface [28]. Thus, inhibitor molecules that possess higher electronegativity would have strong interaction with the metal surface, and therefore, higher inhibition efficiency is expected. The ionization potential and electron affinity are used to obtain the electronegativity of the inhibitors. The controlling parameters can be calculated using the following equation:

$$\chi = \frac{I+A}{2} \tag{7}$$

(c) **Global hardness and softness:** The global hardness (η) is also calculated from the ionization potential and electron affinity values.

$$\eta = \frac{I-A}{2} \tag{8}$$

The global softness (S) is the inverse of global hardness

$$S = 1/\eta \tag{9}$$

Softness is also another important parameter to be considered for the analysis of adsorption capability of the molecule. In the corrosion inhibition study, inhibitors are considered as soft base and the metal surfaces as soft acid [37]. Thus, soft-soft interaction is the most predominant factor for the adsorption of inhibitor. Therefore, those inhibitors that have highest softness value behave as a good corrosion inhibitor.

(d) **The fraction of electron transferred:** When the inhibitor molecule and metallic surface are brought together, until and unless the chemical potential becomes equal, the electron flow will occur from the inhibitor molecule to the metallic atoms. According to Pearson [72], the fraction of electrons transferred from the inhibitor molecule to the metallic surface is given by the following equation:

$$\Delta N = \frac{\chi_M - \chi_{inh}}{2(\eta_M + \eta_{inh})} \tag{10}$$

where χ_M and χ_{inh} are the absolute electronegativity of the metal and the inhibitor molecule, respectively, and η_M and η_{inh} represent the absolute hardness of the metal and the inhibitor molecule, respectively. In order to calculate the fraction of electron transfer, a theoretical value $\chi_{Fe} = 7eV$ is considered for the absolute electronegativity of iron [73–74] and $\eta_{Fe} = 0$, by considering for metallic bulk $I = A$ [75]. Electron transfer from the inhibitor molecule to the metal surface occurs if $\Delta N > 0$ and from the metal surface to the molecule if $\Delta N < 0$ [76,77]. According to Elnga et al. [78], inhibition efficiency increases with increasing electron-donating ability of the molecule at the metal surface if $\Delta N < 3.6$.

In recent past, Kokaji [79] used the work function (ϕ) for the electronegativity of a metal surface as $\chi_{Fe} = 7eV$ is not conceptually correct because it is only associated with the free electron gas Fermi energy of iron where electron-electron interaction is not taken into the consideration. For this reason, currently, researchers have used work function (ϕ) instead of electronegativity of the metal [56, 79, 80]. Therefore, ΔN value calculation is most appropriate with work function (ϕ). By using work function (ϕ) and considering $\eta_M = 0$, Eq. (10) can be written as

$$\Delta N = \frac{\phi - \chi_{inh}}{2(\eta_M + \eta_{inh})} \tag{11}$$

Therefore, those inhibitor molecules having higher ΔN value can easily transfer electrons from the inhibitors to the vacant d-orbital of the metal and can form a strong coordinate bond.

(e) **Dipole moment:** The dipole moment is the only parameter to determine the polarity of the molecule. It is a measure of polarity of the polar covalent bond. The polarity of the molecule is defined as the product of charge of two atoms and distance between the two bonded atoms. The magnitude of μ is [56, 81]:

$$\mu = qR \tag{12}$$

where q is the charge and R is the distance. The SI unit of the dipole moment is coulomb meter (cm) but is commonly reported in the non-SI unit Debye.

In most of the literature, it has been reported that with increasing values of the dipole moment of the inhibitor, inhibition efficiency of the molecule increases [82, 83]. However, some degree of confusion exists wherein several studies report that inhibition efficiency decreases with increasing dipole moment [84, 85]. Therefore, a clear correlation cannot be established between the inhibition efficiency and the dipole moment of the inhibitor.

(f) **Electrophilicity index:** The global electrophilicity index (ω) was first introduced by Parr et al. [86], and it is given by the following equation:

$$\omega = \mu^2 / 4\eta \tag{13}$$

The electrophilicity index mainly measure the capability of the chemical species to accept electrons. The higher the value of ω, the higher is the capability of the molecule to accept electrons. A good nucleophile is therefore considered by the low value of ω and μ and a good electrophile is considered by a high value of ω and μ.

(g) **Fukui indices:** The local reactivity of the chemical species can be analyzed by evaluating Fukui indices [27, 28, 51]. These are the measure of chemical reactivity, indication of the reactive region and the nucleophilic and electrophilic behaviour of the molecule. The Fukui function f_k was defined as the first derivative of the electronic density $\rho(\vec{r})$ with respect to the number of electrons N in a constant external potential $\upsilon(\vec{r})$ [87].

$$f_k = \left(\frac{\partial \rho(\vec{r})}{\partial N} \right)_{\upsilon(\vec{r})}$$

The Fukui functions were written by taking the finite difference approximations as [88, 89]:

$$f_k^+ = q_k(N+1) - q_k(N) \qquad \text{(for nucleophilic attack)} \tag{14}$$

$$f_k^- = q_k(N) - q_k(N-1) \qquad \text{(for electrophilic attack)} \qquad (15)$$

$$f_k^0 = \frac{q_k(N+1) - q_k(N-1)}{2} \qquad \text{(for radical-based attack)} \qquad (16)$$

where q_k is the gross charge of k atom, i.e., the electron density at the point r in space around the molecule. The $q_k(N+1)$, $q_k(N)$, and $q_k(N-1)$ are defined as the charge of the anionic, neutral, and cationic species, respectively.

The nucleophilic and electrophilic attacks are determined by the calculated values of f_k^+ and f_k^- Generally, high value of f_k^+ and f_k^- implies higher capacity of the atom to accept and donate electrons, respectively.

6.4.2.2 Molecular Dynamics (MD) Simulation

MD simulation recently emerged as a modern tool to explore the adsorption behavior of the inhibitor molecule on the metallic surface. MD simulation predicts the most favorable adsorption configuration of surface adsorbed inhibitor molecules. In recent times along with DFT, MD simulation is also carried out for getting better adsorption configuration.

In this simulation, inhibitor molecules, solvent molecules, and corrosive particles like H_3O^+ and Cl^- are constructed from the atomic level, and geometry optimization of their structures are carried out by the smart algorithm which starts with steepest descent method followed by the conjugate gradient method and ends with the Newton method [51]. With the optimized structure of the concerned species, a cubic type simulation box is created with periodic boundary condition in the x, y, and z directions.

For equilibration of the system, MD simulations are performed using canonical ensemble where number of atoms (N), volume (V), and temperature (T) are kept fixed. The interaction of inhibitors on the surface is then simulated by condensed-phase optimized molecular potentials for atomistic simulation studies (COMPASS) force field [28, 89, 90]. COMPASS is an ab initio forcefield where most of the parameters are derived from the ab initio data. COMPASS forcefield has broad coverage in covalent molecules, including most common organic molecules, small inorganic molecules, and polymers [91].

The functional forms of COMPASS forcefield [90, 92, 93] are as follows:

$$E = E_{bond} + E_{angle} + E_{oop} + E_{torsion} + E_{cross} + E_{elec} + E_{il} \quad (17)$$

where E_{bond}, E_{angle}, E_{oop}, $E_{torsion}$, E_{cross}, $E_{elec,}$ and E_{lj} represent the contributions of bond stretching, angle bending, out-of-plane angle coordinates, torsion, cross coupling, electrostatic and van der Waals interactions. Each term of Eq. (17) can be expressed as follows:

$$E_{bond} = \sum_{b}[k_2(b-b_0)^2 + k_3(b-b_0)^3 + k_4(b-b_0)^4] \quad (18)$$

$$E_{angle} = \sum_{\theta}[H_2(\theta-\theta_0)^2 + H_3(\theta-\theta_0)^3 + H_4(\theta-\theta_0)^4] \quad (19)$$

$$E_{torsion} = \sum_{\phi}[V_1[1-\cos(\phi-\phi_0)^2] + V_2[1-\cos(2\phi-\phi_0)^2]$$
$$+ V_3[1-\cos(3\phi-\phi_0)^2]] \quad (20)$$

$$E_{oop} = \sum_{\chi}k_{\chi}\chi^2 \quad (21)$$

$$E_{cross} = \sum_{b}\sum_{b'}F_{bb'}(b-b_0)(b'-b_0') + \sum_{\theta}\sum_{\theta'}F_{\theta\theta'}(\theta-\theta_0)(\theta'-\theta_0')$$
$$+ \sum_{b}\sum_{\theta}F_{b\theta}(b-b_0)(\theta-\theta_0)$$
$$+ \sum_{b}\sum_{\phi}F_{b\phi}(b-b_0)[V_1\cos\phi + V_2\cos 2\phi + V_3\cos 3\phi]$$
$$+ \sum_{b'}\sum_{\phi}F_{b'\phi}(b'-b_0')[V_1\cos\phi + V_2\cos 2\phi + V_3\cos 3\phi]$$
$$+ \sum_{\theta}\sum_{\phi}F_{\theta\phi}(\theta-\theta_0)[V_1\cos\phi + V_2\cos 2\phi + V_3\cos 3\phi]$$
$$+ \sum_{\theta}\sum_{\theta'}\sum_{\phi}k_{\phi\theta\theta'}\cos\phi(\theta-\theta_0)(\theta'-\theta_0') \quad (22)$$

$$E_{elec} = \sum_{i,j} \frac{q_i q_j}{\varepsilon r_{ij}} \tag{23}$$

$$E_{lj} = \sum_{i,j} \varepsilon_{ij} \left[2\left(\frac{r_{ij}^0}{r_{ij}}\right)^9 - 3\left(\frac{r_{ij}^0}{r_{ij}}\right)^6 \right] \tag{24}$$

where b, θ, φ, and χ denote bond lengths, valence angles, torsion angles, and out-of-plane angles, respectively.

The most advantageous aspect of this simulation is that the interaction energy as well as binding energy value between the inhibitor molecule and metallic surfaces can be easily calculated. Higher the negative values of interaction energies, stronger is the interaction between the studied inhibitor molecules and metallic surfaces and thereby higher inhibition efficiency is observed.

The interaction energy between the inhibitor molecules and the metallic surface is calculated by the following equation [51, 94]:

$$E_{\text{interaction}} = E_{\text{total}} - \left(E_{\text{surface} + H_2O + H_3O^+ + Cl^-} + E_{\text{inhibitor}} \right) \tag{25}$$

Here, E_{total} is the total energy of the simulation system, $E_{\text{surface}+H2O+H3O^+ +Cl^-}$ is the energy of the iron surface together with H_2O molecule, H_3O^+ and Cl^- like ions, and $E_{\text{inhibitor}}$ is the energy of the adsorbed inhibitor on the surface.

The binding energy of the inhibitor molecule is calculated as follows:

$$E_{\text{binding}} = -E_{\text{interaction}} \tag{26}$$

The higher binding energy value also suggests a stable and better adsorption of the inhibitor molecule. Therefore, it is also possible to analyze the effectiveness of the inhibitors by the binding energy concepts.

Till now, we have covered both the experimental and theoretical techniques to calculate inhibition efficiency of the inhibitor. Now, we will try to cover the whole picture where N, O, and S donor Schiff bases were used in recent times as corrosion inhibitors. Our emphasis is also to summarize the percentage of inhibition effectiveness obtained from the above-mentioned techniques.

6.5   SCHIFF BASES USED AS CORROSION INHIBITORS

Among the acid solutions, HCl and H_2SO_4 are widely used for different purposes in industry. Hence, our main focus is on these acids. The emphasis herein is to briefly enumerate the different kinds of Schiff base compounds as corrosion inhibitors in these acid solutions. In general, Schiff base molecules contain N, O, and S like heteroatoms in their backbone, and based on the heteroatoms and pi-electrons in the aromatic ring of Schiff base molecules, these are capable to adsorb on the metallic surfaces. Here, the different Schiff base inhibitors are categorized as N, O-donor, N, S-donor, and N, O, S-donor. The following section describes these inhibitors in detail.

6.5.1   N, O-DONOR SCHIFF BASE USED AS INHIBITOR

Recently, Banerjee et al. have reported corrosion inhibition performance of 2-(2-hydroxybenzylideneamino)phenol (L^1), 2-(5-chloro-2-hydroxy-benzylideneamino)phenol (L^2), and 2-(2-hydroxy-5-nitrobenzylideneamino)phenol (L^3) on the mild steel surface in 1 M hydrochloric acid (HCl) solution [28]. Here, the authors have correlated the results obtained from three domains, namely wet chemical experimentation, quantum chemical calculation, and MD simulation. Wet chemical experimentation revealed that corrosion inhibition performance has the following order: $L^3 > L^2 > L^1$. This suggests that $-NO_2$-substituted Schiff base has better protection ability than the –Cl-substituted derivative and unsubstituted Schiff base is the least inhibitor. These experimental findings also corroborated with the theoretical ascertained results. The calculated quantum chemical parameters that best correlated with the inhibition efficiency are E_{HOMO} (highest occupied molecular orbital), E_{LUMO} (lowest unoccupied molecular orbital), ΔE (energy gap), electronegativity (χ), hardness (η), softness (S), and fraction of electron transferred (ΔN). Figure 6.5 shows the distribution of the electron density in the HOMO and LUMO of the inhibitors. From this figure, it is clear that both HOMO and LUMO of the inhibitors is electron rich and are capable to donate and accept electrons. It is also noted that for L^3, one part is responsible for electron donation, while the other part is preferable for electron acceptance. In this way, both the segments of a single inhibitor are playing an important role for the adsorption on the metallic surface. Thus, better the adsorption property, better will be the inhibition efficiency.

In addition to DFT calculations, MD Simulations are carried out to study better insightfulness of the adsorption processes. Figure 6.6 clearly showed that all three-inhibitor molecules are adsorbed on the Fe (1 1 0) surface with almost parallel or flat orientation. This flat orientation is possibly due to the formation of coordination and backbonding between the inhibitor and the metal surface. The interaction energy and binding energy obtained from the MD simulation are also following the order of $L^3 > L^2 > L^1$, which resembles the results obtained from the quantum chemical calculation.

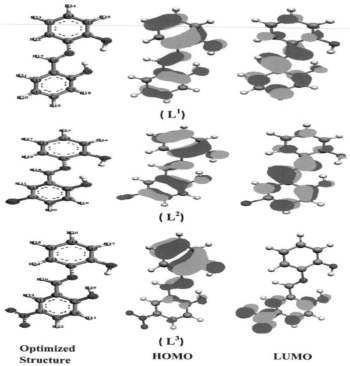

(L¹)

(L²)

(L³)

| Optimized Structure | HOMO | LUMO |

FIGURE 6.5 DFT-derived geometry optimized structure, HOMO, and LUMO plot of three Schiff base molecules at the B3LYP level in the aqueous phase. (Reprinted with permission from Saha, S. K., Dutta, A., Ghosh, P., Sukul, D., & Banerjee, P., (2015). *Phys. Chem. Chem. Phys.*, Adsorption and corrosion inhibition effect of Schiff base molecules on the mild steel surface in 1 M HCl medium: a combined experimental and theoretical approach, *17*, 5679–5690. © 2015 Royal Society of Chemistry.).

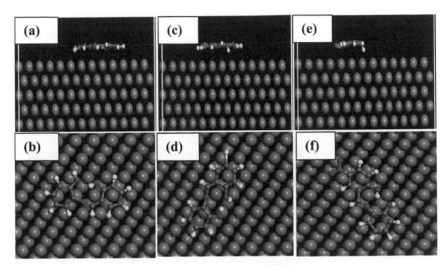

FIGURE 6.6   Equilibrium adsorption configurations of inhibitors L^1 (a and b), L^2 (c and d) and L^3 (e and f) on Fe (1 1 0) surface obtained by MD simulations. Top: top view, Bottom: side view (Reprinted with permission from Saha, S. K., Dutta, A., Ghosh, P., Sukul, D., & Banerjee, P., (2015). *Phys. Chem. Chem. Phys.,* Adsorption and corrosion inhibition effect of Schiff base molecules on the mild steel surface in 1 M HCl medium: a combined experimental and theoretical approach, 17, 5679–5690. © 2015 Royal Society of Chemistry.)

In recent times, we have also performed a theoretical study to correlate the obtained quantum chemical parameters for two mercapto-quinoline Schiff base molecule, namely 3-((phenylimino)methyl)quinoline-2-thiol (PMQ) and 3-((5-methylthiazol-2-ylimino)methyl) quinoline-2-thiol (MMQT) using DFT, with their experimentally obtained inhibition efficiency [27]. By using the ORCA software package, quantum chemical parameters such as E_{HOMO}, E_{LUMO}, ΔE, μ, χ, η, and ΔN were calculated. The MD studies revealed that all the calculated quantum chemical parameters agree well with the obtained inhibition efficiency order, which is MMQT > PMQ. Figure 6.7 shows the HOMO and LUMO orbitals of the inhibitors. From this figure, it is evident that all the inhibitors are electron rich and are capable of donating and accepting electrons with the metal surfaces. Geometry optimized structure of the inhibitors confirms that a steric effect is present in the two studies inhibitor molecules, and more precisely, the steric effect of PMQ is comparatively higher than that of MMQT. The in-house steric effect generated deviates PMQ from the planar structure and is responsible for the less inhibition effect.

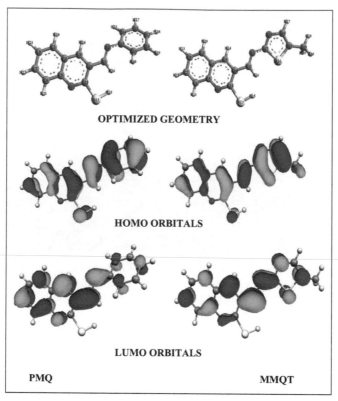

FIGURE 6.7   The optimized geometry: HOMO and LUMO orbitals of PMQ and MMQT at the B3LYP/SV(P), SV/J level of basic set for neutral species in the aqueous phase (Reprinted with permission from Saha, S. K., Ghosh, P., Hens, A., Murmu, N. C., & Banerjee, P., (2015). Density functional theory and molecular dynamics simulation study on corrosion inhibition performance of mild steel by mercapto-quinoline Schiff base corrosion inhibitor. *Phys. E., 66,* 332–341. © 2015 Elsevier.)

Active sites of the inhibitor molecules for adsorption on the metallic surface are also investigated by the Fukui indices. It was clear from the Fukui indices that PMQ and MMQT have many active sites for adsorption on the mild steel surface. Apart from DFT calculation, the obtained interaction energy and binding energy value from MD simulation results also corroborated with the experimental results.

In their work, Duran et al. [95] recently reported that 1,5-bis[2-(2-hydroxybenzylideneamino)phenoxy]-3-oxopentane (D1), 1,5-bis[2-(5-chloro-2-hydroxybenzylideneamino)phenoxy]-3-oxopentane (D2), and 1,5-bis[2-(5-bromo-2-hydroxybenzylideneamino)phenoxy]-3-oxopentane

(D3) behave as a protecting material on the aluminum surfaces in 0.1M HCl medium. It was reported that the inhibition efficiency obtained from the electrochemical impedance spectroscopy and Tafel polarisation measurements follow the order: D3 > D2 > D1. These outcomes imply that in the same N, O donor molecular skeleton, when Br-atom is present as a substituent, the protective ability is higher than that of the Cl-substituted one and Cl-substituted Schiff base is more protective than the unsubstituted one. In this study, the inhibition efficiency of the inhibitors was also cross-checked from the obtained quantum chemical parameters such as E_{HOMO}, E_{LUMO}, and ΔE. A good correlation was obtained between the experimental and theoretical findings.

Gürten et al. [96] fabricated allantoin with 2-hydroxybenzaldehyde to obtain 1-(2-hydroxybenzylidene)-3-(2,5-dioxoimidazolidin-4-yl) urea as Schiff base. Theyinvestigated this compound as corrosion inhibitor for carbon steel in 1M HCl medium. They reported that this molecule behaves as a mixed type inhibitor with 98% efficiency at 5 x 10^{-4} M concentration in 318 K temperature. The adsorption of the inhibitor on the metallic surfaces occurs by the chemisorption process. Thus, HOMO and LUMO play the most crucial role, and they have showed that HOMO and LUMO of the molecule are localized mainly on the phenyl ring as shown in Figure 6.8. This suggests that charge transfer from the HOMO of inhibitor to the metal surface and the opposite trend that is charge transfer from the surface to the LUMO of inhibitor occurs through the phenyl ring of ALS.

Quraishi et al. [97] investigated the effect of N'-(phenylmethylene) isonicotinohydrazide (INHB), N'-(2-hydroxybenzylidene) isonicotinohydrazide (INHS), N'-(furan-2-ylmethylene) isonicotinohydrazide (INHF) and N'-(3-phenylallylidene) isonicotinohydrazide (INHC) on

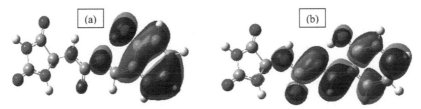

FIGURE 6.8 (a) HOMO and (b) LUMO of ALS. (Reprinted with permission from Gürten, A. A., Keleş, H., Bayol, E., & Kandemirli, F., (2015). The effect of temperature and concentration on the inhibition of acid corrosion of carbon steel by newly synthesized Schiff base *J. Ind. Eng. Chem.*, 27, 68–78. © 2015 Elsevier.

the corrosion of mild steel in 1M HCl medium. They investigated corrosion inhibition performance of the inhibitors by the weight loss, polarization curves, and electrochemical impedance spectroscopy measurement. The inhibition efficiency of Schiff bases followed the order INHC > INHF > INHB > INHS. The higher inhibition performance of INHC is explained on the presence of >C=C< in conjugation with the azomethine group (>C=N–). The extensive delocalized pi-electrons favors its greater adsorption on the mild steel surface. The authors also explained better performance of INHF (93.8%) than INHB (92.9%) is due to the presence of furan heterocyclic ring in the INHF molecule. This is because iron has better coordination affinity toward oxygen- and nitrogen-bearing ligands. The relatively poor performance of INHS may be probably due to the presence of the ortho hydroxyl group that prevents its flat orientation on the metal surface. Quraishi et al. [98] also investigated some new isatin derivative, namely 3-(4-(3-phenylallylideneamino)phenylimino) indolin-2-one (PI) and 3-(4-(4-methoxybenzylideneamino)phenylimino) indolin-2-one (MI), as corrosion inhibitors of mild steel in 20% H_2SO_4 by using the same technique. In all the experimental technique, the inhibition effectiveness order was as follows: PI > MI. In EIS measurement, the decrease in the C_{dl} values and increase in the R_{ct} value are also explained due to the decrease in local dielectric constant and/or an increase in the thickness of the electrical double layer. Parameters such as the E_{HOMO}, E_{LUMO}, energy gap (ΔE), Mulliken atomic charges, and dipole moments were calculated. Good agreement was found between the experimental and theoretical results.

N,N'-bis(*n*-hydroxybenzaldehyde)-1,3-propandiimine (n-HBP) Schiff bases were explored as an effective corrosion inhibitor for steel surface in 1M HCl medium by Danaee et al. [99]. They synthesized three Schiff bases, namely N,N'-bis(2-hydroxybenzaldehyde)-1,3-propandiimine (2-HBP), N,N'-bis(3-hydroxybenzaldehyde)-1,3-propandiimine (3-HBP), and N,N'-bis(4-hydroxybenzaldehyde)-1,3-propandiimine (4-HBP) and studied the effect of hydroxyl group position on the adsorption behavior of the molecule. Comparative studies of these inhibitors showed that the inhibition efficiency follows the order: 3-HBP > 4-HBP > 2-HBP. The obtained inhibition efficiency order is explained in terms of molecular geometry and the nature of their frontier molecular orbitals of the mole-

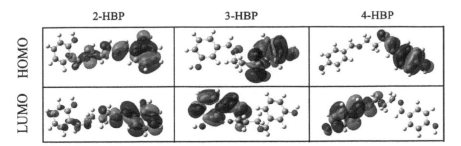

FIGURE 6.9 HOMO and LUMO populations of investigated Schiff bases obtained at the B3LYP/6-31G(d,p) level. (Reprinted with permission from Danaee, I., Ghasemi, O., Rashed, G. R., Avei, M. R., & Maddahy, M. H., (2013). Effect of hydroxyl group position on adsorption behavior and corrosion inhibition of hydroxybenzaldehyde Schiff bases: Electrochemical and quantum calculations. *J. Mole. Struct., 1035,* 247–259. © 2013 Elsevier.)

cule. From the optimized geometry, it is seen that all the molecules are not fully planer. However, if we look at the HOMO and LUMO distribution of the molecule (Figure 6.9), it can be seen that 3-HBP and 4-HBP molecules carry their rich negative centers in small region in comparison to the 2-HBP molecule. This suggests that in 3-HBP and 4-HBP molecules, one part of the molecule is responsible for electron donation and another part is responsible for electron acceptance. This strengthens the bonding between the inhibitor molecule and Fe-surfaces. A good correlation was also observed between the coefficient of E_{HOMO}, E_{LUMO}, ΔE, and the experimental inhibition efficiency.

Farag et al. [100] studied the inhibition effect of N-benzylidene-3-morpholinopropan-1-amine (SB-I), 3-morpholino-N-(pyridin-4-ylmethylene) propan-1-amine (SB-II), and N,N-dimethyl-4-((3-morpholinopropyl-imino) methyl)aniline (SB-III) and its synergistic effect with KI on the corrosion of carbon steel in 0.5M H_2SO_4 solution by the potentiodynamic polarization and electrochemical impedance spectroscopy (EIS) techniques. Potentiodynamic polarization and electrochemical impedance spectroscopy results showed that the efficiency trend of three inhibitors has the following trend: SB-III > SB-II > SB-I. This result implies that the [—N(CH$_3$)$_2$] substitution in benzene ring of SB-III has better prospect of adsorption capability than the pyridine substituent of SB-II and unsubstituted benzene ring of SB-I. Potentiodynamic polarization studies established that SB-I may act as an anodic inhibitor, while SB-II and SB-III act as a mixed-type inhibitor. The authors have also analyzed the synergistic effect with KI, and it was concluded that in presence of KI, the inhibition performance of the Schiff bases improve to more extent.

The effects of 1-(4-methyloxyphenylimino)-1-(phenylhydrazono)-propan-2-one (SB1), 1-(4-methylphenylimino)-1-(phenylhydrazono)-propan-2-one (SB2), 1-(phenylimino)-1 (phenylhydrazono)-propan-2-one (SB3), 1-(4-bromophenylimino)-1-(phenylhydrazono)-propan-2-one (SB4), and 1-(4-chlorophenylimino)-1(phenylhydrazono)-propan-2-one (SB5) on the corrosion of mild steel in 1M HCl medium were investigated by Daoud et al. [101]. The obtained results revealed that at 7.5 x 10^{-5} M concentration, the inhibition efficiency was 95.99%, 92.76%, 91.11%, 90.39%, and 88.31% for SB1, SB2, SB3, SB4, and SB5, respectively. These results showed that the ability of the molecules to adsorb on the mild steel surface depends on the substituent ($-OCH_3$, $-CH_3$, $-H$, $-Br$, and $-Cl$) in the azomethine compounds. Quantum chemical calculations using DFT illustrated that SB1 has a large negative charge on the nitrogen atom of an imino group and the oxygen of an acetyl group (–0.527 e, –0.512 e), which facilitates the better adsorption of SB1 onto the mild steel surface.

The inhibition efficiency of four Schiff base compounds, namely 2-((pyridin-2-ylimino) methyl) phenol (S1), 2-((hexadecylimino) methyl) phenol (S2), 2-((4-hydroxyphenylimino) methyl) phenol (S3), and 1-(4-(2-hydroxybenzylideneamino)phenyl) ethanone (S4) on the carbon steel in 1M HCl was studied by Hegazy et al. [102]. Inhibition efficiency was found to have a strong correlation with the substitutional effect of the inhibitor. The inhibition efficiency order of the investigated inhibitors in 1 M HCl solution was S2 > S3 > S1 > S4. Inhibitor S2 is present in the cationic surfactant form with ammonium ion as the polar group and alkyl chain as the nonpolar group. Consequently, its concentration at the interface is higher than the other studied Schiff base inhibitor, which reflects from its higher inhibition efficiency among the four inhibitors. Among the other three Schiff bases, S3 had the best performance. This has been explained by the authors that the presence of the hydroxyl group (–OH), which is an electron donor, moderately activates electron density on the azomethine ($>C=N$) group. Therefore, higher the binding capability and higher will be inhibition efficiency in comparison to S1 and S4. In S1, no group is attached to the pyridine ring, but inhibitor S3 has a benzene ring attached to the hydroxy group (–OH). Therefore, the inhibitor S3 showed slightly higher inhibition efficiency than the inhibitor S1. Least inhibition efficiency of S4 is explained as the carbonyl group (C=O) present in the inhibitor, which withdraws the electron from the azomethine (>C=N) group, resulting in a weak bonding of azomethine (>C=N) to the metal.

The corrosion inhibition and adsorption properties of (NE)-N-(furan-2-ylmethylidene)-4-({4-[E)-(furan-2-ylmethylidene) amino] phenyl} ethyl) aniline (SB) on the copper surface in 1M HCl medium were studied and investigated by Issaadi et al. [103]. This inhibitor showed 94% inhibition performance. Quantum chemical calculations were also carried out to establish a relationship between the molecular structure and the inhibition effect of the inhibitor. Figure 6.10 shows that the HOMO that two nitrogen atoms (N47 and N48) of azomethine groups and two oxygen atoms (O33 and O41) of furan heterocyclic ring have electron densities with Mulliken

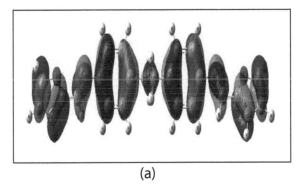

(a)

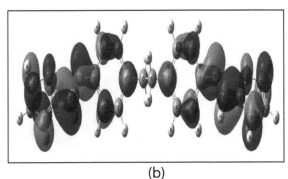

(b)

FIGURE 6.10 Frontier molecule orbitals: (a) HOMO and (b) LUMO density distributions of SB using DFT at the B3LYP/6-31G (d, p). (Reprinted with permission from Issaadi, S., Douadi, T., & Chafaa, S., (2014). Adsorption and inhibitive properties of a new heterocyclic furan Schiff base on corrosion of copper in HCl 1M: Experimental and theoretical investigation. *Appl. Surf. Science.*, *316*, 582–589. © 2014 Elsevier.)

atomic charges of (–0.47), (–0.47), (–0.43), and (–0.43), respectively. With these negative charges, this inhibitor is capable to adsorb on the copper surface by donating the unshared pair of electrons from N and O atoms to the vacant d-orbitals of copper.

The corrosion inhibition performance of N,N'-bis(4-formylphenol)-trimethylenediamine (4-FPTMD) for steel in 1M HCl medium was evaluated by Danaee et al. [104]. The Schiff base consists of two symmetrical parts and contains two benzene rings and two imine groups. The inhibition effectiveness was evaluated by the electrochemical techniques like potentiodyanamic polarization and electrochemical impedance spectroscopy, and the results obtained from this technique show that at 2×10^{-3} M concentration, this inhibitor is capable enough to inhibit corrosion up to 94%. In this work, to support the experimental results and to investigate the relationship between molecular structure of the Schiff base and its inhibition efficiencies, quantum chemical calculations were performed with two different methods: ab initio methods at the Hartreee–Fock (HF) level with the HF/6-31G(d,p) and HF/3-21G basis set and the DFT level with the B3LYP/6-31G(d,p) and

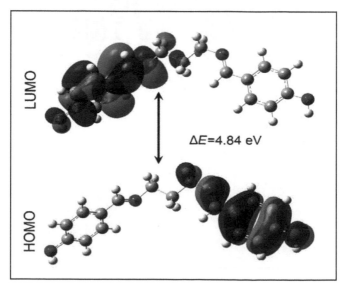

FIGURE 6.11 Molecular orbitals LUMO and HOMO of 4-FPTMD Schiff base obtained at the DFT-B3LYP (6-31G(d,p)) level. (Reprinted from Jafari, H., Danaee, I., Eskandari, H., & Rashvand Avei, M., (2014). Combined Computational and Experimental Study on the Adsorption and Inhibition Effects of N 2 O 2 Schiff Base on the Corrosion of API 5L Grade B Steel in 1 mol/L HCl. *J. Mater. Sci. Technol., 30,* 239–252. © 2014 Elsevier.)

B3LYP/3-21Gbasis sets. All the four different calculation methods yielded very close results. It is also seen from Figure 6.11 that HOMO densities of 4-FPTMD were comparatively distributed over the one side of the molecule and LUMO densities were distributed on the other side of the molecule. This suggested that one side of the molecule is responsible for the donation of electron and the other side is responsible for electron acceptance.

Xi et al. recently investigated the corrosion inhibition effect of a hydrazone derivative, namely 2-(2-{[2-(4-pyridylcabonyl)hydrazono]methyl} phenoxy)acetic acid (PMA), for mild steel surface in the synthetic seawater medium. [105]. A combined experimental and theoretical study was carried out in order to determine inhibition efficiency, reactive sites of the molecule, and mechanism of corrosion inhibition.

The effect of 2-(2 hydroxyphenyl)-2,5-diaza-4,6-dimethyl-8-hydroxy-1,5,7-nonatriene on mild steel in 1.0M HCl and 0.5M H_2SO_4 media was analysed by Mallaiya et al. [106]. The obtained results showed that this inhibitor has more inhibitive capability in the H_2SO_4 medium than in the HCl medium. The inhibitor showed 93% inhibition capability in HCl and 95.7% inhibition capability in H_2SO_4 medium at 600 mg/L concentration.

The adsorption capability of Schiff base nonionic surfactants, namely bis(N,N'-disalicylidene)-1,3-propanediamine-PEG400(PD-400), bis(N,N'-disalicylidene)-1,3-propanediamine-PEG600 (PD-600), bis(N,N'-disalicylidene)-1,8-octanediamine-PEG400 (OD-400), bis(N,N'-disalicylidene)-1,8-octanediamine-PEG600 (OD-600), bis(N,N'-disalicylidene)-1,10-decanediamine-PEG400 (DD-400), and bis(N,N'-disalicylidene)-1,10-decanediamine-PEG600 (DD-600), onto X-65 type tubing steel was studied by Migahed et al. [107]. The results obtained from the polarization, electrochemical impedance spectroscopy, and weight loss measurements revealed that these inhibitors are of mixed type (cathodic/anodic) and the order of inhibition efficiency is DD-600 > DD-400 > OD-600 > PD-600 >OD-400 > PD-400. The results of percentage inhibition efficiency (IE %) of surfactants obtained from EIS, potentiodynamic polarization, and weight loss measurement are in good agreement.

The inhibition effect of N,N'-bis(salicylidene)-1,2-ethylenediamine (Salen) and its reduced form (N,N'-bis(2-hydroxybenzyl)-1,2-ethylenediamine) and a mixture of its starting material, ethylenediamine and salicylaldehyde, on carbon steel in 1 mol L^{-1} HCl solution was studied by corrosion potential measurements, potentiodynamic polarization curves, electrochemical impedance spectroscopy, and spectrophotometry measurements [108]. The experimentally obtained results showed that among all the inhibitors studied, reduced salen has highest inhibition capability. In the acidic media, salen is unstable and undergoes hydrolysis to regenerate its precursor molecules salicylaldehyde and ethylenediamine, but the reduced salen is more stable in the acidic medium.

The interaction and corrosion inhibition properties of four different Schiff bases on mild steel in the HCl medium were investigated by experimental and computational calculation [109]. All the double Schiff bases, namely N,N'-bis(salicylidene)-2-methoxy-phenylmethanediamine (SB1), N,N'-bis(salicylidene)-2-hydroxy-phenylmethanediamine (SB2), N,N'-bis(salicylidene)-4-chloro-phenylmethanediamine (SB3), and N,N'-bis(salicylidene)-phenylmethanediamine (SB4), showed appreciable inhibition efficiency and the order of inhibition efficiency decreased in the order of SB1 ≈ SB2 > SB3 > SB4. Schiff bases act as good inhibitors due to the presence of the –C=N group, planarity, pi-electrons, and lone pairs of electrons present on N atoms. In this study, it was seen that both the –C=N groups present in the inhibitor molecule have steric hindrance that prevent their flat orientation. This non-planar orientation causes these inhibitor molecules (SB1, SB2, SB3 and SB4) to adsorb on the metallic surfaces by slightly tilting over the surface of the metal. The geometry optimized structure obtained from the DFT calculation showed that one of the –C=N groups has lesser steric hindrance and can interact with the steel surface, while in SB3 and SB4, the interaction of both –C=N groups is impossible.

MD simulation based on classical physics under periodic boundary condition has been used to construct a model for the inhibition performance of three Schiff bases, namely 3,5-dibromo salicylaldehyde-2-pyridinecarboxylic acid hydrazide (L1), 3,5-dibromo salicylaldehyde-2-thiol-phenecarboxylic acid hydrazide (L2), and 3,5-dibromo salicylaldehyde-2-aminobenzothiazole (L3), on a mild steel surface [110]. The results illustrate that all the Schiff bases were almost parallel in orientation with respect to the Fe (1 0 0) surfaces, which clearly exemplifies that inhibitor molecules cover a high surface area and forms a compact structured molecular film on the metal surfaces. The obtained interaction energy and binding energy follow the order L2> L1> L3, which agree with the results obtained from the experimental findings. Diffusion coefficient study was also carried out to analyze the effectiveness of the adsorbed inhibitor film, and the study showed that the inhibitor film inhibited diffusion of corrosive ions like Cl⁻ and H_3O^+. The diffusion coefficients followed the order of D(L3) > D(L1) > D(L2) for the Cl⁻corrosive particles and D(L3) > D(L1) > D(L2) for the H_3O^+corrosive particles. Furthermore, calculated quantum chemical parameters such as E_{HOMO}, E_{LUMO}, and ΔE also correlated with the experimentally obtained inhibition efficiency results.

1,8-bis[2-(5-chloro-2-hydroxybenzylideneamino)phenoxy]-3,6-dioxooctane (TC) and 1,8-bis[2-(5-bromo-2-hydroxybenzylideneamino) phenoxy]-3,6-dioxooctane (TB) were explored as a corrosion inhibitor of aluminum in 0.1M HCl by Yurt et al. [111] by using potentiodynamic polarization and electrochemical quartz crystal microbalance measurements. The result obtained from this study showed the inhibition efficiency increases with the increasing concentration of the Schiff bases. Between the two inhibitors, TB has more corrosion resistant capability than TC. This variation in the inhibition efficiency can be explained in light of presence of different electronegative substituents, such as Br and Cl atoms in the molecular structure of the inhibitor. Adsorption ability of the inhibitors onto the metal surface is related with the electronegativity values of substituents. The electronegativity value of Cl (3.16) is higher than that of Br (2.96). Therefore, the presence of Cl atom instead of Br atom in the molecular structure of Schiff base leads to a decrease in charge density on the imine (–C=N) group, oxygen, and benzene ring that are responsible for adsorption. This suggests that chemisorption occurs between the inhibitor and vacant p-orbital of aluminum is more facilitated when Br atom is present in the molecular scaffold.

6.5.2 N, S-DONOR SCHIFF BASE USED AS INHIBITOR

Inhibitory effect of imine compound, 2-(phenylthio)phenyl)-1-(o-tolyl) methanimine (PTM) and its cobalt complex (CoPTM), on low carbon steel (LCS) corrosion in HCl solution was studied using potentiodynamic polarization, electrochemical impedance spectroscopy, scanning electron microscope, and energy-dispersive X-ray spectroscopy methods by KELEŞ et al. [112]. The obtained electrochemical result showed that adsorption of CoPTM was stronger than that of PTM. As the adsorption capability of the CoPTM is stronger than that of the PTM, the inhibition capability of CoPTM is also higher than that of the PTM. Scanning electron microscopy result indicates that in the presence of PTM and CoPTM, the steel surfaces have a better morphology with less corroded features than the surface immersed in 1M HCl solution. It was observed that when the electrode is dipped into the CoPTM mixed solution, it had small corrosion impacts on the steel surface as compared to that when dipped in the PTM containing solution.

The effect of (NE) –N-(thiophen-3-ylmethylidene)-4-({4-[(E)-(thiophen-2-ylmethylidene)amino]phenyl}m-ethyl)aniline on the corrosion of mild steel in HCl and H_2SO_4 solution was investigated using combined experimental studies and theoretical modeling [113]. The experimental result shows that this inhibitor acts as a mixed type inhibitor with predominant anodic effectiveness in HCl and cathodic effectiveness in H_2SO_4 medium. The obtained experimental result affirmed that this inhibitor has nearly similar, i.e., 94% and 95%, inhibition efficiency in HCl and H_2SO_4 medium, respectively. The adsorption of the inhibitor on the metallic surface mainly occurs by the complex type interaction; this means that both chemisorption along with physisorption occur on the surfaces. Quantum chemical calculation, based on DFT is also carried out to determine the reactive sites for the adsorption of the inhibitor. Mulliken charge analysis showed that nitrogen and some carbon atoms of thiophene and benzene rings of the inhibitor have negative charges. These negative charges facilitate the adsorption of the molecule on the mild steel surface.

Solmaz [114] has also investigated the inhibition performance of 5-((E)-4-phenylbuta-1,3-dienylideneamino)-1,3,4-thiadiazole-2-thiol (PDTT) for mild steel corrosion in 0.5M HCl by potentiodynamic polarization, electrochemical impedance spectroscopy, linear polarization resistance, scanning electron microscopy, and atomic force microscopy techniques. The results obtained from these techniques revealed that this inhibitor showed remarkable inhibition effect in the aggressive solution of 0.5M HCl. The author explained that the high inhibition efficiency is due to the blocking of active sites by adsorption of inhibitor molecules on the steel surface.

Two novel Schiff base derivatives were synthesized and used as a novel corrosion inhibitor for mild steel in 1M HCl [115]. The inhibition characteristics of the inhibitors, namely 2-pyridinecarboxaldehyde thiosemicarbazone (2-PCT) and 4-pyridinecarboxaldehyde thiosemicarbazone (4-PCT), were studied in detail following the electrochemical techniques and theoretical calculation methods. The experimental results showed that the inhibition efficiency of 2-PCT is higher than that of 4-PCT. To support the experimental finding, quantum chemical calculation and MD simulation were carried out. The quantum chemical parameters obtained from the DFT and the interaction energy and binding energy obtained from the MD simulation agreed with the results obtained from the experimental findings.

6.5.3   N, O, S-DONOR SCHIFF BASE USED AS INHIBITOR

Döner et al. [116] performed an electrochemical study to analyze the corrosion inhibition effectiveness of 3-[(2-hydroxy-benzylidene)-amino]-2-thioxo-thiazolidin-4-one (HBTT) in 0.5M H_2SO_4 medium. The inhibitor was prepared from 1:1 mol ratio of 3-[(2-hydroxy-benzylidene)-amino]-2-thioxo-thiazolidin-4-one with salicylaldehyde through a condensation reaction in ethanolic media. The potentiodynamic polarization and electrochemical impedance spectroscopy (EIS) measurements revealed that HBTT has 99% corrosion inhibition capability. Potentiodynamic polarization also revealed that HBTT effectively suppresses both anodic and cathodic reactions and can act as a mixed type inhibitor. Scanning electron microscopy confirmed the existence of a protective layer on the mild steel surfaces.

The inhibition performance of 4-(4-bromophenyl)-N'-(2,4-dimethoxy-benzylidene) thiazole-2-carbo-hydrazide (BDTC), 4-(4-bromophenyl)-N'-(4-methoxybenzylidene)thiazole-2-carbohydrazide (BMTC), and 4-(4-bromophenyl)-N'-(4-hydroxybenzylidene) thiazole-2 carbohydrazide (BHTC) against the corrosion of mild steel in 0.5M HCl medium was investigated by weight loss and electrochemical measurement [117]. The inhibition efficiency increased with increasing inhibitor concentration and decreased with an increase in temperature. The obtained inhibition efficiencies of investigated compounds followed the order: BDTC > BHTC > BMTC. The obtained potentiodynamic polarization results indicate that all the studied inhibitors are of mixed type. The higher inhibition efficiency of BDTC is due to the electron donating effect of the two methoxy groups attached to the aromatic ring, which further increases the electron density on the benzene ring and thereby inhibits higher degree of corrosion.

The adsorption and corrosion inhibiting effect of 2-((5-mercapto-1,3,4-thiadiazol-2-ylimino)methyl)phenol (MTMP) on mild steel in 0.5M HCl solution was investigated by weight loss, electrochemical impedance spectroscopy, and potentiodynamic polarization techniques [118]. The obtained results showed that MTMP Schiff base has remarkable inhibition efficiency on the corrosion of mild steel in 0.5M HCl solution. Polarization measurements revealed that MTMP acts as a mixed type corrosion inhibitor with predominant cathodic inhibition. The inhibition efficiency also depends on the concentration of the inhibitor, and in this study, it reached the maximum of 97% at 1.0 mM of MTMP.

A heterocyclic Schiff base furoin thiosemicarbazone (FTSC) was explored as a corrosion inhibitor of mild steel in 1M HCl by Jacob et al. [30]. The obtained experimental outcome revealed that FTSC provides good corrosion inhibition efficiency even at low concentrations for mild steel in HCl medium, and its inhibition efficiency is higher than that of its parent amine, thiosemicarbazide molecule. Furthermore, Tafel polarization studies showed that FTSC acts as a mixed type inhibitor.

Cheng et al. [119] assessed the corrosion inhibition of copper in chloride solution by using triazolyl-acylhydrazone derivatives, namely salicylal-[5-(p-methyl)-phenyl-4-amino-(1,2,4-triazolyl)-2-thiol]-acylhydrazone (STA), anisalicylal-[5-(p-methyl)-phenyl-4-amino-(1,2,4-triazolyl)-2-thiol]-acylhydrazone (ATA), and vanillin-[5-(p-methyl)-phenyl-4-amino-(1,2,4-triazolyl)-2-thiol]-acylhydrazone (VTA). The results obtained from the polarization, electrochemical impedance spectroscopy, and weight loss measurements revealed inhibition efficiency increasing in the following order: VTA > STA > ATA. The adsorption of the inhibitor molecule on the copper surface mainly occurs by both chemisorption and physisorption; however, chemisorption of inhibitors on the copper surface plays the dominant role in corrosion inhibition.

6.6 CONCLUSION

In summary, extensive research studies have been carried out on the Schiff base molecules as promising inhibitors to minimize metallic corrosion. It is shown that Schiff base molecules are one of the most efficient corrosion inhibitors for the prevention of metallic corrosion. Their inhibition efficiency increases with increasing molecular concentration and time of immersion. Among the different N, O, S donor Schiff bases, N, O donor Schiff bases are the most frequently used ones due to their robustness as well as low-cost starting material. To analyze the corrosion inhibition effectiveness of the inhibitors, experimental and theoretical analyses were carried out by scientist and researchers worldwide. It was found that in many cases, experimental techniques are costly, time consuming, and unable to predict the mechanism of corrosion inhibition clearly. Keeping in mind these shortcomings, this chapter has mainly focused on the utilization of modern quantum chemical methods in the form of DFT to describe the adsorption capability of the inhibitors on the metallic surfaces in the solution medium. It is evident that DFT is a powerful tool with unique capability for performing practical directive calculations on inhibitors containing complex compounds and/or many body systems such as large organic molecules, etc. DFT offers the facility to design the molecule at the molecular level, including the choice and location of various functional groups as directed by the experimental work. The outcome obtained from the DFT calculations, such as energy of HOMO (E_{HOMO}), energy of LUMO (E_{LUMO}), energy gap (ΔE), electronegativity (χ), hardness (η), softness (S), electrophilicity index (ω), and fraction of electron transfer (ΔN), are very useful parameters to determine the inhibition efficiency of the inhibitors. Apart from these parameters, the distribution of the electron density in the frontier molecular orbitals is also very useful to analyze which segment of a concerned molecule is responsible for the flow and relay of electrons with the metal surfaces. In addition, MD simulation is another powerful tool to investigate the adsorption phenomenon of the inhibitor molecule on the desired metallic surfaces. The obtained interaction and binding energy are also helpful to predict the binding capability of the molecule. Therefore conclusively, DFT along with MD simulation is a highly useful computational technique to measure and provide a quick survey and perfect picture of the inhibition pathway and in determining the interactions between the inhibitor molecules and the concerned metal surfaces.

ACKNOWLEDGMENT

Authors are thankful to Dr. Pijush Pal Roy, the Director of CSIR-Central Mechanical Engineering Research Institute, Durgapur, India for his immense support. SKS would like to acknowledge the Department of Science and Technology (DST), New Delhi, India for his DST INSPIRE Fellowship.

KEYWORDS

- acid inhibition
- adsorption mechanism
- DFT calculation
- Fukui indices
- MD simulation
- Schiff base inhibitor

REFERENCES

1. Keles, H., Keles, M., Dehri, I., & Serindag, O., (2008). The inhibitive effect of 6-amino-m-cresol and its Schiff base on the corrosion of mild steel in 0.5 M HCl medium. *Mater. Chem. Phys., 112,* 173–179.
2. Wang, H., Fan, H., & Zheng, J., (2003). Corrosion inhibition of mild steel in hydrochloric acid solution by a mercapto-triazole compound. *Mater. Chem. Phys., 77,* 655–661.
3. El-Maksoud, S. A. A., & Fouda, A. S., (2005). Some pyridine derivatives as corrosion inhibitors for carbon steel in acidic medium. *Mater. Chem. Phys., 93,* 84–90.
4. Aljourani, J., Raeissi, K., & Golozar, M. A., (2009). Benzimidazole and its derivatives as corrosion inhibitors for mild steel in 1M HCl solution. *Corros. Sci., 51,* 1836–1843.

5. Zheludkevich, M. L., Yasakau, K. A., Poznyak, S. K., & Ferreira, M. G. S., (2005). Triazole and thiazole derivatives as corrosion inhibitors for AA2024 aluminium alloy. *Corros. Sci., 47*, 3368–3383.

6. Obot, I. B., Obi-Egbedi, N. O., & Umoren, S. A., (2009). The synergistic inhibitive effect and some quantum chemical parameters of 2, 3-diaminonaphthalene and iodide ions on the hydrochloric acid corrosion of aluminium. *Corros. Sci., 51*, 276–282.

7. Hosseini, M. G., Ehteshamzadeh, M., & Shahrabi, T., (2009). Protection of mild steel corrosion with Schiff bases in 0.5 MH 2 SO 4 solution. *Electrochim. Acta., 52*, 3680–3685.

8. Zhang, K., Xu, B., Yang, W., Yin, X., Liu, Y., & Chen, Y., (2015). Halogen-substituted imidazoline derivatives as corrosion inhibitors for mild steel in hydrochloric acid solution. *Corros. Sci., 90*, 284–295.

9. Ortega-Toledo, D. M., Gonzalez-Rodriguez, J. G., Casales, M., Martinez, L., & Martinez-Villafañe, A., (2011). CO_2 corrosion inhibition of X-120 pipeline steel by a modified imidazoline under flow conditions. *Corros. Sci., 53*, 3780–3787.

10. Finšgar, M., (2013). EQCM and XPS analysis of 1, 2, 4-triazole and 3-amino-1, 2, 4-triazole as copper corrosion inhibitors in chloride solution. *Corros. Sci., 77*, 350–359.

11. Lebrini, M., Traisnel, M., Lagrenée, M., Mernari, B., & Bentiss, F., (2008). Inhibitive properties, adsorption and a theoretical study of 3, 5-bis (n-pyridyl)-4-amino-1, 2, 4-triazoles as corrosion inhibitors for mild steel in perchloric acid. *Corros. Sci., 50*, 473–479.

12. Tourabi, M., Nohair, K., Traisnel, M., Jama, C., & Bentiss, F., (2013). Electrochemical and XPS studies of the corrosion inhibition of carbon steel in hydrochloric acid pickling solutions by 3,5-bis (2-thienylmethyl)-4-amino-1, 2, 4-triazole. *Corros. Sci., 75*, 123–133.

13. Kosari, A., Moayed, M. H., Davoodi, A., Parvizi, R., Momeni, M., Eshghi, H., & Moradi, H., (2014). Electrochemical and quantum chemical assessment of two organic compounds from pyridine derivatives as corrosion inhibitors for mild steel in HCl solution under stagnant condition and hydrodynamic flow. *Corros. Sci., 78*, 138–150.

14. Li, X., Xie, X., Deng, S., & Du, G., (2014). Two phenylpyrimidine derivatives as new corrosion inhibitors for cold rolled steel in hydrochloric acid solution. *Corros. Sci., 87*, 27–39.

15. Gao, H., Li, Q., Dai, Y., Luo, F., & Zhang, H. X., (2010). High efficiency corrosion inhibitor 8-hydroxyquinoline and its synergistic effect with sodium dodecylbenzenesulphonate on AZ91D magnesium alloy. *Corros. Sci., 52*, 1603–1609.

16. Wang, D., Yang, D., Zhang, D., Li, K., Gao, L., & Lin, T., (2015). Electrochemical and DFT studies of quinoline derivatives on corrosion inhibition of AA5052 aluminium alloy in NaCl solution, *Appl. Surf. Sci., 357*, 2176–2183.

17. Awad, M. K., (2004). Semiempirical investigation of the inhibition efficiency of thiourea derivatives as corrosion inhibitors. *J. Electroanal. Chem., 567*, 219–225.

18. Quraishi, M. A., Ansari, F. A., & Jamal, D., (2003). Thiourea derivatives as corrosion inhibitors for mild steel in formic acid. *Mater. Chem. Phys., 77*, 687–690.

19. Sasikumar, Y., Adekunle, A. S., Olasunkanmi, L. O., Bahadur, I., Baskar, R., Kabanda, M. M., Obot, I. B., & Ebenso, E. E., (2015). Experimental, quantum chemical and Monte Carlo simulation studies on the corrosion inhibition of some alkyl imidazolium ionic liquids containing tetrafluoroborate anion on mild steel in acidic medium. *J. Mole. Liquids., 211*, 105–118.

20. Tang, Y., Zhang, F., Hu, S., Cao, Z., Wu, Z., & Jing, W., (2013). Novel benzimidazole derivatives as corrosion inhibitors of mild steel in the acidic media. Part I: Gravimetric, electrochemical, SEM and XPS studies. *Corros. Sci., 74*, 271–282.

21. Aljourani, J., Raeissi, K., & Golozar, M. A., (2009). Benzimidazole and its derivatives as corrosion inhibitors for mild steel in 1M HCl solution. *Corros. Sci., 51*, 1836–1843.

22. Yadav, M., Kumar, S., Sinha, R. R., Bahadur, I., & Ebenso, E. E., (2015). New pyrimidine derivatives as efficient organic inhibitors on mild steel corrosion in acidic medium: electrochemical, SEM, EDX, AFM and DFT studies. *J. Mol. Liquids., 211*, 135–145.

23. Yıldız, R., (2015). An electrochemical and theoretical evaluation of 4, 6-diamino-2-pyrimidinethiol as a corrosion inhibitor for mild steel in HCl solutions. *Corros. Sci., 90*, 544–553.

24. Abdallah, M., Helal, E. A., & Fouda, A. S., (2006). Aminopyrimidine derivatives as inhibitors for corrosion of 1018 carbon steel in nitric acid solution. *Corros. Sci., 48*, 1639–1654.

25. Saranya, J., Sounthari, P., Parameswari, K., & Chitra, S., (2016). Acenaphtho [1, 2-b] quinoxaline and acenaphtho [1,2-b] pyrazine as corrosion inhibitors for mild steel in acid medium. *Measurement, 77*, 175–186.

26. Li, X., Deng, S., & Fu, H., (2011). Three pyrazine derivatives as corrosion inhibitors for steel in 1.0 MH 2 SO 4 solution. *Corros. Sci., 53*, 3241–3247.

27. Saha, S. K., Ghosh, P., Hens, A., Murmu, N. C., & Banerjee, P., (2015). Density functional theory and molecular dynamics simulation study on corrosion inhibition performance of mild steel by mercapto-quinoline Schiff base corrosion inhibitor. *Phys. E., 66*, 332–341.

28. Saha, S. K., Dutta, A., Ghosh, P., Sukul, D., & Banerjee, P., (2015). *Phys. Chem. Chem. Phys.,* Adsorption and corrosion inhibition effect of Schiff base molecules on the mild steel surface in 1 M HCl medium: a combined experimental and theoretical approach, *17*, 5679–5690.

29. Ansaria, K. R., Quraishia, M. A., & Singh, A., (2014). Schiff's base of pyridyl substituted triazoles as new and effective corrosion inhibitors for mild steel in hydrochloric acid solution. *Corros. Sci., 79*, 5–15.

30. Jacob, K. S., & Parameswaran, G., (2010). Corrosion inhibition of mild steel in hydrochloric acid solution by Schiff base furoin thiosemicarbazone. *Corros. Sci., 52*, 224–228.

31. Leçe, H. D., Emregül, K. C., & Atakol, O., (2008). Difference in the inhibitive effect of some Schiff base compounds containing oxygen, nitrogen and sulfur donors. *Corros. Sci., 50*, 1460–1468.

32. Issaadi, S., Douadi, T., Zouaoui, A., Chafaa, S., Khan, M. A., & Bouet, G., (2011). "Novel thiophene symmetrical Schiff base compounds as corrosion inhibitor for mild steel in acidic media. *Corros. Sci., 53*, 1484–1488.

33. Ansari, K. R., Quraishi, M. A., & Singh, A., (2014). Schiff's base of pyridyl substituted triazoles as new and effective corrosion inhibitors for mild steel in hydrochloric acid solution. *Corros. Sci., 79*, 5–15.

34. Satyanarayana, M. G. V., Himabindu, V., Kalpana, Y., Kumar, M. R., & Kumar, K., (2009). Part 1: In-silico studies as corrosion inhibitor and its experimental investigation on mild steel in wet-lab. *J. Mol. Struc. (Theochem), 912*, 113–118.

35. Muster, T. H., Hughes, A. E., Furman, S. A., Harvey, T., Sherman, N., Hardin, S., Corrigan, P., Lau, D., Scholes, F. H., White, P. A., Glenn, M., Mardel, J., Garcia, S. J., & Mol, J. M. C., (2009). A rapid screening multi-electrode method for the evaluation of corrosion inhibitors. *Electrochim.Acta., 54*, 3402–3411.

36. Yan, Y., Wang, X., Zhang, Y., Wang, P., & Zhang, J., (2013). Theoretical evaluation of inhibition performance of purine corrosion inhibitors. *Molecular Simulation, 39*, 1034–1041.

37. Obot, I. B., & Gasem, Z. M., (2014). Theoretical evaluation of corrosion inhibition performance of some pyrazine derivatives. *Corros. Sci., 83*, 359–366.

38. Walker, M. L., (1994). *Method and Composition for Acidizing Subterranean Formations*, in: US Patent 5,366,643, Halliburton Company, Duncan, Okla.

39. Baddini, A. L. Q., Cardoso, S. P., Hollauer, E., & Gomes, J. A. C. P., (2007). Statistical analysis of a corrosion inhibitor family on three steel surfaces (duplex, super-13 and carbon) in hydrochloric acid solutions. *Electrochim. Acta, 53*, 434–446.

40. Hill, D. G., & Jones, A., (2003). An engineered approach to corrosion control during matrix acidizing of HTHP sour carbonate reservoir, *Corrosion*, pp. 03121.

41. Hill, D. G., & Romijn, H., (2000). Reduction of risk to the marine environment from oilfield chemicals: environmentally improved acid corrosion inhibition for well stimulation, *Corrosion*, pp. 00342.

42. Nešic, S., & Sun, W., (2010). 2.25-Corrosion in Acid Gas Solutions, in: J. A. R. Tony (Ed.), *Shreir's Corrosion*, Elsevier, Oxford, pp. 1270–1298.

43. Badr, G. E., (2009). The role of some thiosemicarbazide derivatives as corrosion inhibitors for C-steel in acidic media. *Corros. Sci., 51*, 2529–2536.

44. Panossian, Z., Almeida, N. L. D., Sousa, R. M. F. D., Pimenta, G. D. S., & Marques, L. B. S., (2012). *Corros. Sci., 58*, 1–11.

45. Brondel, D., Edwards, R., Hayman, A., Hill, D., Mehta, S., & Semerad, T., (1994). Corrosion in the oil industry. *Oilfield Rev., 6*, 4–18.

46. Quraishi, M., & Jamal, D., (2000). Fatty acid triazoles: novel corrosion inhibitors for oil well steel (N-80) and mild steel. *J. Am. Oil Chem. Soc., 77*, 1107–1111.

47. Popov, B. N., (2015). Chapter 14–Corrosion Inhibitors. *Corrosion Engineering.*, 581–597.

48. Finšgar, M., & Jackson, J., (2014). Application of corrosion inhibitors for steels in acidic media for the oil and gas industry: a review. *Corros. Sci., 86*, 17–41.

49. Smith, C. F., Dollarhide, F. E., & Byth, N. B., (1978). Acid corrosion inhibitors-are we getting what we need?. *J. Petrol. Technol., 30*, 737–746.

50. Sastri, V. S., (2011). *Green Corrosion Inhibitors*, Wiley, June.

51. Saha, S. K., & Banerjee, P., (2015). A theoretical approach to understand the inhibition mechanism of steel corrosion with two aminobenzonitrile inhibitors. *RSC Adv., 5*, 71120–71130.

52. Deng, S., Li, X., & Fu, H., (2011). Two pyrazine derivatives as inhibitors of the cold rolled steel corrosion in hydrochloric acid solution. *Corros. Sci., 53*, 822–828.

53. Li, X., Deng, S., & Fu, H., (2011). Three pyrazine derivatives as corrosion inhibitors for steel in 1.0 MH 2 SO 4 solution. *Corros. Sci., 53*, 3241–3247.

54. Olasunkanmi, L. O., Obot, I. B., Kabanda, M. M., & Ebenso, E. E., (2015). Some quinoxalin-6-yl derivatives as corrosion inhibitors for mild steel in hydrochloric acid: Experimental and theoretical studies. *J. Phys. Chem. C., 119*, 16004–16019.

55. Obot, I. B., & Obi-Egbedi, N. O., (2010). Theoretical study of benzimidazole and its derivatives and their potential activity as corrosion inhibitors. *Corros. Sci. 52*, 657–660.

56. Obot, I. B., Macdonald, D. D., & Gasem, Z. M., (2015). Density functional theory (DFT) as a powerful tool for designing new organic corrosion inhibitors. Part 1: An overview. *Corros. Sci., 99*, 1–30.

57. Saha, S. K., Ghosh, P., Chowdhury, A. R., Samanta, P., Murmu, N. C., Lohar, A. K., & Banerjee, P., (2014). Corrosion control of chrome steel ball in nitric acid medium using Schiff base ligand and corresponding metal complexes: a combined experimental and theoretical study. *Can. Chem. Trans., 2*, 381.

58. Geerlings, P., De Proft, F., & Langenaeker, W., (2003). Conceptual density functional theory. *Chem. Rev., 103*, 1793–1874.

59. Nagy, A., (1998). Density functional theory and application to atoms and molecules. *Phys. Rev., 298*, 1–79.

60. Seminario, J. M., & Politzer, P., (1995). *Modern Density Functional Theory: A Tool for Chemistry*, Elsevier, Amsterdam.

61. Nalewajski, R. F., (1996). *Topics in Current Chemistry: Density Functional Theory*, Springer, Berlin.

62. Labanowski, J., & Andzelm, J., (1995). *Theory and Applications of Density Functional Approaches to Chemistry*, Springer, Berlin.

63. Springborg, M., (1997). *DFT Methods in Chemistry and Material Science*, Wiley, New York,.

64. Chong, D. P., (1995). *Recent Advances in Density Functional Methods*, World Scientific, Singapore,.

65. Yang, W., Levy, M., & Trickey, S., (1998). Special issue: Symposium on density functional and applications (part I of II): introduction. *Int. J. Quant. Chem., 69*, 227–227.

66. Geerlings, P., De Proft, F., & Martin, J. M. L., (1996). In recent developments and applications of modern density functional theory, in: J. M. Seminario (Ed.), *Theoretical and Computational Chemistry*, Elsevier, Amsterdam, *4*, pp. 773.

67. Fukui, K., (1982). Role of Frontier Orbitals in Chemical Reactions, *Science, 218*, 747.

68. Parr, R. G., & Yang, W., (1989). *Density Functional Theory of Atoms and Molecules*, Oxford University Press, Oxford.

69. Saha, S. K., Hens, A., Chowdhury, A. R., Lohar, A. K., Murmu, N. C., & Banerjee, P., (2014). Molecular dynamics and density functional theory study on corrosion inhibitory action of three substituted pyrazine derivatives on steel surface. *Can. Chem. Trans., 2*, 489–503.

70. Guoa, L., Zhub, S., Zhanga, S., Hea, Q., & Li, W., (2014). Theoretical studies of three triazole derivatives as corrosion inhibitors for mild steel in acidic medium. *Corros. Sci., 87*, 366–375.

71. Lukovits, I., Kalman, E., & Zucchi, F., (2001). Corrosion inhibitors-correlation between electronic structure and efficiency. *Corrosion., 57*, 3–8.

72. Pearson, R. G., (1963). Hard and soft acids and bases. *J. Am. Chem. Soc., 85*, 3533–3539.

73. Musa, A. Y., Jalgham, R. T. T., & Mohamad, A. B., (2012). Molecular dynamic and quantum chemical calculations for phthalazine derivatives as corrosion inhibitors of mild steel in 1M HCl. *Corros. Sci., 56*, 176–183.

74. Musa, A. Y., Kadhum, A. A. H., Mohamad, A. B., & Takriff, M. S., (2011). Molecular dynamics and quantum chemical calculation studies on 4,4-dimethyl-3-thiosemicarbazide as corrosion inhibitor in 2.5 M H_2SO_4. *Mater. Chem. Phys., 129*, 660–665.

75. Shokry, H., (2014). Molecular dynamics simulation and quantum chemical calculations for the adsorption of some Azo-azomethine derivatives on mild steel. *J. Mol. Struct., 1060*, 80–87.

76. Kokalj, A., (2010). Is the analysis of molecular electronic structure of corrosion inhibitors sufficient to predict the trend of their inhibition performance. *Electrochim. Acta., 56*, 745–755.

77. Kovacevic, N., & Kokalj, A., (2011). Analysis of molecular electronic structure of imidazole-and benzimidazole-based inhibitors: a simple recipe for qualitative estimation of chemical hardness. *Corros. Sci., 53*, 909–921.

78. Awad, M. K., Mustafa, M. R., & Elnga, M. M. A., (2010). Computational simulation of the molecular structure of some triazoles as inhibitors for the corrosion of metal surface. *J. Mole. Struct.(Theochem)., 959*, 66–74.

79. Kokaji, A., (2012). *Chem. Phys., 393*, 1.

80. Cao, Z., Tang, Y., Cang, H., Xu, J., Lu, G., & Jing, W., (2014). Novel benzimidazole derivatives as corrosion inhibitors of mild steel in the acidic media. Part II: Theoretical studies. *Corros. Sci., 83*, 292–298.

81. Atkins, P., & De Paula J., (2006). *Atkins Physical Chemistry*, eighth ed., Oxford, New York.

82. Sahin, M., Gece, G., Karci, F., & Bilgic, S., (2008). Experimental and theoretical study of the effect of some heterocyclic compounds on the corrosion of low carbon steel in 3.5% NaCl medium. *J. Appl. Electrochem., 38*, 809–815.

83. Quraishi, M. A., & Sardar, R., (2003). Hector bases–a new class of heterocyclic corrosion inhibitors for mild steel in acid solutions. *J. Appl. Electrochem., 33*, 1163–1168.

84. Khaled, K. F., Babic-Samardzija, N. K., & Hackerman, N., (2005). Theoretical study of the structural effects of polymethylene amines on corrosion inhibition of iron in acid solutions. *Electrochim.Acta., 50*, 2515–520.

85. Bereket, G., Hur, E., & Ogretir, C., (2002). Quantum chemical studies on some imidazole derivatives as corrosion inhibitors for iron in acidic medium. *J. Mol. Struct. (Theochem.), 578*, 79–88.

86. Parr, R. G., Sventpaly, L., & Liu, S., (1999). Electrophilicity Index. *J. Am. Chem. Soc., 121*, 1922–1924.

87. Proft, F. D., Martin, J. M. L., & Geerlings, P., (1996). Calculation of molecular electrostatic potentials and Fukui functions using density functional methods. *Chem. Phys. Lett., 256,* 400–408.

88. Contreras, R. R., Fuentealba, P., Galvan, M., & Perez, P., (1999). A direct evaluation of regional Fukui functions in molecules. *Chem. Phys. Lett., 304,* 405–413.

89. Dutta, A., Saha, S. K., Banerjee, P., & Sukul, D., (2015). Correlating electronic structure with corrosion inhibition potentiality of some bis-benzimidazole derivatives for mild steel in hydrochloric acid: Combined experimental and theoretical studies. *Corros. Sci., 98,* 541–550.

90. Sun, H., (1998). COMPASS: an ab initio force-field optimized for condensed-phase applications overview with details on alkane and benzene compounds. *J. Phys. Chem. B., 102,* 7338–7364.

91. Chakraborty, T., Hens, A., Kulashresth, S., Murmu, N. C., & Banerjee, P., (2015). Calculation of diffusion coefficient of long chain molecules using molecular dynamics. *Physica. E., 69,* 371–377.

92. Sun, H., Ren, P., & Fried, J. R., (1998). The COMPASS force field: parameterization and validation for phosphazenes. *Comput. Theor. Polym. S., 8,* 229–246.

93. Bunte, S. W., & Sun, H., (2000). Molecular modeling of energetic materials: the parameterization and validation of nitrate esters in the COMPASS force field. *J. Phys. Chem. B., 104,* 2477–2489.

94. Materials Studio 6.1 Manual, (2007). Accelrys, Inc., San Diego, CA.

95. Şafak, S., Duran, B., Yurt, A., & Türkoğlu, G., (2012). Schiff bases as corrosion inhibitor for aluminium in HCl solution. *Corros. Sci., 54,* 251–259.

96. Gürten, A. A., Keleş, H., Bayol, E., & Kandemirli, F., (2015). The effect of temperature and concentration on the inhibition of acid corrosion of carbon steel by newly synthesized Schiff base *J. Ind. Eng. Chem., 27,* 68–78.

97. Ahamad, I., Prasad, R., & Quraishi, M. A., (2010). Thermodynamic, electrochemical and quantum chemical investigation of some Schiff bases as corrosion inhibitors for mild steel in hydrochloric acid solutions. *Corros. Sci., 52,* 933–942.

98. Ansari, K. R., & Quraishi, M. A., (2015). Experimental and quantum chemical evaluation of Schiff bases of isatin as a new and green corrosion inhibitors for mild steel in 20% H 2 SO 4. *J. Taiwan. Inst. Chem. Eng., 54,* 145–154.

99. Danaee, I., Ghasemi, O., Rashed, G. R., Avei, M. R., & Maddahy, M. H., (2013). Effect of hydroxyl group position on adsorption behavior and corrosion inhibition of hydroxybenzaldehyde Schiff bases: Electrochemical and quantum calculations. *J. Mole. Struct., 1035,* 247–259.

100. Farag, A. A., & Hegazy, M. A., (2013). Synergistic inhibition effect of potassium iodide and novel Schiff bases on X65 steel corrosion in 0.5 M H 2 SO 4. *Corros. Sci., 74,* 168–177.

101. Hamani, H., Douadi, T., Al-Noaimi, M., Issaadi, S., Daoud, D., & Chafaa, S., (2014). Electrochemical and quantum chemical studies of some azomethine compounds as corrosion inhibitors for mild steel in 1M hydrochloric acid. *Corros. Sci., 88,* 234–245.

102. Hegazy, M. A., Hasan, A. M., Emara, M.M., Bakr, M. F. & Youssef, A. F., (2012). Evaluating four synthesized Schiff bases as corrosion inhibitors on the carbon steel in 1 M hydrochloric acid, *Corros. Sci., 65,* 67–76.

103. Issaad, S., Douadi, T., & Chafaa, S., (2014). Adsorption and inhibitive properties of a new heterocyclic furan Schiff base on corrosion of copper in HCl 1M: Experimental and theoretical investigation. *Appl. Surf. Science., 316,* 582–589.

104. Jafari, H., Danaee, I., Eskandari, H., & Rashvand Avei, M., (2014). Combined Computational and Experimental Study on the Adsorption and Inhibition Effects of N 2 O 2 Schiff Base on the Corrosion of API 5L Grade B Steel in 1 mol/L HCl. *J. Mater. Sci. Technol., 30,* 239–252.

105. Liu, B., Xi, H., Li, Z., & Xia, Q., (2012). Adsorption and corrosion-inhibiting effect of 2-(2-{[2-(4-Pyridylcarbonyl) hydrazono] methyl} phenoxy) acetic acid on mild steel surface in seawater. *Appl. Surf. Science., 258,* 6679–6687.

106. Mallaiya, K., Subramaniam, R., Srikandan, S. S., Gowri, S., Rajasekaran, N., & Selvaraj, A., (2011). Electrochemical characterization of the protective film formed by the unsymmetrical Schiff's base on the mild steel surface in acid media. *Electrochim. Acta., 56,* 3857–3863.

107. Migahed, M. A., Farag, A. A., Elsaed, S. M., Kamal, R., Mostfa, M., & Abd El-Bary, H., (2011). Synthesis of a new family of Schiff base nonionic surfactants and evaluation of their corrosion inhibition effect on X-65 type tubing steel in deep oil wells formation water.Mater. *Chem. Phys., 125,* 125–135.

108. Silva, A. B., D'Elia, E., & Gomes, J. A. C. P., (2010). Carbon steel corrosion inhibition in hydrochloric acid solution using a reduced Schiff base of ethylenediamine. *Corros. Sci., 52,* 788–793.

109. Soltani, N., Behpour, M., Ghoreishi, S. M., & Naeimi, H., (2010). Corrosion inhibition of mild steel in hydrochloric acid solution by some double Schiff bases. *Corros. Sci., 52,* 1351–1361.

110. Xie, S. W., Liu, Z., Han, G. C., Li, W., Liu, J., & Chen, Z. C., (2015). *Comp. Theor. Chem. 1063,* 50.

111. Yurt, A., & Aykın, Ö., (2011). Diphenolic Schiff bases as corrosion inhibitors for aluminium in 0.1 M HCl: Potentiodynamic polarisation and EQCM investigations. *Corros. Sci., 53,* 3725–3732.

112. Keles, H., Emir, D. M., & Keles, M., (2015). A comparative study of the corrosion inhibition of low carbon steel in HCl solution by an imine compound and its cobalt complex. *Corros. Sci., 101,* 19–31.

113. Daoud, D., Douadi, T., Issaadi, S., & Chafaa, S., (2014). Adsorption and corrosion inhibition of new synthesized thiophene Schiff base on mild steel X52 in HCl and H 2 SO 4 solutions. *Corros. Sci., 79,* 50–58.

114. Solmaz, R., (2010). Investigation of the inhibition effect of 5-((E)-4-phenylbuta-1, 3-dienylideneamino)-1, 3, 4-thiadiazole-2-thiol Schiff base on mild steel corrosion in hydrochloric acid. *Corros. Sci., 52,* 3321–3330.

115. Xu, B., Yang, W., Liu, Y., Yin, X., Gong, W., & Chen, Y., (2014). Experimental and theoretical evaluation of two pyridinecarboxaldehyde thiosemicarbazone compounds as corrosion inhibitors for mild steel in hydrochloric acid solution. *Corros. Sci., 78,* 260–268.

116. Doner, A., Sahin, E. A., Kardas, G., & Serindag, O., (2013). *Corros. Sci., 66,* 278–284.

117. Kumar, C. B. P., & Mohana, K. N., (2014). Corrosion inhibition efficiency and adsorption characteristics of some Schiff bases at mild steel/hydrochloric acid interface. *J. Taiwan. Inst. Chem. Eng., 45,* 1031–1042.

118. Solmaz, R., Altunbas, E., & Kardas, G., (2011). Adsorption and corrosion inhibition effect of 2-((5-mercapto-1, 3, 4-thiadiazol-2-ylimino) methyl) phenol Schiff base on mild steel.Mater. *Chem. Phys., 125,* 796–801.
119. Tian, H., Cheng, Y. F., Li, W., & Hou, B., (2015). Triazolyl-acylhydrazone derivatives as novel inhibitors for copper corrosion in chloride solutions. *Corros. Sci., 100,* 341–352.

CHAPTER 7

CONCEPTUAL DENSITY FUNCTIONAL THEORY AND ITS APPLICATION TO CORROSION INHIBITION STUDIES

IME BASSEY OBOT,[1] SAVAŞ KAYA,[2] and CEMAL KAYA[2]

[1]Centre of Research Excellence in Corrosion, Research Institute, King Fahd University of Petroleum and Minerals, Dhahran 31261, Kingdom of Saudi Arabia

[2]Cumhuriyet University, Faculty of Science, Department of Chemistry, Sivas, 58140, Turkey

CONTENTS

ABSTRACT

Corrosion scientists have identified new corrosion inhibitor molecules either by incrementally changing the structures of existing inhibitors or by testing hundreds of compounds in the laboratory, these experimental means are often very expensive and time-consuming. Developments and major advances in computer hardware and softwares have opened the door for powerful use of theoretical chemistry in corrosion inhibition research. Using Conceptual Density Functional Theory (CDFT), corrosion scientists are now able to predict electronic, molecular and adsorption properties of corrosion inhibitor molecules. Density Functional Theory (DFT) offers the facility to design, at the molecular level, the molecular structure, including the choice and location of various functional groups and to build in selected aromaticity, as dictated by experimental feedback from studies on existing inhibitors. This chapter is structured as follows: introduction, corrosion mechanisms of steel, brief introduction to corrosion inhibitors, brief introduction to DFT, conceptual DFT parameters related to corrosion inhibition, an illustrative example of the application of conceptual DFT to corrosion inhibition studies, and conclusion.

7.1   INTRODUCTION

Chemical/electrochemical reactions occur at the interface between the corrosive medium and the surface of the metal, which result in metal dissolution into the solution and formation of corrosion product. The annual direct cost of metallic corrosion in US in 1998 was estimated to be 3.1% of the gross domestic product (GDP) or approximately $300 billion. A significant milestone in the effect of corrosion on the US economy occurred in 2013 when the total cost of corrosion in the US exceeded $1 trillion annually for the first time [1]. The annual cost of corrosion worldwide is estimated at $2.2 trillion (2010), which is about 3% of the world's GDP of $73.33 trillion [2]. One of the most effective alternatives for the protection of metallic surfaces against corrosion is to use organic inhibitors containing nitrogen, oxygen, sulfur, and aromatic ring in their molecular structure [3–5]. The existing data reveal that most organic inhibitors act by adsorption on the metal surface. This is influenced by the nature and surface charge of the metal, the type of aggressive electrolyte, and the chemical structure of inhibitors [6, 7].

Experimental methods are useful in the discovery of new corrosion inhibitors. However, they are generally expensive and time-consuming and are deficient in studying inhibition mechanism at the three-dimensional atomic level. The use of quantum chemical calculations, especially the density functional theory (DFT), as a tool in the design of effective corrosion inhibitors at the molecular level has received a lot of attention in recent times [8–12]. This is supported by ongoing computer hardware and software advances, which have opened the door for the powerful use of theoretical chemistry (computational chemistry) in corrosion inhibition research. Through DFT calculations, the corrosion inhibition efficiencies of molecules are associated with quantum chemical parameters such as the energies of the highest occupied molecular orbital (E_{HOMO}) and the lowest unoccupied molecular orbital (E_{LUMO}), HOMO–LUMO energy gap (ΔE), chemical hardness (η), softness (S), electronegativity (χ), and electrophilicity (ω). A recent comprehensive review by us on the use of DFT as a tool in the design of corrosion inhibitors is available in the literature and the references therein [13].

The main goal of this chapter is to explain how conceptual DFT could be applied to corrosion inhibition studies. The chapter is arranged as follows: introduction, corrosion mechanisms of steel, brief introduction to corrosion inhibitors, brief introduction to DFT, conceptual DFT parameters related to corrosion inhibition, an illustrative example of the application of conceptual DFT to corrosion inhibition studies, and conclusion.

7.2 CORROSION MECHANISMS OF STEEL

There are different mechanisms by which steel corrodes depending on the corrosive media. Some of the corrosive media of industrial importance, such as acid, alkali, carbon dioxide (CO_2), and hydrogen sulfide (H_2S), are discussed below.

The corrosion process of steel in aqueous environments involves two reactions [14], namely the oxidation reaction at the anode

$$M_o \rightarrow M^{n+} + ne^- \text{ anodic reaction} \tag{1}$$

and the reduction reaction at the cathode. The three possible cathodic reactions are

$$2H_3O^+ + 2e^- \rightarrow 2H_2O + H_2 \text{ (acid solutions)} \tag{2}$$

$$O_2 + 4H^+ + 4e^- \rightarrow 2H_2O \text{ (acid solutions)} \tag{3}$$

$$O_2 + 2H_2O + 4e^- \rightarrow 4OH^- \text{(alkaline solutions)} \tag{4}$$

When the corrosive medium contains other species such as a ferric ion and an acid such as nitric acid, the following cathodic reactions may be encountered:

$$Fe^{3+} + e^- \rightarrow Fe^{2+} \tag{5}$$

$$3H^+ + NO_3^- + 2e^- \rightarrow HNO_2 + H_2O \tag{6}$$

When several cathodic reactions are possible, the cathodic reaction leading to the largest corrosion current is considered to be the principle cathodic electronation reaction corresponding to the given potential. This observation is confirmed by the increased corrosion rate of iron in aerated solutions as compared to that in deoxygenated solutions.

The reaction in deaerated acid solutions can be written as

$$2H_3O^+ + 2e^- \rightarrow 2H_2O + H_2 \tag{7}$$

and the reaction in oxygenated solutions can be written as

$$O_2 + 4H^+ + 4e^- \rightarrow 2H_2O \tag{8}$$

The type of acid also appears to affect the corrosion rate, as is the case with nitric acid because of the following reaction:

$$3H^+ + NO_3^- + 2e^- \rightarrow HNO_2 + H_2O \tag{9}$$

For internal corrosion in oil pipeline containing CO_2, the basic corrosion reactions are presented as follows:

CO_2 dissolves in the presence of water, forming carbonic acid, which is corrosive to carbon steel, as shown in Eq. (10) [15]:

$$CO_2 \text{ (g)} + H_2O \text{ (l)} \rightarrow H_2CO_3 \text{ (aq)} \tag{10}$$

Several mechanisms have been proposed for the dissolution of steel in aqueous, deareated CO_2 solutions. The main corrosion process can be summarized by three cathodic (Eq. (10a), Eq. (10b), and Eq. (10c)) and one anodic Eq. (11) reactions:

$$2H_2CO_3 + 2e^- \rightarrow H_2 + HCO_3^- \tag{10a}$$

$$2HCO_3^- + 2e^- \rightarrow H_2 + 2CO_3^{2-} \tag{10b}$$

$$2H^+ + 2e^- \rightarrow H_2 \tag{10c}$$

$$Fe \rightarrow Fe^{2+} + 2e^- \tag{11}$$

Due to these processes, a corrosion layer is formed on the steel surface. The properties of this layer and its influence on the corrosion rate are important factors to be taken into account when studying the corrosion of steels in CO_2 aqueous solutions. Iron carbonate, $FeCO_3$, plays an important role in the formation of protective layers. Its formation can be explained using Eq. (12), Eq. (13a), and Eq. (13b). Because of its low solubility, $FeCO_3$ precipitates out of solution ($pK_{sp} = 10.54$ at 25° C):

$$Fe + CO_3^{2-} \rightarrow FeCO_3 \tag{12}$$

$$Fe^{2+} + 2HCO_3^- \rightarrow Fe(CO_3)_2 \tag{13a}$$

$$Fe(CO_3)_2 \rightarrow FeCO_3 + CO_2 + H_2O \tag{13b}$$

The corrosion layer formed on carbon and low alloy steels is also composed of cementite (Fe_3C). It is generally agreed that Fe_3C is cathodic to ferrite in CO_2 environment, and this finding is confirmed by the corrosion of ferrite.

Hydrogen sulfide easily dissolves in water. The solubility is related to partial pressure and temperature. The produced water of a sulfide-containing gas well has high acidity, which may cause serious corrosion. The dissolved hydrogen sulfide may be rapidly ionized. The dissociation reaction is as follows [16]:

$$H_2S \rightarrow HS^- + H^+ \tag{14}$$

$$HS^- \rightarrow S^{2-} + H^+ \tag{15}$$

Hydrogen ion is a strong depolarizer. After capturing electrons from a steel surface, it is reduced to a hydrogen atom. This process is known as cathodic reaction. The iron that loses an electron reacts with sulfide ion to generate iron sulfide. This process is known as anodic reaction. The iron as anode fastens the dissolution reaction. Thus, corrosion is caused. The electrochemical equations mentioned earlier are as follows:

$$Anode\ reaction: \qquad Fe \rightarrow Fe^{2+} + 2e^{-} \qquad (16a)$$

$$Cathode\ reaction: \qquad 2H^{+} + 2e^{-} \rightarrow 2H \qquad (16b)$$

$$Anode\ product \qquad Fe^{2+} + S^{2-} \rightarrow FeS \qquad (16c)$$

$$Total\ reaction \qquad Fe + H_2S \rightarrow FeS + 2H \qquad (16d)$$

7.3 CORROSION INHIBITORS

A great number of scientific studies have been devoted to the subject of corrosion inhibitors. However, most of what is known has grown from trial and error experiments, both in the laboratories and in the field. Rules, equations, and theories to guide inhibitor development or use are very limited. By definition, a corrosion inhibitor is a chemical substance that, when added in small concentration to an environment, effectively decreases the corrosion rate. Corrosion can be controlled by using an appropriate technique; however, it is nearly impossible to prevent corrosion completely [17]. Corrosion inhibitors currently applied for steel protection in aqueous environment are either inorganic or organic in nature. Due to environmental concerns, the use of inorganic inhibitors such as chromates, molybdates, phosphonates, nitrites, and silicates to control the corrosion of steel and its alloys has come under severe criticism. Thus, in the face of growing global demand for effective inhibitors that are environmental friendly, the focus will increasingly shift to organic inhibitors. There are different techniques for controlling corrosion, one of which is the use of anticorrosion compounds. The mode of action of these compounds is either by adsorbing on the metal surface or by reacting with some impurities in the system that may cause corrosion. Based on the way they function, inhibitors used can be any of the following types: anodic, cathodic, passivating, vapor phase, film-forming, and neutralizing inhibitors [18].

The role of inhibitors in oil extraction and processing industries is of paramount importance; in fact, they are given the first consideration for protection against corrosion. The use of corrosion mitigation is regarded as one of the most economically wise and practical technique for corrosion protection. The application of suitable corrosion inhibitors can be incorporated with the use of lower grade steels, and it is of better capital economy in comparison with the use of expensive high-grade alloys in the same condition [19].

It is a general assumption that these compounds perform the inhibitive action through the process of adsorption. Adsorption is an adhesive process where a substance (ions, atoms, or molecules of a gas, liquid, or dissolved solid) called adsorbate adheres to a surface (adsorbent). The efficiency of an organic inhibitor is dependent on its adsorption property on the metal surface. Further, the adsorption of a particular inhibitor is dependent on its physical and chemical properties: functional groups, aromaticity, steric effect, electronic arrangement of its molecules, density of electrons at the donor atoms, and characteristic of the π orbital of the donating electrons. It is also established that the nature of the metal, pH, composition, microstructure, temperature, and electrochemical potential between metal-solution interfaces affect adsorption and hence the efficiency of the inhibitor [20]. According to the type of forces involved, the adsorption phenomenon can either be (a) physisorption, (b) chemisorption, or combination of both (a) and (b) [14]:

(a) Physisorption involves electrostatic interaction between the electric charge at the surface of the metal (adsorbent) and ionic charges or dipoles of the inhibitor (adsorbate). In physisorption, heat of absorption is low, and therefore, this type of adsorption is unstable at high temperature.

(b) Chemisorption involves charge sharing or electron transfer from the inhibitor molecules to the metal surface. A coordinate type of bond is formed as a result of this interaction between the metal adsorbent and the π-electrons or unshared electron pairs of the inhibitor adsorbate. In this mode of adsorption, either the anodic or the cathodic reaction or both are reduced from the adsorption of inhibitor on the corresponding active sites. Unlike physisorption which is characterized by low bond energy, chemisorption energy is higher and is therefore stable at higher temperatures. The adsorption process, regardless of whether physical or chemical, results in the formation of a protective hydrophobic surface film that causes the removal of water molecule at the metal-solution interface by the inhibitor molecules. This serves as a barrier to the metal dissolution in the electrolyte [21]:

$$Org_{(sol)} + nH_2O_{(ads)} \rightarrow Org_{(ads)} + nH_2O_{(sol)} \tag{17}$$

When inhibitor molecules get adsorbed on the surface of the metal, they retard or stifle the corrosion process by:

- Changing the rate of anodic-cathodic reaction.
- Retarding the rate of diffusion from reactants to the surface of the metal.
- Decreasing the electrical resistance of the metal surface.

There are several factors to be considered for choosing an appropriate corrosion inhibitor even if it is efficient:

(a) Availability
(b) Cost
(c) Toxicity
(d) Environmental friendliness

7.4 BRIEF THEORY OF DENSITY FUNCTIONAL THEORY (DFT)

The basis of the DFT has been described in many excellent articles and books [22–25]. Accordingly, we will provide only a brief summary here and will do so by minimizing the mathematics as much as possible. The recent impact of DFT in the development of quantum electrochemistry is considerable and can be linked to achievements at the end of the 1980s when gradient-corrected and hybrid functional methods were introduced. Based on the well-known Hohenberg–Kohn theorems (Kohn was awarded a Nobel Prize in physics in 1964 for his work on DFT in the same year that Pople was also awarded a Nobel Prize in the same field), DFT focuses on the electron density, $\rho(r)$, itself as the carrier of all information in the molecular (or atomic) ground state, rather than on the single electron wave function, which is one per electron. Because the electron density arises from the collective contributions of all electrons, there is considerable simplification of the many-bodied Schrödinger equation because of the reduction in the complexity (number of degrees of freedom) of the system. In summary, the Hohenberg–Kohn theorem establishes that the ground state of an electronic system is just a functional of the electronic density. In principle, one only needs the knowledge of the density to calculate all the properties of a system.

The famous Hohenberg–Kohn theorem states that "apart from a trivial additive constant, the electronic density $\rho(r)$ determines the external (i.e., due to the nuclei) potential $v(r)$." On the other hand, $\rho(r)$ determines the number of electrons, N, of a system by integration:

$$N = \int p(r)dr \qquad (18)$$

and hence $\rho(r)$ determines v and N. Accordingly, $\rho(r)$ also determines the Hamiltonian of an N electron system, and thereby the energy, E. Otherwise stated, E is a functional of $\rho(r)$ or of N and $v(r)$.

$$E = E[p(r)] \qquad (19)$$

or

$$E = E[N, v(r)] \qquad (20)$$

The molecular Hamiltonian, H_{op}, can be written in the Born-Oppenheimer approximation, neglecting relativistic effects, as:

$$H_{op} = -\sum_i^N \frac{1}{2}\nabla_i^2 - \sum_A^N \sum_i^N \frac{Z_A}{r_{iA}} + \sum_{i<j}^N \sum_j^N \frac{1}{r_{ij}} + \sum_{B<A}^n \sum_A^n \frac{Z_A Z_B}{R_{AB}} \qquad (21)$$

Here, summation of i and j run over electrons and the summations of A and B run over all nuclei; r_{ij}, r_{iA}, and R_{AB} denote electron-electron, electron-nuclei, and internuclei distances, respectively. Because H_{op} determines the energy of the system by Schrödinger's equation:

$$H_{op}\Psi = E\Psi \tag{22}$$

ψ being the electronic wave function, $\rho(r)$ ultimately determines the system's energy and all other ground state electronic properties.

A generalized DFT expression can be written as follows:

$$E_{DFT}[\rho] = T_S[\rho] + E_{ne}[\rho] + J[\rho] + E_{xc}[\rho] \tag{23}$$

where T_s is the kinetic energy functional (S denotes that the kinetic energy is obtained from a Slater determinant), E_{ne} is the electron-nuclear attraction functional, J is the Coulomb part of the electron-electron repulsion functional, and E_{xc} represents the exchange correlation functional. The dependence of each of these terms on the electron density, ρ, is represented by ρ in brackets following each term.

7.5   CONCEPTUAL DFT PARAMETERS RELATED TO CORROSION INHIBITION

Conceptual DFT developed by Robert G Parr and others has elucidated several molecular parameters/descriptors that are widely used in corrosion inhibition studies [26–28]. Conceptual DFT parameters can be applied in the structure and performance research of corrosion inhibitors to make faster design and evaluation. This method, which is an effective way to conduct research on complex systems at the molecular, atomic, and even electronic level, can provide specific information about molecular structure, electron distribution, and the adsorption process of corrosion inhibitors onto metal and oxide surfaces. Some of the electronic properties afforded by conceptual DFT reactivity indices calculations include molecular orbitals (HOMO and LUMO orbitals), HOMO energy (E_{HOMO}), LUMO energy (E_{LUMO}), energy gap ($E_{LUMO} - E_{HOMO}$), ionization potential (I), electron affinity (A), electrophilicity index (ω), hardness (η), softness (S), electronegativity (χ), fraction of electrons transferred to the metal (ΔN), energy change during charge transfer to the metal (ΔE), Fukui functions, and local softness. These electronic properties can in theory be used to design new effective organic inhibitors with high efficiencies in a variety of corrosive environments

Within the framework of the conceptual DFT, the major equations and derivations for the reactivity parameters useful for corrosion inhibition studies can be summarized as follows [29–35]:

The chemical potential is defined as the negative of electronegativity:

$$\chi = -\mu = -\left(\frac{\partial E}{\partial N}\right)_{v(r)} \tag{24}$$

and hardness (η) is defined as:

$$\eta = \frac{1}{2}\left(\frac{\partial \mu}{\partial N}\right)_{v(r)} = \frac{1}{2}\left(\frac{\partial^2 E}{\partial N^2}\right)_{v(r)} \tag{25}$$

Using the finite difference approximation, global hardness can be approximated as:

$$\eta = \left(\frac{I - A}{2}\right) \tag{26}$$

$$\mu = -\left(\frac{I + A}{2}\right) \tag{27}$$

where I and A are the first vertical ionization potential and electron affinity, respectively, of the chemical system.

Using Koopmans' theorem, and by setting $I = -E_{HOMO}$ and $A = -E_{LUMO}$, hardness can be expressed as:

$$\mu = \frac{E_{LUMO} + E_{HOMO}}{2} \tag{28}$$

and

$$\eta = \frac{E_{LUMO} - E_{HOMO}}{2} \tag{29}$$

where E_{HOMO} is the energy of highest occupied molecular orbital and E_{LUMO} is the energy of the lowest unoccupied molecular orbital.

The global softness (S) is the inverse of global hardness and is given as:

$$S = \frac{1}{\eta} = \left(\frac{\partial N}{\partial \mu}\right)_{v(r)} \tag{30}$$

The global electrophilicity index (ω) is given by:

$$\omega = \frac{\mu^2}{4\eta} \tag{31}$$

The Fukui function, which measures reactivity in a local sense, is by far the most important local reactivity index. By using a scheme of finite difference approximations, this procedure condenses the values around each atomic site into a single value that characterizes the atom in the molecule. With this approximation, the condensed Fukui function becomes:

$$f_k^+ = q_k(N+1) - q_k(N) \qquad \text{(for nucleophilic attack)} \tag{32a}$$

$$f_k^- = q_k(N) - q_k(N-1) \qquad \text{(for electrophilic attack)} \tag{33b}$$

where q_k is the gross charge of atom k in the molecule, i.e., the electron density at the point r in space around the molecule. N corresponds to the number of electrons in the molecule, $N+1$ corresponds to a singly charged anion with an electron added to the LUMO of the neutral molecule, and $N-1$ corresponds to singly charged cation with an electron removed from the HOMO of the neutral molecule.

Morell et al. [36] have proposed a local reactivity descriptor (LRD) called the dual descriptor (DD) represented by:

$$f^{(2)}(r) \equiv \Delta f(r) \tag{34}$$

It is another useful function to reveal reactive sites. DD allows one to obtain simultaneously the preferable sites for nucleophilic attacks ($\Delta f(r) > 0$) and preferable sites for electrophilic attacks ($\Delta f(r) < 0$) into the system at the point r. DD has proven to be much more efficient than Fukui function as a local reactivity indicator because of its power to distinguish those sites of true nucleophilic and electrophilic behavior. The formal definition of the DD $\Delta f(r)$ that is closely related to Fukui function is:

$$\Delta f(r) = f^+(r) - f^-(r) \tag{35}$$

The local softness $s(r)$ can be defined as:

$$s(r) = \left(\frac{\partial \rho(r)}{\partial \mu} \right)_{v(r)} \tag{36}$$

Eq. (36) can also be written as:

$$s(r) = \left(\frac{\partial \rho(r)}{\partial N} \right)_{v(r)} \times \left(\frac{\partial N}{\partial \mu} \right)_{v(r)} = f(r)S \tag{37}$$

Atomic softness values can easily be calculated as follows:

$$s_k^+ = [q_k(N+1) - q_k(N)]S \tag{38}$$

$$s_k^- = [q_k(N) - q_k(N-1)]S \tag{39}$$

$$s_k^o = \frac{S[q_k(N+1) - q_k(N-1)]}{2} \tag{40}$$

Some conceptual DFT molecular descriptors used to model chemical reactivity and molecular characterization/design of organic materials for metallic protection [37] are listed below.

Conceptual DFT descriptors (Global)
Energy of the highest occupied molecular orbital E_{HOMO}
Energy of the lowest unoccupied molecular orbital E_{LUMO}
Ionization potential: Removing an electron from a molecular system X (X $\rightarrow$ X$^+$ +e$^-$) $IP(\approx -E_{HOMO})$
Electron affinity: Attaching an additional electron to a molecular system X (X+e$^-$$\rightarrow$ X$^-$) $EA(\approx -E_{LUMO})$

Chemical potential, defined as the change in electron energy E upon change in the total number of electrons N

$$\mu = \left(\frac{\delta E}{\delta N} \right)_v$$

Absolute electronegativity $\chi = -\mu \approx -1/2(E_{HOMO} + E_{LUMO})$

Molecular hardness, defined as the change in the chemical potential μ upon change in the total number of electrons N

$$\eta = -\left(\frac{\delta \mu}{\delta N} \right)_v \approx -(E_{HOMO} - E_{LUMO})$$

Molecular softness $S = \dfrac{1}{2\eta}$

Molecular polarizability (note that molecules arrange themselves toward a state of minimum polarizability and maximum hardness) α

Electrophilicity index $\omega = \dfrac{\mu^2}{2\eta} = \dfrac{\chi^2}{2\eta}$

Charge distribution

Net atomic charges (atom r) $QA(r)$

Polar surface area, describing the spatial surface density distribution PSA

Molecular dipole moment μ

Site-specific molecular descriptors (Local)

Electrophilic Fukui function, defined as the change in electron density ρ at atom r upon addition of electrons to the system (N = number of electrons

$$f^+(r) = \left(\frac{\delta \rho(r)}{\delta N} \right)_{v(r)}^+ \approx \rho_{N+1}(r) - \rho_N(r)$$

Nucleophilic Fukui function, defined as the change in electron density ρ at atom r upon removal of electrons from the system (N = number of electrons

$$f^-(r) = \left(\frac{\delta \rho(r)}{\delta N} \right)_{v(r)}^- \approx \rho_N(r) - \rho_{N-1}(r)$$

Local electrophilicity index $\omega(r) = \omega \times f^+(r)$

Local nucleophilicity index $\omega(r) = \omega \times f^-(r)$

Reactivity-selectivity descriptor or dual descriptor: $f'(r)$ measures reactivity toward nucleophilic and $f(r)$ toward electrophilic attacks; therefore, electrophilic sites are identified by $\Delta f(r) > 0$. ρLUMO and ρHOMO are the electron densities of the LUMO and HOMO orbitals, respectively.

$$\Delta f(r) = f^+(r) - f^-(r) \quad \text{or} \quad \Delta f(r) \approx \rho_{LUMO}(r) - \rho_{HOMO}(r)$$

7.6 AN ILLUSTRATIVE EXAMPLE OF THE APPLICATION OF CONCEPTUAL DFT TO CORROSION INHIBITION STUDIES

7.6.1 DESIGN STRATEGY

The strategy adopted here is to modify 2-mercaptobenzimidazole (MBI) with a methyl group to generate 2-(methylthio) benzimidazole (MTBI). The modified molecules are subsequently screened and ranked based on computational modeling using reactivity parameters derived from conceptual DFT. The computational results are compared with a reference molecule MBI, which has shown great promise as an excellent corrosion inhibitor for steel even in high turbulent flow conditions from literature results. This procedure is fast and reduces tedious laboratory time and resources, which could be spend in qualifying molecules as corrosion inhibitors.

7.6.2 COMPUTATIONAL DETAILS

DFT calculations were carried out using the Gaussian-09 program [38]. The exchange–correlation was treated using hybrid, B3LYP functionals. A full optimization was performed using the 6-311G ++ (d, p) basis sets. This basis set is well known to provide accurate geometries and electronic properties for a wide range of organic compounds [39]. Moreover, the 6-311G++ (d,p) basis set was used because it gives, according to convergency tests, results comparable to those obtained from plane-wave calculations; the differences in energies and bond lengths between the 6-311G++ (d,p) and plane-wave basis sets are ~0.01 eV and <0.01 Å, respectively [39]. A vibrational analysis was carried out for each optimized molecule to ensure that there was no imaginary frequency in the optimization steps. This ensures that the optimized molecules have reached a minimum energy in the potential energy surface. It is well known that the phenomenon of electrochemical corrosion takes place in the aqueous phase. Therefore, to ensure the accuracy of the data, the solvent effect has to be taken into account in the calculation. In this study, self-consistent reaction field (SCRF) methods were performed using the polarized continuum method (PCM) as a model for the solvent. In this model, the solvent is treated as an expanse of dielectric media and the solute as a trapped molecule in a cavity surrounded by the solvent [40].

7.6.3 RESULTS AND DISCUSSION

The molecular and optimized structures of the investigated inhibitors are shown in Figure 7.1. The quantum chemical parameters that influence the electronic interaction between steel and inhibitor molecules, such as E_{HOMO}, E_{LUMO}, the energy gap (ΔE), the dipole moment (D), and the lipophilicity ($logP$) of the inhibitors, are presented in Table 7.1. The localization of the frontier molecular orbitals (HOMO and LUMO) of the benzimidazole inhibitors studied is also presented in Figure 7.2.

HOMO and LUMO are widely used to predict the adsorption centers of inhibitor molecules. E_{HOMO} indicates the tendency of the molecule to donate electrons. The higher the value of E_{HOMO}, the greater is the ability of that molecule to donate electrons. E_{LUMO} indicates the propensity of a molecule to accept electrons. The lower the value of E_{LUMO}, the greater is the ability of that molecule to accept electrons [41]. The energy gap, ΔE, is an important parameter that indicates the reactivity tendency of organic molecule for the metal surface [42]. As ΔE decreases, the reactivity of the molecule increases, leading to an increase in adsorption on the metal surface.

According to the quantum chemical results obtained from Table 7.1, the E_{HOMO} of the inhibitors increased in the following order: MTBI > MBI, E_{LUMO} decreased in the following order: MBI > MTBI, while the order of the energy gap, ΔE, decreased as follows: MTBI < MBI. It is clear from the above ranking that if the ordering of E_{HOMO}, E_{LUMO}, and ΔE are considered, MTBI is expected to be a more effective corrosion inhibitor than MBI. The dipole

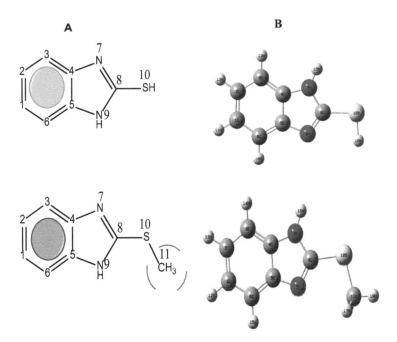

FIGURE 7.1   Molecular and optimized structures of (A) MBI and (B) MTBI.

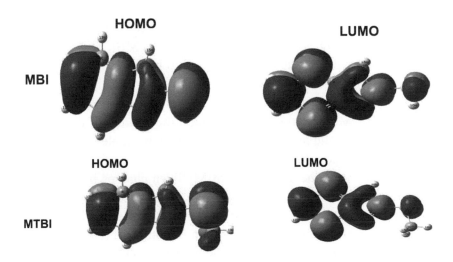

FIGURE 7.2   Plots of the HOMO and LUMO orbital distribution for MBI and MTBI.

TABLE 7.1   Quantum Chemical Parameters for MBI and MTBI

Properties	MBI	MTBI
Total Energy (eV)	−21094.14	−22245.86
E_{HOMO} (eV)	−6.257	−6.090
E_{LUMO} (eV)	−0.817	−0.887
Energy gap (ΔE) (eV)	5.440	5.203
Dipole moment (D)	3.622	2.978
Log P	1.98	2.23

moment, which indicates molecular polarity, was also computed. Results from Table 7.1 indicate that MTBI has a lower dipole moment than MBI. In the literature, there is no consensus on the influence of the dipole moment on corrosion inhibition. Some authors have suggested that an increase in dipole moment leads to a decrease in inhibition, while others think that a decrease in dipole moment can lead to an increase in inhibition [43].

Lipophilicity (*logP*) is a measure of the tendency of a molecule to prefer an oil-like environment to an aqueous one. High and more positive *logP* values indicate decreasing solubility of a molecule in an aqueous environment. As documented in the literature, *logP* is an indication of the hydrophobicity of a molecule, which is related to the mechanism of formation of the protective layer on the metal surface [44]. It is clear from Table 7.1 that MTBI has a higher value of *logP* than MBI. This means that MTBI is more hydrophobic and is expected to bind stronger to the steel surface than MBI. Both MTBI and MBI have log $P < 3$, indicating their environmentally acceptable profile [45]. As can be seen in Figure 7.2, the HOMO and LUMO electron densities are on the entirety of the inhibitor molecules. This could make it easier for the two molecules to adsorb in a parallel manner onto a steel surface.

Figure 7.3 shows the distribution of Mulliken charges from the analysis of the Mulliken population analysis in the two-benzimidazole molecules investigated. It is evident that the benzimidazole ring and S atoms are the major active sites for the adsorption of both inhibitors onto a steel surface. However, the presence of the methyl group on MTBI makes electrons more available on the benzimidazole ring for bonding with a steel

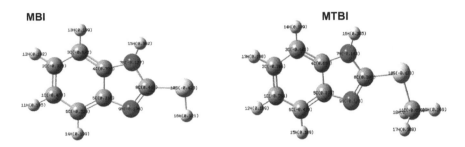

FIGURE 7.3   Atomic Mulliken charge plots for MBI and MTBI.

surface. This is evident in the high negative charges of the S atom in MTBI (–0.438) when compared to MBI (–0.428).

7.7   CONCLUSIONS

Conceptual DFT can offer the facility to design, at the molecular level, the molecular structure, including the choice and location of various functional groups and to build in selected aromaticity, as dictated by experimental feedback from studies on existing inhibitors. Several reactivity parameters derived from conceptual DFT can be correlated with corrosion inhibition activity as discussed in this chapter. We have shown that conceptual DFT parameters can be used as a reliable approach to screen and select potential organic corrosion inhibitors prior to experimental validation.

ACKNOWLEDGMENTS

The authors would like to acknowledge the support received from King Abdulaziz City for Science and Technology (KACST) for funding this work under the National Science Technology Plan (NSTIP) grant No. 13-ADV1737-04. Also, the support provided by the Deanship of Scientific Research (DSR) and the Center of Research Excellence in Corrosion (CORE-C), at King Fahd University of Petroleum & Minerals (KFUPM) is gratefully acknowledged.

KEYWORDS

- benzimidazole
- conceptual DFT
- corrosion inhibition
- global reactivity descriptors
- local reactivity descriptors
- steel

REFERENCES

1. "Cost of Corrosion to exceed $1 Trillion in the United States in 2013-G2MT Labs-*The Future of Materials Conditions Assessment*," http://www.g2mtlabs.com/2011/06/nace-cost-of-corrosion-study-update/.
2. Al Hashem, A., (2011). "Corrosion in the Gulf Cooperation Council (GCC) states: Statistics and Figures," *in proceedings of the Corrosion*, UAE, Abu Dhabi, UAE.
3. Obot, I. B., Ebenso, E. E., Obi-Egbedi, N. O., Afolabi, A. S., & Gasem, Z. M., (2012). Experimental and theoretical investigations of adsorption characteristics of itraconazole as green corrosion inhibitor at a mild steel/hydrochloric acid interface. *Res. Chem. Intermed.*, *38*, 1761–1779.
4. Khaled, K. F., (2008). Molecular simulation, quantum chemical calculations and electrochemical studies for inhibition of mild steel by triazoles. *Electrochim. Acta.*, *53*, 3484–3492.
5. Awad, M. K., Mustafa, M. R., & Abo Elnga, M. M., (2010). Computational simulation of the molecular structure of some triazoles as inhibitors for the corrosion of metal surface. *J. Mol. Struct. (THEOCHEM)*, *959*, 66–74.
6. Obot, I. B., Ebenso, E. E., & Kabanda, M. M., (2013). Metronidazole as environmentally safe corrosion inhibitor for mild steel in 0.5 M HCl: Experimental and theoretical investigation. *J. Environ. Chem. Eng.*, *1*, 431–439.
7. Obot, I. B., Obi-Egbedi, N. O., (2010). Theoretical study of benzimidazole and its derivatives and their potential activity as corrosion inhibitors, *Corros. Sci.*, *52*, 657–660.
8. Obi-Egbedi, N. O., Obot, I. B., El-Khaiary, M. I., Umoren, S. A., & Ebenso, E. E., (2012). Computational simulation and statistical analysis on the relationship between corrosion inhibition efficiency and molecular structure of some phenanthroline derivatives on mild steel surface. *Int. J. Electrochem. Sci.*, *7*, 5649–5675.
9. Obot, I. B., & Obi-Egbedi, N. O., (2010). Adsorption properties and inhibition of mild steel in sulphuric acid solution by ketoconazole: experimental and theoretical investigation, *Corros. Sci.*, *52*, 198–204.
10. Obot, I. B., Obi-Egbedi, N. O., & Umoren, S. A., (2009). The synergistic inhibitive effect and some quantum chemical parameters of 2,3-diaminonapthalene and iodide ions on the hydrochloric acid corrosion of aluminium. *Corros. Sci.*, *51*, 276–282.

11. Khaled, K. F., (2011). Molecular modeling and electrochemical investigations of the corrosion inhibition of nickel using some thiosemicarbazone derivatives. *J. Appl. Electrochem., 41*,423–433.

12. Obot, I. B., Umoren, S. A., Gasem, Z. M., Suleiman, R., & El Ali, B., (2015). Theoretical prediction and electrochemical evaluation of vinylimidazole and allylimidazole as possible green corrosion inhibitors for carbon steel in 1 M HCl. *J. Ind. Eng. Chem., 21*,1328–1339.

13. Obot, I. B., Macdonald, D. D., & Gasem, Z. M., (2015). Density functional theory (DFT) as a powerful tool for designing new organic corrosion inhibitors. Part 1: An overview, *Corros. Sci., 99*, 1–30.

14. Sastri, V. S., (2012). *Corrosion and Electrochemistry*. In green corrosion chemistry and engineering opportunities and challenges; Sharma, S. K. Ed., Wiley-VCH Verlag GmbH & Co. KGaA: Germany 33–69.

15. Nesic, S., (2007). Key issues related to modelling of internal corrosion of oil and gas pipelines–A review. *Corros. Sci., 49*, 4308–4338.

16. Popoola, L. T., Grema, A. S., Latinwo, G. K., Gutti, B., & Balogun, A. S., (2013). Corrosion problems during oil and gas production and its mitigation. *Int. J. Ind. Chem., 4*, 35–49.

17. Brondel, D., Edwards, R., Hayman, A., Hill, D., & Semerad, T., (1994). Corrosion in the Oil Industry, *Oilfield Rev., 6*, 4–18.

18. Rajeev, P., Surendranathan, A. O., & Murthy, Ch. S. N., (2012). Corrosion mitigation of the oil well steels using organic inhibitors-A review. *J. Mater. Environ. Sci., 3*, 856–869.

19. Horsup, D. I., Clark, J. C., Binks, B. P., Fletcher, P. D. I., & Hicks, J. T., (2010). The fate of oilfield corrosion inhibitors in multiphase systems, *Corrosion., 66*, 1–14.

20. Nazari, M. H., Allahkaram, S. R., & Kermani, M. B., (2010). The effects of temperature and pH on the characteristics of corrosion product in CO_2 corrosion of grade X70 steel. *Mater & Des., 31*, 3559–3563.

21. Khaled, K. F., (2003). The inhibition of benzimidazole derivatives on corrosion of iron in 1 M HCl solutions, *Electrochim. Acta., 48*, 2493–2503.

22. Geerlings, P., De Proft, F., & Langenaeker, W., (2003). Conceptual density functional theory. *Chem. Rev., 103*,1793–1874.

23. Nagy, A., (1998). Density functional theory and application to atoms and molecules. *Phys. Rev., 298*, 1–79.

24. Seminario, J. M., & Politzer, P., (1995). *Modern Density Functional Theory: A Tool for Chemistry.* Elsevier: Amsterdam, Volume 2, p. 404.

25. Nalewajski, R. F., (1996). *Topics in Current Chemistry: Density Functional Theory,* Springer, Berlin, Volume 183, pp. 1–24.

26. Parr, R. G., Donnelly, R. A., Levy, M., & Palke, W. E., (1978). Electronegativity: The Density Functional Viewpoint. *J. Chem. Phys., 68*, 3801–3807.

27. Parr, R. G., & Pearson, R. G., (1983). Absolute hardness: companion parameter to absolute electronegativity. *J. Am. Chem. Soc., 105*, 7512–7516.

28. Parr, R. G., & Yang, W., (1989). *Density Functional Theory of Atoms and Molecules*, Oxford University Press: New York, USA.

29. Chermette, H., (1999). Chemical reactivity indexes in density functional theory. *J. Comp. Chem., 20*,129–154.

30. Parr, R. G., & Chattaraj, P. K., (1991). Principle of maximum hardness. *J. Am. Chem. Soc., 113,* 1854–1855.

31. Iczkowski, R. P., & Margrave, J. L., (1961). Electronegativity, *J. Am. Chem. Soc., 83,* 3547–3551.

32. Koopmans, T., (1933). Ordering of wave functions and Eigen-energies to the individual electrons of an atom. *Physica., 1,*104–113.

33. Yang, W., & Parr, R. G., (1985). Hardness, softness and the Fukui function in the electronic theory of metals and catalysis. *Proc. Natl. Acad. Sci., 82,* 6723–6726.

34. Parr, R. G., Sventpaly, L., & Liu, S., (1999). Electrophilicity Index. *J. Am. Chem. Soc., 121,* 1922–1924.

35. Fukui, K., (1982). Role of Frontier orbitals in chemical reactions. *Science., 218,* 747–754.

36. Morell, C., Grand, A., & Toro-Labbe, A., (2005). A new dual descriptor for chemical reactivity. *J. Phys. Chem. A., 109,* 205–212.

37. Obot, I. B., (2014). Recent advances in computational design of organic materials for corrosion protection of steel in aqueous media, In: Aliofkhazraei M (ed). *Developments in Corrosion Protection, INTECH, Croatia, 123*–51.

38. Frisch, M. J., et al, (2009). Gaussian 09, Revision D.01, Gaussian, Inc., Wallingford CT.

39. Kovacevic, N., & Kokalj, A., (2011). DFT study of interaction of azoles with Cu(111) and Al(111) Surfaces: Role of azole nitrogen atoms and dipole-dipole interactions, *J. Phys. Chem.C., 115,* 24189–24197.

40. Fu, J., Li, S., Wang, Y., Liu, X., & Lu, L., (2011). Computational and electrochemical studies on the inhibition of corrosion of mild steel by L-Cysteine and its derivatives. *J. Mater. Sci., 46,* 3550–3559.

41. Chidiebere, M. A., Ogukwe, C. E., Oguzie, K. L., Eneh, C. N., & Oguzie, E. E., (2012). Corrosion inhibition and adsorption behavior of punicagranatum extract on mild steel in acidic environments: Experimental and theoretical studies. *Ind. Eng. Chem. Res., 51,* 668–677.

42. Obot, I. B., Obi-Egbedi, N. O., & Eseola, A. O., (2011). Anticorrosion potential of 2-Mesityl-1H-imidazo[4,5-f][1,10]-phenanthroline on mild steel in sulfuric acid solution: Experimental and theoretical study. *Ind. Eng. Chem. Res., 50,* 2098–2110.

43. Guo, L., Dong, W., & Zhang, S., (2014). Theoretical challenges in understanding the inhibition mechanism of copper corrosion in acid media in the presence of three triazole derivatives. *RSC Adv.,4,* 41956–41967.

44. Al-Sabagh, A. M., Nasser, N. M., Farag, A. A., Migahed, M. A., Eissa, A. M. F., & Mahmoud, T., (2013). Structure effect of some amine derivatives on corrosion inhibition efficiency for carbon steel in acidic media using electrochemical and quantum theory methods,' Egypt. *J. Petrol., 22,* 101–116.

45. Martin, R. L., Alink, B. A., McMahon, J. A., & Weare, R., (1999). Improvement of environmental properties of corrosion inhibitors, *SPE International*, pp. 50796, Houston, Texas, SPE, 735.

CHAPTER 8

PHASE DESCRIPTION OF REACTIVE SYSTEMS

ROMAN F. NALEWAJSKI

Department of Theoretical Chemistry, Jagiellonian University, Gronostajowa 2, 30-387 Cracow, Poland

CONTENTS

ABSTRACT

The equilibrium states of molecular systems extrermize the system resultant entropy combining the *classical* (probability) and *nonclassi-*

The following notation is adopted: A denotes a *scalar*, A is the row/column *vector*, $\mathbf{A}$ represents a square or rectangular *matrix*, and dashed symbol $\hat{A}$ stands for the quantum-mechanical operator of the physical property, A. The logarithm of the Shannon information measure is taken as an arbitrary but fixed base: log = log2 corresponds to the information content measured in *bits* (binary digits), while log = ln expresses the amount of information in *nats* (natural units), 1 nat = 1.44 bits.

cal (phase/current) information contributions. Such *phase*-transformed states are explored in both the bimolecular reactive complex R = A----B and its acidic (A) and basic (B) reactants. The isolated subsystems A^0 and B^0 exhibit the equilibrium distributions $\{\rho_\alpha^0 = \rho_\alpha[N_\alpha^0, v_\alpha], \alpha = A, B\}$ for their initial (integer) numbers of electrons $\{N_\alpha^0\}$ and external potentials $\{v_\alpha\}$ due to the fragment constituent nuclei. The *intra*-reactant equilibria of the "promoted" subsystems in the polarized reactive system $R_n^+ = (A^+|B^+)$, consisting of geometrically rigid but electronically relaxed densities $\{\rho_\alpha^+ = \rho_\alpha^+[N_\alpha^0, v_R]\}$ of the mutually *closed* (nonbonded) reactants in the combined external potential $v_R = v_A + v_B$, determine the initial state for the subsequent B→A charge transfer (CT): $N_{CT} = N_A^*$ $- N_A^0 = N_B^0 - N_B^* > 0$. This electron flow establishes the final, *inter*-reactant equilibrium in $R_b^* = (A^*|B^*)$ as a whole, combining the resultant densities $\{\rho_\alpha^* = \rho_\alpha^*[N_\alpha^*, v_R]\}$ of the geometrically "frozen" but mutually *open* (bonded) fragments: $\rho_R = \rho_R[N_R, v_R] = \Sigma_\alpha \rho_\alpha^*$, $N_R = N_A^* + N_B^* = N_A^0 + N_A^0$. The continuity equations are summarized and the *in situ* descriptors of CT processes between such polarized complementary subsystems in their internal equilibrium states are examined. The classical and non-classical contributions to the resultant entropy/information measures are partitioned into their additive and nonadditive components. It is argued that for the internal equilibria in polarized reactants, the nonvanishing *in situ* CT derivatives of the system resultant entropies are due to their nonadditive contributions alone.

8.1   INTRODUCTION

The probability-based classical information theory (CIT) [1–6] of molecular systems has been successfully applied to explore molecular electron distributions, density pieces attributed to atoms-in-molecules (AIM), patterns of chemical bonds, and preferences in reactivity phenomena, e.g., [7–10]. The CIT concepts have been used to explore the information distribution in molecules, identify the entropic principles of the molecular electronic structure, determine the most objective bonded atoms, and extract multiplicities of chemical bonds and their covalent/ionic composition from electronic communications in molecules. The nonadditive Fisher information [11] has been linked to electron localization function (ELF) [12–16] of modern density functional theory (DFT) [17–19] and provides an efficient contragradience (CG) probe [20] for localizing chemical bonds. The orbital communication theory (OCT) [9, 10, 21–32] has identified the *bridge*-bonds [9, 10, 33–39] originating from the cascade propagations of information between AIM, involving intermediate orbitals.

In entropic theories of molecular electronic structure, one ultimately requires [10, 40–45] the *quantum* generalizations of the familiar Fisher [1] and Shannon [2] descriptors of CIT, which are appropriate for the *complex* probability amplitudes (wavefunctions) of quantum mechanics (QM). In this quantum information theory (QIT) [10], the state *probability* distribution reflects its *modulus* factor and generates the classical entropy/information components, while the wavefunction *phase* or its gradient (electronic current) density gives rise to the nonclassical complement in the corresponding resultant measure. The classical information terms, conceptually rooted in DFT, probe the entropic content of the *incoherent* (disentangled) local events, while their nonclassical supplements provide the information contribution due to the mutual *coherence* (entanglement) of such local events, which is inherent in full QM description. In modern DFT the Harriman-Zumbach-Maschke(HZM) construction [10, 46–48] of Slater determinants yielding the specified electron density provides a convenient framework for separating the modulus and phase aspects of general electronic states. The overall QIT concepts allow one to distinguish the information content of states generating the same electron density but differing in their current patterns.

The quantum extension of the classical Fisher information has been proposed [10, 40–45] using its association with the average kinetic energy of electrons and the resultant *global* entropy has been inferred using the known relation between densities of the complementary classical Fisher and Shannon measures. The expectation value of the (*non*-Hermitian) *complex*-entropy operator [10, 45] generates the probability and phase contributions in the resultant measure as its real and imaginary parts. This extension satisfies the requirement that the known dependence between densities of the classical Shannon (*uncertainty*-information) and Fisher (*determinicity*-information) components also relates to the nonclassical components of the resultant measures.

Generalized entropic principles of QIT determine the *equilibrium* states of molecules and their constituent fragments. Two types of such entropic rules have been explored, for the *phase*-maxima of the nonclassical entropy alone and for extremum of the resultant measure. The former determines the so-called *vertical* equilibrium state, while the latter generates the system *horizontal* equilibrium, with the optimum "thermodynamic" phase being related to the logarithm of the state probability density. Both the global and gradient measures in the *resultant*-entropy rules have been shown to give rise to the same optimum *phase* solution.

In a bimolecular chemical reaction both the system electron distribution and its geometry relax, when reactants interact chemically with each other at a finite separation between them. Such displacements ultimately determine the equilibrium electron distribution for the current value of the reaction progress-variable, e.g., the intrinsic reaction coordinate. Within the electron-*following* perspective of the familiar Born-Oppenheimer (BO) approximation, each (open) reactant responds to the presence of the reaction partner by changing its electron density and the effective average numbers of electrons. This perturbation first induces the equilibrium *polarization* (P) of the mutually *closed* reactants and eventually affects a (fractional) charge transfer (CT) between them, when a hypothetical barrier for the flow of electrons between the two mutually *open* subsystems is finally lifted. The spontaneous responses of molecules to displacements in their external potential and/or the average number of electrons are all grounded in DFT and explain gross features of reactivity preferences [49–51]. Such perturbation-response relations have been formulated in the DFT-based reactivity theory called Charge Sensitivity Analysis (CSA) [49, 50] or Conceptual DFT [51].

The classical concepts of the chemical *electronegativity* [49–55], *hardness* [49–51, 56, 57], *softness* or *Fukui function* (FF) [49–51, 58], and *electrophilicity/nucleophilicity* indices [59, 60] help to understand the complex phenomenon of chemical reactivity. Its adequate description calls for quantities measuring both the electronic responses of subsystems to external perturbation created by the presence of the other reactant [49, 50] and their coupling to the system geometry [61–66]. The information-theoretic (IT) descriptors provide the *entropy representation* of the reaction mechanism and reveal the whole complexity of the process [67–69]. The

Minimum-Energy-Path sections of the entropy/information functionals, in both the position and momentum spaces, have uncovered the presence of additional features revealing the chemically important regions, where the bond-*breaking* and bond-*forming* processes actually occur. The "classical" language of DFT, focusing solely on displacements in the electron distribution accompanying breaking and forming of chemical bonds in chemical reactions, rather than on the concomitant changes in the system *many*-electron wave function, loses the phase aspect of reactivity phenomena. It is retained in quantum probability amplitudes resulting from the Superposition Principle of QM [70] and reflected by the "nonclassical" entropy/information contributions [10, 40–45] and molecular phase/current channels [10, 71] of QIT.

In the present analysis, we explore the *in situ* reactivity descriptors [49, 50, 72] related to CT processes between subsystems of the bimolecular acid-base complex, representing the polarized reactants in their (internal) *phase*-equilibrium states. We begin with an overview of the basic degrees-of-freedom of such quantum systems, summarize the entropy/information components of their electronic states at hypothetical reaction stages, and examine the optimum ("horizontal") phases marking both the *phase*-equilibrium in the reactive system as a whole and the internal equilibria in its acid/base subsystems. We shall also identify the additive/nonadditive components of the resultant information measures and explore their contributions to populational derivatives describing the CT phenomena.

8.2 DENSITY AND CURRENT ATTRIBUTES OF QUANTUM STATES

We first consider the simplest case of *one*-electron ($N=1$) system. The electron density $\rho(r) = Np(r)$ then also represents the probability distribution $p(r;t) \equiv |\psi(r;t)|^2$ generated by the squared modulus of the system wavefunction $\psi(r;t) = \langle r|\psi(t)\rangle$ describing a general quantum state $|\psi(t)\rangle$. We adopt the usual BO approximation of the fixed nuclear positions. In this molecular scenario, one then envisages a single electron moving in the external potential $v(r)$ due to the "frozen" nuclei, described by the Hamiltonian

$$\hat{H}(r) = -\left(\hbar^2 / 2m\right)\nabla^2 + v\left(r\right) = \hat{T}(r) + v\left(r\right). \tag{1}$$

It combines the external potential $v(r)$ due to the nuclei in their fixed positions and the kinetic-energy operator $\hat{T}(r) = \hat{p}(r)^2/(2m) = -(\hbar^2/2m)\nabla^2$, where $\hat{p}(r)$ denotes the momentum operator in position representation: $\hat{p}(r) = -i\hbar\nabla$. Its eigensolutions,

$$\hat{H}(r)\varphi_s\left(r\right) = E_s\varphi_s\left(r\right), \qquad s = 0, 1, \ldots, \tag{2}$$

represent amplitudes $\{\varphi_s(r) = \langle r|\varphi_s\rangle\}$ of the stationary electronic states at specified time t:

$$\begin{aligned}\psi_s\left(r,t\right) &= \langle r|\psi_s(t)\rangle = \varphi_s(r)\exp\left[-i\left(E_s/\hbar\right)t\right]\\ &\equiv \varphi_s(r)\exp\left[-i\omega_s t\right] \equiv \varphi_s(r)\exp\left[i\phi_s(t)\right].\end{aligned} \tag{3}$$

They correspond to the sharply specified electronic energies $\{E_k\}$, with the lowest ($s = 0$) eigenvalue marking the system ground state, and the stationary (*time*-independent) probability distribution, the expectation value of the probability-density operator $\hat{\rho}(r) = |r\rangle\langle r|$:

$$p_s\left(r;t\right) = \langle \psi_s(t)|\hat{\rho}(r)|\psi_s(t)\rangle = |\varphi_s\left(r\right)|^2 = p_s\left(r\right). \tag{4}$$

Each (nondegenerate) stationary state $\psi_s(r, t) = \varphi_s(r)\exp[i\varphi_s(t)]$ exhibits *time*-independent modulus factor $\varphi_s(r) = \langle r|\varphi_s\rangle$ and purely *time*-dependent phase $\varphi_s(t)$, i.e., the exactly vanishing *spatial* phase component $\chi_s(r) = 0$ in the resultant phase $\Phi_s(r; t) = \chi_s(r) + \varphi_s(t)$. It generates the stationary probability distribution $p_s(r, t) = p_s(r) = \varphi_s(r)^2$ and vanishing current density given by the expectation value of its quantum-mechanical operator

$$\hat{j}(r) = \frac{1}{2m}[\hat{\rho}(r)\hat{p} + \hat{p}\,\hat{\rho}(r)], \tag{5}$$

$$\begin{aligned}j_s\left(r\right) &= \langle\varphi_s|\hat{j}(r)|\varphi_s\rangle\frac{\hbar}{2mi}[\phi_s^*(r)\nabla\phi_s(r) - \phi_s(r)\nabla\phi_s^*(r)]\\ &= \frac{\hbar}{m}p_s(r)\,\nabla\varphi_s(r) = 0.\end{aligned} \tag{6}$$

The *state* dynamics is determined by the Schrödinger equation (SE),

$$i\hbar[d|\psi_s(t)\rangle/dt] = \hat{H}|\psi_s(t)\rangle \tag{7}$$

which implies the associated derivative of the expectation value $\langle\psi_s(t)|\hat{A}_s|\psi_s(t)\rangle \equiv \langle A\rangle_\psi$:

$$\partial\langle A\rangle_\psi/\partial t = (i/\hbar)\langle[\hat{H},\hat{A}]\rangle_\psi, \tag{8}$$

determined by the commutator $[\hat{H},\hat{A}] = \hat{H}\hat{A} - \hat{A}\hat{H}$. The Schrödinger dynamics of the particle wavefunction,

$$\psi(r,t) \equiv \langle r|\psi(t)\rangle = R(r,t)\exp[i\Phi(r,t)], \tag{9}$$

$$(i\hbar)^{-1}\hat{H}(r)\psi(r,t) = \partial\psi(r,t)/\partial t, \tag{10}$$

and its Hermitian-conjugate give rise to the sourceless form of the continuity equation for the electronic probability distribution $p(r,t) = |\psi(r,t)|^2 = R(r,t)^2$:

$$\partial p(r,t)/\partial t = -\nabla\cdot j(r,t)$$

or

$$\sigma_p(r,t) \equiv dp(r,t)/dt = \partial p(r,t)/\partial t + \nabla\cdot j(r,t) = 0, \tag{11}$$

which also implies the conservation in time of the wavefunction/probability normalization:

$$d\langle\psi(t)|\psi(t)\rangle/dt = d[\int p(r,t)\,dr]/dt = \int[dp(r,t)/dt]\,dr = 0. \tag{12}$$

Therefore, the time dependence of the probability density in the *fixed* monitoring volume element, reflected by the *partial* derivative $\partial p(r,t)/\partial t$, is due to the probability outflow determined by the divergence of the current density: $-\nabla\cdot j(r,t)$. The *probability*-continuity also implies the vanishing *total* time derivative $\dot{p}(r,t) \equiv \sigma_p(r,t) = 0$ (the particle probability source), which expresses the time rate of change of the particle density in an infinitesimal volume element flowing with the particle.

The SE also implies the following *time*-dependence of the wavefunction *phase*-component,

$$\partial\Phi(\mathbf{r},t)/\partial t = [\hbar/(2m)]\{R(\mathbf{r},t)^{-1}\Delta R(\mathbf{r},t) - [\nabla\Phi(\mathbf{r},t)]^2\} - v(\mathbf{r})/\hbar, \qquad (13)$$

which generates the associated *time*-rate of *phase*-density $\pi(\mathbf{r},t) = \Phi(\mathbf{r},t)^2$:

$$\partial\pi(\mathbf{r},t)/\partial t = 2\Phi(\mathbf{r},t)\,\partial\Phi(\mathbf{r},t)/\partial t. \qquad (14)$$

In general (complex) quantum states, the wavefunction phase is evolving with time, changing the *phase*-related contributions to resultant descriptors of the overall information content. This time rate of change in the state phase is seen to be determined by spatial variations of both the modulus (R) and phase (Φ) components of molecular electronic states as well as by the shape of the external potential (v).

To summarize, the wavefunction *modulus* factor R, the classical amplitude of the particle probability function $p = R^2$, and the state *phase* component Φ, or its gradient $\nabla\Phi$ determining the current density

$$\mathbf{j}(\mathbf{r},t) = \langle\psi(t)|\,\hat{\mathbf{j}}(\mathbf{r})|\psi(t)\rangle = (\hbar/m)p(\mathbf{r},t)\nabla\Phi(\mathbf{r},t) \equiv p(\mathbf{r},t)\mathbf{V}(\mathbf{r},t), \qquad (15)$$

where the current-per-particle $\mathbf{V} = \mathbf{j}/p$ reflects the effective velocity of the probability fluid, constitute two fundamental "degrees-of-freedom" in the quantum IT treatment of electronic states:

$$\psi \Leftrightarrow (R, \Phi) \Leftrightarrow (p, \mathbf{j}).$$

Finally, the corresponding total and partial *time*-derivatives of the probability current give:

$$\sigma_j(\mathbf{r},t) = d\mathbf{j}(\mathbf{r},t)/dt = \mathbf{V}(\mathbf{r},t)\sigma_p(\mathbf{r},t) + p(\mathbf{r},t)\,d\mathbf{V}(\mathbf{r},t)/dt \quad \text{and}$$
$$\partial\mathbf{j}(\mathbf{r},t)/\partial t = (i/\hbar)\,\langle[\hat{H}, \hat{\mathbf{j}}(\mathbf{r})]\rangle_\psi$$
$$= \mathbf{V}(\mathbf{r},t)\,\partial p(\mathbf{r},t)/\partial t + p(\mathbf{r},t)\,\partial\mathbf{V}(\mathbf{r},t)/\partial t$$
$$= -\mathbf{V}(\mathbf{r},t)\nabla\cdot\mathbf{j}(\mathbf{r},t) + p(\mathbf{r},t)\,\partial\mathbf{V}(\mathbf{r},t)/\partial t. \qquad (16)$$

For example, in the stationary (bonded) state of Eq. (3), when $\Phi_s(\mathbf{r},t) = \varphi_s(t)$ and hence $\mathbf{j}_s = \mathbf{V}_s = \partial\mathbf{V}_s/\partial t = \mathbf{0}$, $d\mathbf{j}_s/dt = \partial\mathbf{j}_s/\partial t = \mathbf{0}$.

8.3 RESULTANT INFORMATION MEASURES AND *PHASE-EQUILIBRIA*

In QIT [10, 40–45], one should distinguish the entropic content of states exhibiting the same electron distribution but differing in composition of their currents. Obviously, the CIT treatment of molecular systems, which uses the classical entropic concepts to probe the quantum probability distribution alone, fails to do so. The relevant quantum extension of the information measures, which explicitly recognize the density of the state phase or its electronic current, calls for the nonclassical complements to the classical information quantities, because the latter have been designed to probe solely the state probability distribution. A similar generalization of the classical (probability) communication systems is required in the quantum extension of the communication treatment of entropic bond multiplicities [10, 71].

The average Fisher's measure [1] of the classical *gradient*-information content in the probability density $p(r) = |\psi(r)|^2 = \varphi(r)^2$ of a *single*-particle state $|\psi\rangle$ described by the wavefunction

$$\psi(r) = \langle r|\psi\rangle = \varphi(r)\exp[i\chi(r)], \tag{17}$$

is reminiscent of von Weizsäcker's [73] inhomogeneity correction to the kinetic energy functional in the Thomas–Fermi theory,

$$I[p] = \int [\nabla p(r)]^2/p(r)\,dr = \int p(r)\,[\nabla \ln p(r)]^2\,dr \equiv \int p(r)\,I_p(r)\,dr$$

$$= 4\int [\nabla \varphi(r)]^2\,dr \equiv I[\varphi]. \tag{18}$$

This classical descriptor characterizes an effective "narrowness" (determinicity, sharpness) of the particle probability distribution. It follows from its amplitude form $I[\varphi]$ that it measures a length of the *modulus*-gradient.

The complementary Shannon's [2] entropy descriptor,

$$S[p] = -\int p(r)\ln p(r)\,dr \equiv \int p(r)\,S_p(r)\,dr \equiv -2\int \varphi^2(r)\ln \varphi(r)\,dr \equiv S[\varphi], \tag{19}$$

reflects the average "spread" (indeterminacy, uncertainty) of the random position variable. It also provides the amount of information received, when this uncertainty is removed by an appropriate particle localization experiment: $F[p] \equiv S[p]$. The densities-per-electron of these probability (modulus) functionals, $I_p(r)$ and $S_p(r)$, respectively, satisfy the classical relation

$$I_p(r) = [\nabla S_p(r)]^2. \tag{20}$$

The resultant entropy/information descriptors of the quantum state $|\psi\rangle$ combine these familiar classical contributions and the associated nonclassical supplements due to the state spatial phase or the probability current it generates:

$$\begin{aligned}
I[\psi] &\equiv -4\langle\psi|\nabla^2|\psi\rangle \equiv \langle\psi|\hat{I}|\psi\rangle = 4\int|\nabla\psi(r)|^2\,dr \equiv \int p(r)\,I(r)\,dr \\
&= I[p] + 4\int p(r)\,[\nabla\chi(r)]^2\,dr \equiv \int p(r)\,[I_p(r) + I_\chi(r)]\,dr \\
&\equiv I[p] + I[\chi] \equiv I[p,\chi] \\
&= I[p] + \left(\frac{2m}{\hbar}\right)^2 \int\left[\frac{j^2(r)}{p(r)}\right]dr = \int p(r)\,[I_p(r) + I_j(r)]\,dr \\
&\equiv I[p] + I[j] \equiv I[p,j],
\end{aligned} \tag{21}$$

$$\begin{aligned}
\tilde{I}[\psi] &\equiv I[p] - I[\chi] \equiv \tilde{I}[p] + \tilde{I}[\chi] \equiv \tilde{I}[p\,\chi] \\
&= \int p(r)\,[\tilde{I}_p(r) + \tilde{I}_\chi(r)]\,dr \equiv \int p(r)\tilde{I}(r)\,dr,
\end{aligned} \tag{22}$$

$$\begin{aligned}
S[\psi] &\equiv -\langle\psi|\ln p + 2\chi|\psi\rangle \equiv \langle\psi|\hat{S}|\psi\rangle = \int\psi(r)^*\,\hat{S}(r)\psi(r)\,dr \\
&\equiv \int p(r)\,S(r)\,dr \\
&= \int p(r)[S_p(r) - 2\chi(r)]\,dr \equiv \int p(r)[S_p(r) + S_\chi(r)]\,dr \\
&\equiv S[p] + S[\chi] \equiv S[p,\chi].
\end{aligned} \tag{23}$$

It should be observed that the phase-entropy terms are nonpositive, $S[\chi] \leq 0$ and $\tilde{I}[\chi] \leq 0$, while the gradient *phase*-information, proportional to the kinetic energy due to electronic current, is nonnegative: $I[\chi] = I[j] \geq 0$. Above, these information components have been expressed as expectation values of the corresponding multiplicative "operators" in the position representation, which also measure the functional resultant density-per-electron:

i)   of the gradient *determinicity*-information $I[\psi]$,

$$I(r) = [\nabla \ln p(r)]^2 + 4[\nabla \chi(r)]^2, \tag{24}$$

ii) of the *indeterminicity*-information (*gradient entropy*) $\tilde{I}[\psi]$,

$$\tilde{I}(r) = [\nabla \ln p(r)]^2 - 4[\nabla \chi(r)]^2, \tag{25}$$

iii) and of the global entropy $S[\psi]$:

$$S(r) = -[\ln p(r) + 2\chi(r)]. \tag{26}$$

The resultant gradient-*information*, the expectation value of the Hermitian operator

$$\hat{I} = -4\nabla^2 = (8m/\hbar^2), \hat{T}, \tag{27}$$

is proportional to the particle average kinetic energy $T[\psi] = \langle \psi | \hat{T} | \psi \rangle$, corresponding to the quantum-mechanical operator $\hat{T}(r) = -[\hbar^2 / (2m)]\Delta$ of Eq. (1),

$$I[\psi] = \langle \psi | \hat{I} | \psi \rangle - (8m/\hbar^2)\, T[\psi], \tag{28}$$

and characterizes the state overall gradient-*deterministic* aspect.

The nonclassical terms in densities-per-electron of the resultant gradient *information* and global entropy obey Eq. (20),

$$I_\chi(r) = [\nabla S_\chi(r)]^2, \tag{29}$$

while the nonclassical gradient and global *entropy*-densities satisfy the modified relation:

$$\tilde{I}_\chi(r) = -[\nabla S_\chi(r)]^2. \tag{30}$$

A proper explanation of this apparent change of sign calls for the *complex*-entropy concept [45]:

$$S[\psi] = \langle \psi | -2\ln \psi | \psi \rangle \equiv \langle \psi | \hat{S} | \psi \rangle = S[p] + i S[\chi] \equiv S[p] + S[\chi]. \tag{31}$$

It corresponds to the density-per-electron,

$$S(r) = S_p(r) + iS_\chi(r) \equiv S_p(r) + S_\chi(r) = -[\ln p(r) + 2i\chi(r)] , \qquad (32)$$

with the *phase*-component $S_\chi(r)$ being attributed to its imaginary part. The complex entropy is applicable to (complex) amplitudes (wavefunctions) of QM and provides a natural generalization of the familiar classical measure of the entropy content in probability distribution. The *non*-Hermitian *entropy* (uncertainty) operator $\hat{S} = -2\ln\psi$ generates the probability $S[p]$ and phase $S[\chi]$ components of the complex resultant measure $S[\chi]$ as its real and imaginary parts, respectively.

To summarize, the Hermitian *information* (determinicity) operator $\hat{I} = -4\nabla^2 = (8m/\hbar^2)\,\hat{T}$ gives rise to the *real* expectation value of the state resultant Fisher-type information content $I[\psi] = \langle\psi|\hat{I}|\psi\rangle$, while the *non*-Hermitian *entropy* (indeterminicity) operator $\hat{S} = -2\ln\psi$ generates the complex average quantity $S[\psi] = \langle\psi|\hat{S}|\psi\rangle$. The classical and nonclassical gradient analogs then follow from the same type of a mutual relation between the information and entropy densities:

$$\tilde{I}_p(r) = [\nabla S_p(r)]^2 \quad \text{and} \quad \tilde{I}_\chi(r) = [\nabla S_\chi(r)]^2 = [i\nabla S_\chi(r)]^2 = -[\nabla S_\chi(r)]^2. \quad (33)$$

The molecular (horizontal) *phase*-equilibria mark the extrema of both the global and gradient measures of the resultant entropy. The relevant information principle is directly suggested by the universal characteristic of the preceding mutual relation, which links the resultant Shannon and Fisher measures of the information content in quantum states. Let us assume, for reasons of simplicity, the particle *strong*-stationary state $\psi(r, t) = \varphi(r)\exp[i\phi(t)] \equiv \psi(\varphi, \phi)$, in which the (spatial) phase and current identically vanish: $\chi(r) = 0$ and $j[\varphi] = 0$. In order to determine the optimum ("thermodynamic") spatial phase $\chi_{eq.}(r)$ for the given probability distribution $p(r) = \varphi(r)^2$, $\chi_{eq.}(r) \equiv \chi_{eq.}[p; r]$, in the equilibrium state

$$\psi_{eq.}(r, t) = \varphi_{eq.}(r)\exp[i\phi(t)],$$

$$\varphi_{eq.}(r) = \varphi(r)\exp[i\chi_{eq.}(r)] \equiv \varphi_{eq.}[\varphi, \chi_{eq.}; r], \qquad (34)$$

one searches for the maximum of the state resultant gradient entropy with respect to its phase "variable" $\chi_{eq.}(r)$:

$$\delta_\chi \tilde{I}[\varphi_{eq.}] = 0 \implies \chi_{eq.}(r) = -(1/2)\ln p(r). \tag{35}$$

This equilibrium "horizontal" phase is seen to be related to the negative logarithm of probability density $p(r)$. It generates the state (*time*-independent) current density,

$$j_{eq.}(r, t) = (\hbar/m)p(r)\nabla\chi_{eq.}(r) = -[\hbar/(2m)]\nabla p(r) = j_{eq.}(r), \tag{36}$$

and hence a finite (stationary) probability source:

$$\sigma_p^{eq.}(r,t) = \partial p(r)/\partial t + \nabla \cdot j_{eq.}(r) = \nabla \cdot j_{eq.}(r) = \sigma_p^{eq.}(r) = -[\hbar/(2m)]\Delta p(r). \tag{37}$$

It should be observed, however, that–by Gauss' theorem–this equilibrium local source of probability density does not affect its overall normalization:

$$\int \sigma_p^{eq.}(r)\,dr = -[\hbar/(2m)]\int \nabla \cdot \nabla p(r)\,dr = -[\hbar/(2m)]\iint_{S\to\infty}[\nabla p(r)]_n\,dS = 0. \tag{38}$$

To summarize, the extremum of resultant entropy identifies a "thermodynamic" analogue $\psi_{eq.}(r, t)$ of the quantum state $\psi(r, t) = \langle r|\psi(t)\rangle$, determined by the optimum spatial phase $\chi_{eq.}(r, t) = -(1/2)\ln p(r, t)$:

$$\psi_{eq.}(r, t) = \psi(r, t)\exp[i\chi_{eq.}(r, t)] = R(r, t)\exp\{i[\phi(r, t) + \chi_{eq.}(r, t)]\}$$
$$\equiv R(r, t)\exp[i\Phi_{eq.}(r, t)]. \tag{39}$$

This equilibrium state represents the (unitary) *phase*-transform of the original quantum state $|\psi(t)\rangle$:

$$|\psi_{eq.}(t)\rangle = \hat{U}[p;r]|\psi(t)\rangle, \quad \hat{U}[p;r] = \exp[i\chi_{eq.}(r, t)], \quad \hat{U}\hat{U}^{-1} = \hat{U}\hat{U}^\dagger = 1. \tag{40}$$

In this phase transformation of the system wavefunction, the probability distribution is conserved,

$$p_{eq.}(r, t) = |\psi_{eq.}(r, t)|^2 = |\psi(r, t)|^2 = R(r, t)^2 = p(r, t), \tag{41}$$

while the probability current of the original state ψ,

$$j(r, t) = \langle\psi|\hat{j}(r)|\psi\rangle = (\hbar/m)p(r, t)\nabla\Phi(r, t), \tag{42}$$

is modified in the equilibrium state $\psi_{eq.}$ by term proportional to the negative gradient of $p(r, t)$,

$$
\begin{aligned}
j_{eq.}(r, t) &= \langle \psi_{eq.} | \hat{j}(r) | \psi_{eq.} \rangle = (\hbar/m)\, p(r, t) \nabla[\phi(r, t) + \chi_{eq.}(r, t)] \\
&= j(r, t) - [\hbar/(2m)]\, \nabla p(r, t),
\end{aligned}
\tag{43}
$$

which modifies the original current-divergence $\nabla \cdot j$ of the probability continuity relation by a correction term proportional to the Laplacian of probability density:

$$
\nabla \cdot j_{eq.} = \nabla \cdot j - [\hbar/(2m)]\, \Delta p.
\tag{44}
$$

Therefore, the *phase*-transformation generating the horizontal equilibrium state does not affect the *time*-derivative of state probability distribution of Eq. (41),

$$
(\partial p_{eq.}/\partial t) \equiv -\nabla \cdot j_{eq.} + \sigma_p{}^{eq.} = (\partial p/\partial t) = -\nabla \cdot j,
\tag{45}
$$

where the finite probability source $\sigma_p{}^{eq.} = dp_{eq.}/dt$ in equilibrium state is proportional to the probability Laplacian:

$$
\sigma_p{}^{eq.} = (\partial p_{eq.}/\partial t) + \nabla \cdot j_{eq.} = \sigma_p - [\hbar/(2m)]\, \Delta p = - [\hbar/(2m)]\, \Delta p.
\tag{46}
$$

For example, in equilibrium states derived from the stationary states of Eq. (2), for which $p_s = p_s(r)$ and $\partial p_s/\partial t = \nabla \cdot j_s = 0$, one finds the *time*-independent probability source:

$$
\sigma_{s,p}{}^{eq.}(r) = \nabla \cdot j_s{}^{eq.}(r) = - [\hbar/(2m)]\, \Delta p_s(r).
\tag{47}
$$

8.4   REACTION STAGES IN ACID-BASE SYSTEMS

Consider a chemical interaction between reactants in a bimolecular reactive system $R = A\text{----}B$ composed of the complementary *acidic* (A) and *basic* (B) subsystems and comprising together $N_R = N_A + N_B = N$ electrons. It is customary in the theory of chemical reactivity to examine several hypothetical stages of a chemical process in such CT systems, involving either the mutually *closed* (nonbonded, disentangled) or *open* (bonded, entangled) reactants [7, 49, 50, 74, 75].

The *Separated-Reactant Limit* (SRL) $R^\infty = A^0 \xleftrightarrow{\infty} B^0 \equiv \{\alpha^0(\infty)\}$ of the *isolated* (disentagled) or *dissociated* (entangled) species represents a collection of the infinitely separated, (noninteracting, mutually non-bonded) free species A^0 and B^0 exhibiting their respective ground-state densities $\{\rho_\alpha^0 = \rho_\alpha^0[N_\alpha^0, v_\alpha]\}$, $\alpha = A, B$, the equilibrium distributions for the subsystem initial (integer) numbers of electrons $\{N_\alpha^0\}$, $N_A^0 + N_B^0 = N$, and the external potentials $\{v_\alpha\}$ of reactants, due to their own nuclei. These reference electron densities also define the reactant probability distributions, the shape factors of isolated fragments $\{\alpha^0\}$,

$$p_\alpha^0(r) = \rho_\alpha^0(r)/N_\alpha^0, \qquad \int p_\alpha^0(r)\,dr = 1. \qquad (48)$$

The equilibrium states of the disentangled, free reactants represent their *phase*-transformed ground states corresponding to the fragment equilibrium "thermodynamic" phases:

$$\chi_{eq.}(\alpha^0) = -(\tfrac{1}{2})\ln p_\alpha^0(r) = \chi_{eq.}[p_\alpha^0, r], \qquad \alpha = A, B. \qquad (49a)$$

Their entangled analogs, with a common molecular ancestor, which interacted in the past and then have been separated [75], exhibit the equilibrium phase related to the density of the combined system:

$$\chi_{eq.}(\alpha^0+\beta^0) = -(\tfrac{1}{2})\ln p^0(r; \infty) = \chi_{eq.}[p^0(\infty); r],$$
$$p^0(\infty) = p_A^0 + p_B^0, \qquad \alpha = A, B. \qquad (49b)$$

Two *intermediate* stages of *interacting* species are introduced at a finite separation and given mutual orientation of the geometrically "frozen" reactants in $R(R_{A—B}) = A \xleftarrow{R_{A-B}} B$ specified by the external potential of R as a whole, $v_R(R_{A—B}) = v_A(R_A) + v_B(R_B) \equiv v(R_{A—B})$, for the current values of the relative reaction "coordinates" $R_{A—B}$:

The *nonbonded* (*n*) *promolecular* reference $R_n^0 = A^0----B^0 = (A^0|B^0)$ corresponds to the electronically and geometrically "frozen" (mutually *closed*) free subsystems separately exhibiting the same distributions as in R^∞, either brought from infinity to their current mutual positions $R_{A—B}$ in a molecule (disentangled reactants).

The "frozen" electron distributions of free reactants determine the overall promolecular density:

$$\rho_R^0(\mathbf{R}_{A-B}) = \rho_A^0(\mathbf{R}_{A-B}) + \rho_B^0(\mathbf{R}_{A-B}) \equiv N_R\, p_R^0(\mathbf{R}_{A-B})$$
$$= N_A p_A^0(\mathbf{R}_{A-B}) + N_B p_B^0(\mathbf{R}_{A-B}) \tag{50}$$

or its probability factor:

$$p_R^0(\mathbf{R}_{A-B}) = (N_A/N_R)\, p_A^0(\mathbf{R}_{A-B}) + (N_B/N_R)\, p_B^0(\mathbf{R}_{A-B})$$
$$\equiv P_A p_A^0(\mathbf{R}_{A-B}) + P_B p_B^0(\mathbf{R}_{A-B}), \tag{51}$$

$$\int \rho_R^0(\mathbf{r})\,d\mathbf{r} = N_R^0 = N_A^0 + N_B^0 = N \quad \text{and} \quad \int p_R^0(\mathbf{r})\,d\mathbf{r} = P_A + P_B = 1.$$

It generates the promolecular phase characterizing both the entangled free reactants at molecular positions in R_b^0: at molecular positions in the "bonded" promolecule $R_b^0 = (A^0|B^0)$:

$$\chi_{eq.}(A^0|B^0) = -\,(\tfrac{1}{2})\ln p_R^0(\mathbf{R}_{A-B}). \tag{52}$$

One thus interprets ρ_R^0 as representing the *bonded* (*b*) promolecular system $R_b^0 = A^0-B^0 = (A^0|B^0)$ consisting of the mutually open ("frozen") free reactants. In this global approach to the promolecular "equilibrium," one ascribes the *horizontal*-phase of Eq. (52), due to the overall promolecular density ρ_R^0, to the promolecule as a whole and both its fragments. Indeed, when there is no dividing "wall" separating the two reactants in the "bonded" system, as symbolized by the *broken* vertical line in R_b^0, each *open* fragment can be regarded as effectively extending over the whole system, thus accquiring itself the global promolecular distribution. Therefore, this local "thermodynamic" phase of the whole promolecular complex also characterizes its mutually open subsystems.

One observes, however, that in this "frozen" (nonequilibrium) electron distribution, the local chemical potentials are not equalized:

$$\mu^0(\mathbf{r}) \equiv \delta E_v[\rho]/\delta\rho(\mathbf{r})\big|_R^0 \neq \mu^0(\mathbf{r'}) \neq \ldots \neq \mu^0(\mathbf{r'});$$

here, the density functional $E_v[\rho]$ for the average electron energy [17, 18],

$$E[\psi] = \langle\psi|\hat{H}|\psi\rangle \equiv \int\rho(r)\,v(r)\,dr + F[\rho] \equiv E_v[\rho]$$
$$= V_{ne}[\rho] + (T[\rho] + V_{ee}[\rho]), \qquad (53)$$

contains the external potential interaction contribution $V_{ne}[\rho]$ and the sum $F[\rho]$ of electron kinetic $T[\rho] = T[\psi[\rho]]$ and repulsion ($V_{ee}[\rho]$) energies.

The disentangled (independent) reactants in $R_n^{\,0}$ can be formally linked to the product of the ground states $\{\psi_\alpha^0 = \psi_\alpha^0[N_\alpha^0, v_\alpha]\}$ in separate subsystems, $\psi_R^0(A^0|B^0) = \psi_A^0(A)\,\psi_B^0(B)$, which generates the joint probabilities $P_R^0(A^0\wedge B^0) = P_A^0(A)\,P_B^0(B)$. Therefore, the effective subsystem probabilities $\{P(\alpha^0) = \mathrm{tr}_{\beta\neq\alpha}P_R^0(A^0\wedge B^0) = P_\alpha^0(\alpha)\}$ still reflect the frozen electron distribution of the free reactants, giving rise to the equilibrium phases of Eq. (49a).

The nonbonded *polarization* (P) stage of $R_n^{\,+} = (A^+|B^+) = A^+$----$B^+$ for the same relative position $\mathbf{R}_{A-B}$ of the geometrically "frozen" subsystems consists of the disentangled (mutually *closed*), electronically relaxed reactants. It is characterized by the "promoted" subsystem densities $\{\rho_\alpha^+ = \rho_\alpha'[N_A^0, N_B^0, v_R] \equiv N_\alpha^0 p_\alpha'\}$, equilibrium distributions for their initial (integer) numbers of electrons, and the overall external potential of $R_n^{\,+}$ as a whole, which combine into the overall distribution of the whole polarized complex $R_n^{\,+}$:

$$\rho_R^+ = \rho_A^+ + \rho_B^+ \equiv N_R p_R^+, \quad \int\rho_R^+(r)\,dr = N_R \quad \text{or}$$
$$p_R^+ = P_A^0 p_A^+ + P_B^0 p_B^+, \quad \int p_R^+(r)\,dr = 1. \qquad (54)$$

The internal equilibria in each subsystem of $R_n^{\,+}$ imply the equalization of the reactant *local* chemical potentials at the fragment *global* level,

$$\mu_\alpha^+(r) = \mu_\alpha^+[N_A^0, N_B^0, v_R] = \partial E[N_A, N_B, v_R]/\partial N_\alpha|_R^+ = \mu_\alpha^+[\rho_A^+, \rho_B^+]. \qquad (55)$$

The mutually closed, polarized reactants thus exhibit the *internally* equalized but different chemical potentials of both fragments, $\mu_A^+ \neq \mu_B^+$ and thermodynamic phases (χ_A^+, χ_B^+) related to the subsystem polarized probability distributions,

$$\{\chi_{eq.}(\alpha^+) = -(\tfrac{1}{2})\ln p_\alpha^+ \equiv \chi_\alpha^+\}, \qquad (56)$$

marking their separate (*intra*-reactant) equilibria in the mutually non-bonded but interacting subsystems.

At the *bonded* P-stage of $R_b^+ = A^+ - B^+ = (A^+|B^+)$, one formally attributes to each entangled subsystem the *global* horizontal phase corresponding to the combined electron distribution of Eq. (54), describing the whole polarized reactive system R^+ and representing its overall phase intensity:

$$\chi_{eq.}(R_b^+) = \chi_{eq.}[N_A^0, N_B^0, v_R] = -(\tfrac{1}{2})\ln p_R^+ \equiv \chi_R^+. \tag{57}$$

This global thermodynamic phase also characterizes each externally open (bonded) reactant fragment in R_b^+, now effectively extending over the whole system. One again notes, however, that this electron distribution of the nonequilibrium open subsystems generates the nonequalized local chemical potentials.

Finally, the *inter*-reactant equilibrium state of the geometrically "frozen" but mutually *open* (bonded, electronically relaxed) reactants in $R_b = A^* - B^* = (A^*|B^*)$, i.e., the ground state of the reactive system as a whole, is characterized by equalization of the local chemical potential at the system global level,

$$\mu^*(r) = \partial E[N, v_R]/\partial N|_R^* = \mu_R[N_R, v_R]/= \mu_R[\rho_R]. \tag{58}$$

In this equilibrium state of the electronically relaxed complex, the effective subsystem densities extend over the whole reactive system: $\rho(\alpha_b) = \rho_R$, $\alpha = A, B$. The stationary electron density of the whole reactive system,

$$\rho_R(R_{A-B}) \equiv N_R p_R, \int p_R(r)\,dr = 1, \tag{59}$$

generates the equilibrium phase of R_b as a whole:

$$\chi_{eq.}(R_b) = \chi_{eq.}[N_R, v_R] = -(\tfrac{1}{2})\ln p_R(r) \equiv \chi_R. \tag{60}$$

This *global phase*-intensity also characterizes the externally open (bonded, entangled) subsystems in R_b: $\chi_{eq.}(\alpha_b) = \chi_R$.

One could also formally envisage the nonbonded case of the disentagled subsystems in $R_n = A^* ---- B^* = (A^*|B^*)$ of the same overall electron density, for which the *partial* equilibrium phases of subsystems are related to pieces $\{\rho_\alpha^* = N_\alpha^* p_\alpha^*\}$ of $\rho_R = \Sigma_\alpha \rho_\alpha^*$:

$$\chi_{eq.}(\alpha^*) = -(\tfrac{1}{2})\ln p_\alpha^* = \chi_{eq.}[p_\alpha^*] \equiv \chi_\alpha^*, \quad \alpha = A, B. \tag{61}$$

In such a collection of the mutually nonbonded (disentangled) molecular subsystems, one thus formally introduces the reactant densities

$$\{\rho_\alpha^* = \rho_\alpha^*[N_A^*, N_B^*, v_R]\}, \quad \rho_A^* + \rho_B^* = \rho_R[N_R, v_R], \tag{62}$$

which already exhibit effects of the (fractional) B→A CT:

$$N^{CT} = N_A^* - N_A^0 = N_B^0 - N_B^* > 0. \tag{63}$$

It equalizes the chemical potentials of subsystems:

$$\mu_A^*[N_A^*, N_B^*, v_R] = \mu_B^*[N_A^*, N_B^*, v_R] = \mu_R[N_R, v_R]. \tag{64}$$

In DFT, the optimum wavefunctions for the specified electron density of N electrons, $\rho = Np$, can be constructed using the HZM scheme [10, 46–48]. Such functions appear in Levy's [19] constrained-search definition of the universal density functional for the sum of the electron kinetic and repulsion energies. This DFT framework introduces the complete set of the density-conserving Slater determinants build using equidensity orbitals (EO) of the *plane*-wave type:

$$\{\varphi_k(r) = R(r)\exp[i\Phi_k(r)]\}. \tag{65}$$

The equilibrium EO [10, 40, 42, 43, 46] adopts an equal, density-dependent modulus part $R(r) = p(r)^{1/2}$ and the spatial phase function composed of the "orthogonality" $[F_k(r)]$ and "thermodynamic" $[\chi(r)]$ parts:

$$\Phi_k(r) = k \cdot f(r) + \chi(r) \equiv F_k(r) + \chi(r), \tag{66}$$

with the density-dependent vector function $f(r) = f[p; r]$, common to all EO and linked to the Jacobian of the $r \rightarrow f(r)$ transformation. The optimum orbitals for the specified ground-state probability distribution $p_0(r) = R_0(r)^2$,

$$\varphi_k(r) = [p_0(r)]^{1/2} \exp\{i[k \cdot f[p_0; r] + \chi(r)]\} \equiv R_0(r) \exp(i\Phi_k[p_0; r]) \equiv \varphi_k[p_0; r],$$
$$\tag{67}$$

are thus specified by the "orthogonality" phase $F_k[p_0; r] = k \cdot f[p_0; r]$, with the *wave*-vector (reduced momentum) $k = k[p_0]$ and the density-dependent vector field $f_0(r) = f[p_0; r]$ resulting from the ordinary variational principle for the minimum electron energy of the familiar self-consistent field (SCF) theories, e.g., in the Hartree–Fock (HF) or Kohn–Sham (KS) methods.

The "thermodynamic" phase function $\chi(r)$, common to all equilibrium EO, subsequently results from the subsidiary maximum entropy principle. More specifically, the optimum form of this contribution in the resultant phase component $\Phi_k(r)$ results from the extremum principle of the resultant entropy for the given ground-state density $\rho_0 = Np_0$ and probability distribution p_0, determined at an earlier energy optimization stage. The equilibrium horizontal phase of these orbitals,

$$\chi_{eq.}(r) = - (\tfrac{1}{2}) \ln p_0(r) \equiv \chi_{eq.}[p_0; r],$$
$$\tag{68}$$

then generates the resultant current of N electrons,

$$j[\chi_{eq.}[p_0]; r] = (N\hbar/m)p(r)\nabla\chi_{eq.}(r) = - [\hbar/(2m)]\nabla\rho_0(r),$$
$$\tag{69}$$

reflecting the negative density gradient.

8.5   CHARGE TRANSFER DERIVATIVES OF ELECTRONIC DESCRIPTORS

The phase approach to molecular equilibria introduces a new class of IT-sensitivities of reactants, which can be used to characterize their entropic propensities in chemical reactions. In *classical* CSA, only the energy derivatives with respect to the external potential (v) and number of electrons (N) of the effective charge distributions in subsystems and in R as a whole have been examined [49, 50]. The *phase*-equilibria additionally call for the associated energy conjugates of the *nonclassical* (phase/current) state parameters, which we shall address in this section. In the *acid-base* complexes of the preceding section, such descriptors of reactants can be subsequently combined with the relevant *in-situ* properties characterizing the B→A CT [72]. We shall express such quantities of the equilibrium polarized subsystems in terms of their charge and phase/current sensitivities. The nonclassical IT descriptors of the polarized reactants will be subsequently combined with the resultant quantities describing the whole reactive system.

Consider again the mutually *closed* (nonbonded), polarized reactants in $R_n^+ = (A^+|B^+)$. Their equilibrium phase distributions $\{\chi_\alpha^+(r)\}$ of Eq. (54) determine the subsystem currents,

$$j_\alpha^+ = (\hbar/m)\, \rho_\alpha^+ \nabla \chi_\alpha^+ = -[\hbar/(2m)]\rho_\alpha^+ \nabla \ln p_\alpha^+ = -[\hbar/(2m)]\nabla \rho_\alpha^+,\ \alpha = A, B, \tag{70}$$

proportional to the negative gradient of the equilibrium electron density in the polarized fragment. These subsystem phases also determine the resultant phase χ_R^+ of Eq. (56) and the net current in R_n^+:

$$j_R^+ = j_A^+ + j_B^+ = -[\hbar/(2m)](\nabla \rho_A^+ + \nabla \rho_B^+)$$
$$= -[\hbar/(2m)]\nabla \rho_R^+ = (\hbar/m)\rho_R^+ \nabla \chi_R^{\,i}. \tag{71}$$

The CT derivative of the resultant phase of the entangled polarized reactants [Eq. (57)] then reads:

$$\partial \chi_R^+(r)/\partial N_{CT} = \partial \chi_R^+(r)/\partial N_{CT} = \Sigma_{\alpha=A,B}\, [\partial \chi_R^+(r)/\partial N_\alpha](\partial N_\alpha/\partial N_{CT})$$
$$= \Sigma_{\beta=A,B}\, \Sigma_{\alpha=A,B}\, [\partial \chi_\beta^+(r)/\partial N_\alpha](\partial N_\alpha/\partial N_{CT})$$
$$= \{[\partial \chi_A^+(r)/\partial N_A] - [\partial \chi_A^+(r)/\partial N_B]\}$$
$$- \{[\partial \chi_B^+(r)/\partial N_B] - [\partial \chi_B^+(r)/\partial N_A]\}$$
$$\equiv \phi_A^{CT}(r) - \phi_B^{CT}(r) \equiv \Phi^{CT}(r), \tag{72}$$

where Φ^{CT} stands for the *in situ* FF descriptor of R_n^+ as a whole, given by the difference of effective FF indices $\{\phi_\alpha^{CT}\}$ of reactants. By analogy to the FF matrix of molecular fragments in CSA,

$$\mathbf{f}^+(r) = \{f_{\alpha,\beta}(r) = \partial \rho_\beta^+(r)/\partial N_\alpha\}, \tag{73}$$

one introduces in QIT the matrix of analogous *phase* derivatives calculated for the fixed external potential due to the nuclei determined by the "frozen" molecular geometry,

$$\partial\chi_\beta^+(r)/\partial N_\alpha = -(1/2)\,\partial[\ln p_\beta^+(r)]/\partial N_\alpha = -(1/2)\,\partial\ln[\rho_\beta^+(r)/N_\beta]/\partial N_\alpha$$

$$= -(1/2)\,\{[f_{\alpha,\beta}(r)/\rho_\beta^+(r)] - (\delta_{\alpha,\beta}/N_\beta)\} \equiv \phi_{\alpha,\beta}(r);\ \alpha,\ \beta \in \{A, B\}, \qquad (74)$$

determined by the equilibrium-*phase* derivatives $\phi_{\alpha,\beta}(r)$ measuring the corresponding indices-per-electron. One recalls that the *in situ* FF [72],

$$f_\alpha^{CT}(r) = \partial\rho_\alpha(r)/\partial N_{CT} = f_{\alpha,\alpha}(r) - f_{\beta,\alpha}(r);\ \alpha,\ \beta\neq\alpha \in \{A, B\}, \qquad (75)$$

combines into the corresponding global CT derivative

$$F^{CT}(r) \equiv f_A^{CT}(r) - f_B^{CT}(r), \qquad (76)$$

which represents the population sensitivity of R_n^+ with respect to the internal CT between reactants.

The *in situ* differentiation of the resultant current j_R^+ [Eq. (71)] in the equilibrium (polarized, entangled) reactants similarly gives

$$\partial j_R^+(r)/\partial N_{CT} = -[\hbar/(2m)]\,\partial[\nabla\rho_R^+(r)]/\partial N_{CT}$$
$$= -[\hbar/(2m)]\,\nabla[\partial\rho_R^+(r)/\partial N_{CT}] \equiv -[\hbar/(2m)]\,\nabla F^{CT}(r), \qquad (77)$$

where we have observed that the two differentiations with respect to different variables, $\partial/\partial N_{CT}$ and $\nabla = \partial/\partial r$, commute with one another. This CT derivative of electronic current is thus seen to be determined by the gradient of the *in situ* FF in the reactant resolution:

$$F^{CT}(r) = \partial\rho_R^+(r)/\partial N_{CT} = \Sigma_{\alpha=A,B}\,\Sigma_{\beta=A,B}\,[\partial\rho_\beta^+(r)/\partial N_\alpha]\,(\partial N_\alpha/\partial N_{CT})$$

$$= [f_{A,A}(r) - f_{B,A}(r)] - [f_{B,B}(r) - f_{A,B}(r)] \equiv f_A^{CT}(r) - f_B^{CT}(r). \qquad (78)$$

The electronic current $j[\rho] = (\hbar/m)\,\rho\,\nabla\chi[\rho]$ for the equilibrium phase $\chi[\rho] = -(\tfrac{1}{2})\ln(\rho/N)$ gives the following derivative with respect to the external inflow/outflow of electrons:

$$\partial j/\partial N = (\hbar/m)\,\{(\partial\rho/\partial N)\,\nabla\chi[\rho] + \rho\,\partial(\nabla\chi[\rho])/\partial N\}$$
$$= (\hbar/m)\,\{(\partial\rho/\partial N)\,\nabla\chi[\rho] + \rho\,\nabla(\partial\chi[\rho]/\partial N)\}$$
$$\equiv (\hbar/m)\,(f\nabla\chi + \rho\,\nabla\phi) = -[\hbar/(2m)]\,\nabla f, \qquad (79)$$

where the global FF descriptor

$$f(r) = \partial \rho(r)/\partial N. \tag{80}$$

Thus, the FF gradient is seen to determine the overall populational sensitivity of electronic current in the given electron distribution. One similarly determines the populational derivative of the equilibrium "thermodynamic" phase,

$$\partial \chi[\rho]/\partial N = -(1/2)] [(f/\rho) - (1/N)], \tag{81}$$

the global analog of Eq. (74).

Consider next the global equilibrium in $R_n = A^*----B^* = (A^*|B^*)$, with both (mutually closed) fragments already displaying the B→A CT: $\rho_A^* + \rho_B^* = \rho_R$, $\{\int \rho_\alpha^* dr = N_\alpha^*\}$. This optimum flow of electrons brings about the chemical potential equalization in R_n, $\mu_A^*[\rho_A^*] = \mu_B^*[\rho_B^*] = \mu_R[\rho_R]$, but the nonbonded (disentangled) status of both subsystems is reflected by their different equilibrium phases [Eq. (61)]. The fragment chemical potentials $\{\mu_\alpha\}$ and elements of the hardness matrix $\eta_R{}' = \{\eta_{\alpha,\beta}\}$ represent populational derivatives, calculated for the fixed external potential v reflecting molecular geometry, of the system ensemble-average electronic energy $E_v(\{N_\beta\})$ [49–57]. These quantities are defined by the corresponding partial derivatives of the *grand*-ensemble average value of the system electron energy with respect to average electron populations $\{N_\alpha\}$ on subsystems:

$$\mu_\alpha \equiv \partial E_v(\{N_\gamma\})/\partial N_\alpha, \quad \eta_{\alpha,\beta} = \partial^2 E_v(\{N_\gamma\})/\partial N_\alpha \partial N_\beta = \partial \mu_\alpha/\partial N_\beta. \tag{82}$$

The associated global properties of R similarly involve differentiation with respect to the overall number of electrons in the whole reactive system:

$$\mu_R = \partial E_v(N_R)/\partial N_R, \quad \eta_R = \partial^2 E_v(N_R)/\partial N_R^2 = \partial \mu_R/\partial N_R. \tag{83}$$

The optimum amount of the fractional N_{CT} of Eq. (63) is then determined by the difference in chemical potentials of the (equilibrium) polarized reactants in R_n^+, i.e., its effective CT gradient,

$$\mu_{CT} = \partial E_v(N_{CT})/\partial N_{CT} = \mu_A^+ - \mu_B^+ < 0, \tag{84}$$

and the *in situ* hardness (η_{CT}) or softness (S_{CT}) for this process,

$$\eta_{CT} = \partial \mu_{CT}/\partial N_{CT} = \eta_{A,A} + \eta_{B,B} - \eta_{A,B} - \eta_{B,A} = S_{CT}^{-1}, \tag{85}$$

representing the CT Hessian and its inverse, respectively. The interreactant CT,

$$N_{CT} = -\mu_{CT} S_{CT}, \tag{86}$$

then generates the second-order stabilization energy:

$$E_{CT} = \mu_{CT} N_{CT}/2 = -(\mu_{CT})^2 S_{CT}/2 < 0. \tag{87}$$

The stationary characteristic of the overall electron distribution in $R_b^* = R$ implies that FF descriptors of the mutually open (bonded, entangled) reactants in the global equilibrium distribution $\rho_R = \rho_A^* = \rho_B^*$ give rise to the vanishing CT differences of Eq. (75):

$$f_\alpha^{CT}(N_A^*, N_B^*) = f_{\alpha,\alpha}(N_A^*, N_B^*) - f_{\beta,\alpha}(N_A^*, N_B^*) \equiv f_{\alpha,\alpha}^* - f_{\beta,\alpha}^* = 0 \text{ or}$$

$$f_{\alpha,\alpha}^* = f_{\beta,\alpha}^* = \partial \rho_R/\partial N_R = f_R. \tag{88}$$

Thus, in the global equilibrium state of R as a whole, the diagonal FF locally equalizes with its *off*-diagonal ("cross") complement.

8.6 POPULATIONAL DERIVATIVES OF ENTROPY/INFORMATION MEASURES

Let us now examine the populational derivatives of information measures themselves. One first observes that a functional $G[\rho]$ of the electron density $\rho(r) = Np(r)$ can be regarded as an explicit function of the system overall number of electrons N and a functional of the density *shape*-factor $p(r)$, which is itself a function of N, $p(r) = p[N, r]$: $G[\rho] = G[p(N), N] \equiv G(N)$. Accordingly, properties depending on this probability distribution alone are only implicitly dependent on N: $F[p] = F[p(N)] = F(N)$. Thus, the gradient information $I[p]$ [Eq. (18)] and Shannon entropy $S[p]$ [Eq. (19)] fall into the second category, while their electron density analogs,

$$I[\rho] = \int [\nabla\rho(r)]^2/\rho(r)\,dr = N\,I[p] \equiv I[p(N), N] \quad \text{and}$$

$$S[\rho] = -\int \rho(r)\ln\rho(r)\,dr = N(S[p] - \ln N) \equiv S[p(N), N], \qquad (89)$$

determine the classical information measures depending both explicitly and implicitly on the overall number of electrons.

The electronic FF, measuring the populational derivative of electron density, has both explicit and implicit components:

$$f(r) = \partial\rho(r)/\partial N = p(r) + N[\partial p(r)/\partial N], \qquad \int f(r)\,dr = 1, \quad \text{or}$$

$$g(r) \equiv [\partial p(r)/\partial N] = [f(r) - p(r)]/N, \qquad \int g(r)\,dr = 0, \qquad (90)$$

while its equilibrium phase $\chi_{eq}[p(r), r]$ analog involves the *shape*-FF factor $g(r)$:

$$\phi_{eq.}(r) \equiv \partial\chi_{eq.}[p(N), r]/\partial N = [\partial\chi_{eq.}(r)/\partial p(r)]\,g(r) = -[2p(r)]^{-1}g(r). \quad (91)$$

The populational derivatives of the classical measures of the information content in electron density are locally FF-weighted:

$$\partial I[\rho]/\partial N = \int \frac{\delta I[\rho]}{\delta\rho(r)}f(r)\,dr \equiv \int i_\rho(r)\,f(r)\,dr, \qquad i_\rho = (\rho^{-1}\rho)^2 - 2\rho^{-1}\Delta\rho;$$

$$\partial S[\rho]/\partial N = \int \frac{\delta I[\rho]}{\delta\rho(r)}f(r)\,dr \equiv \int s_\rho(r)\,f(r)\,dr = -\int f(r)\ln\rho(r)\,dr - 1,$$

$$s_\rho = -(\ln\rho)\,\Delta\rho - 1. \qquad (92)$$

The nonclassical *phase*-supplements $I[\chi] = -\tilde{I}[\chi]$ [Eqs. (21) and (22)] and $S[\chi]$ [Eq. (23)] in the resultant measures of the information content are similarly weighted by the *phase*-FF:

$$\partial I[\chi]/\partial N = \int \frac{\delta I[\chi]}{\delta\chi(r)}\phi(r)\,dr \equiv \int i_\chi(r)\,\phi(r)\,dr = -\partial\tilde{I}[\chi]/\partial N,$$

$$i_\chi = -8[\nabla p\cdot\nabla\chi - p\,\Delta\chi];$$

$$\partial S[\chi]/\partial N = \int \frac{\delta S[\chi]}{\delta\chi(r)}\phi(r)\,dr \equiv \int s_\chi(r)\,\phi(r)\,dr, \qquad s_\chi = -2p. \qquad (93)$$

It should be observed that in the global phase-equilibrium, for $\chi_{eq.}[p(N),$
$r] = -(1/2)\ln p(r)$, the magnitudes of the classical and nonclassical entropies equalize,

$$-S[\chi] = S[p] \quad \text{and} \quad -\tilde{I}[\chi] = I[p], \tag{94}$$

so that the resultant entropies in the equilibrium state $\psi_{eq.} = \psi \exp(i\chi_{eq.})$
identically vanish, $S[\psi_{eq.}] = \tilde{I}[\psi_{eq}] = 0$, while the resultant gradient information $I[\psi_{eq.}] = 2I[p]$. This also implies

$$\partial S[\psi_{eq.}]/\partial N = \partial \tilde{I}[\psi_{eq.}]/\partial N = 0 \quad \text{and}$$

$$\partial I[\psi_{eq.}]/\partial N = 2\,\partial I[p]/\partial N = 2\int i_p(r)\,g(r)\,dr, \quad i_p = (p^{-1}\nabla p)^2 - 2p^{-1}\Delta p. \tag{95}$$

Consider next the nonbonded (disentangled) subsystems in $R_n^+ =$
$(A^+|B^+)$. The entropy/information functionals of subsystems, $G[\rho_\alpha]$, $G = S$,
I or $\tilde{I}$, determine their additive parts

$$G^{add.}[\rho_A^+, \rho_B^+] = G[\rho_A^+] + G[\rho_B^+], \tag{96}$$

while the system overall electron distribution of Eq. (53) defines its total
entropy/information contents,

$$G[\rho_A^+ + \rho_B^+] = G^{total}[\rho_A^+, \rho_B^+] = G^{add.}[\rho_A^+, \rho_B^+] + G^{nadd.}[\rho_A^+, \rho_B^+], \tag{97}$$

and their nonadditive parts:

$$G^{nadd.}[\rho_A^+, \rho_B^+] = G^{total}[\rho_A^+, \rho_B^+] - G^{add.}[\rho_A^+, \rho_B^+] = G[\rho_R^+] - G[\rho_A^+] - G[\rho_B^+]. \tag{98}$$

A similar partition also applies to the classical/nonclassical components of these information measures. Consider first an illustrative example of the additive/nonadditive partition of contributions to the resultant
global entropy [Eq. (23)]:

$$S[\psi_R^+] = S^{class.}[\psi_R^+] + S^{nclass.}[\psi_R^+] \equiv S_{total}^{class.}[p_A^+, p_B^+] + S_{total}^{nclass.}[\chi_A^+, \chi_B^+]$$

$$\equiv S_{total}[p_A^+, p_B^+; \chi_A^+, \chi_B^+] = S_{add.}[\psi_R^+] + S_{nadd.}[\psi_R^+],$$

$$(99)$$

$$S_{add.}[\psi_R^+] = S_{add.}^{class.}[p_A^+, p_B^+] + S_{add.}^{nclass.}[\chi_A^+, \chi_B^+],$$

$$S_{nadd.}[\psi_R^+] = S_{nadd.}^{class.}[p_A^+, p_B^+] + S_{nadd.}^{nclass.}[\chi_A^+, \chi_B^+].$$

Its classical part

$$
\begin{aligned}
S^{class.}[\rho_R^+] &= -\int p_R^+ \ln p_R^+ \, dr = S[p_R^+] \\
&= -\int (P_A^0 p_A^+ + P_B^0 p_B^+) \ln (P_A^0 p_A^+ + P_B^0 p_B^+) dr \\
&\equiv S_{total}^{class.}[p_A^+, p_B^+],
\end{aligned}
$$

$$(100)$$

exhibits the following components:

$$
\begin{aligned}
S_{add.}^{class.}[p_A^+, p_B^+] &= P_A^0 S[p_A^+] + P_B^0 S[p_B^+] \\
&= -P_A^0 \int p_A^+ \ln p_A^+ dr - P_B^0 \int p_B^+ \ln p_B' \, dr,
\end{aligned}
$$

$$S_{nadd.}^{class.}[p_A^+, p_B^+] = S_{total}^{class.}[p_A^+, p_B^+] - S_{add.}^{class.}[p_A^+, p_B^+]. \qquad (101)$$

A related division of the nonclassical part of the resultant global entropy,

$$
\begin{aligned}
S^{nclass.}[\rho_R^+] &= -2\int p_R^+ \chi_R^+ \, dr = S[\chi_R^+] \\
&= -2\int (P_A^0 p_A^+ P_B^0 p_B^+)(\chi_A^+ + \chi_B^+) \, dr \\
&= S_{total}^{nclass.}[\chi_A^+, \chi_B^+],
\end{aligned}
$$

$$(102)$$

gives:

$$
\begin{aligned}
S_{add.}^{nclass.}[\chi_A^+, \chi_B^+] &= P_A^0 S[\chi_A^+] + P_B^0 S[\chi_B^+] \\
&= -2(P_A^0 \int p_A^+ \chi_A^+ dr + P_B^0 \int p_B^+ \chi_B^+ dr) \\
&= P_A^0 \int p_A^+ \ln p_A^+ dr + P_B^0 \int p_B^+ \ln p_B^+ dr, \\
S_{nadd.}^{nclass.}[\chi_A^+, \chi_B^+] &= S_{total}^{nclass.}[\chi_A^+, \chi_B^+] - S_{add.}^{nclass.}[\chi_A^+, \chi_B^+] \\
&= P_A^0 \int p_A^+ \ln p_B^+ dr + P_B^0 \int p_B^+ \ln p_A^+ dr.
\end{aligned}
$$

$$(103)$$

Next, let us consider the resultant gradient information of Eq. (21):

$$
\begin{aligned}
I[\psi_R^+] &= I^{class.}[\psi_R^+] + I^{nclass.}[\psi_R^+] \equiv I_{total}^{class.}[p_A^+, p_B^+] + I_{total}^{nclass.}[\chi_A^+, \chi_B^+] \\
&= I_{total}[p_A^+, p_B^+; \chi_A^+, \chi_B^+] = I_{add.}[\psi_R^+] + I_{nadd.}[\psi_R^+],
\end{aligned}
$$

$$I_{add.}[\psi_R^{+}] = I_{add.}^{class.}[p_A^{+}, p_B^{+}] + I_{add.}^{nclass.}[\chi_A^{+}, \chi_B^{+}],$$

$$I_{nadd.}[\psi_R^{+}] = I_{nadd.}^{class.}[p_A^{+}, p_B^{+}] + I_{nadd.}^{nclass.}[\chi_A^{+}, \chi_B^{+}]. \tag{104}$$

The additive/nonadditive division of its overall classical contribution [11, 20, 75],

$$\begin{aligned}
I^{class.}[\psi_R^{+}] = I[p_R^{+}] &= \int (\nabla p_R^{+})^2/p_R^{+}\, dr \\
&= \int (P_A^{0}\nabla p_A^{+} + P_B^{0}\nabla p_B^{+})^2/(P_A^{0}p_A^{+} + P_B^{0}p_B^{+})\, dr \\
&= I_{total}^{class.}[p_A^{+}, p_B^{+}],
\end{aligned} \tag{105}$$

reads:

$$\begin{aligned}
I_{add.}^{class.}[p_A^{+}, p_B^{+}] &= P_A^{0}I[p_A^{+}] + P_B^{0}I[p_B^{+}] \\
&= P_A^{0}\int (\nabla p_A^{+})^2/p_A^{+}\, dr + P_B^{0}\int (\nabla p_B^{+})^2/p_B^{+}\, dr,
\end{aligned}$$

$$I_{nadd.}^{class.}[p_A^{+}, p_B^{+}] = I_{total}^{class.}[p_A^{+}, p_B^{+}] - I_{add.}^{class.}[p_A^{+}, p_B^{+}]. \tag{106}$$

The nonclassical part of the resultant gradient information

$$\begin{aligned}
I^{nclass.}[\psi_R^{+}] = I[\chi_R^{+}] &= 4\int p_R^{+}(\nabla \chi_R^{+})^2\, dr \\
&= 4\int (P_A^{0}p_A^{+} + P_B^{0}p_B^{+})(\nabla \chi_A^{+} + \nabla \chi_B^{+})^2\, dr \\
&= I_{total}^{nclass.}[\chi_A^{+}, \chi_B^{+}],
\end{aligned} \tag{107}$$

partitions as follows:

$$I_{total}^{nclass.}[\chi_A^{+}, \chi_B^{+}] = I_{add.}^{nclass.}[\chi_A^{+}, \chi_B^{+}] + I_{nadd.}^{nclass.}[\chi_A^{+}, \chi_B^{+}],$$

$$\begin{aligned}
I_{nadd.}^{nclass.}[\chi_A^{+}, \chi_B^{+}] &= P_A^{0}I[\chi_A^{+}] + P_B^{0}I[\chi_B^{+}] \\
&= 4[P_A^{0}\int p_A^{+}(\nabla \chi_A^{+})^2\, dr + P_B^{0}\int p_B^{+}(\nabla \chi_B^{+})^2\, dr],
\end{aligned}$$

$$\begin{aligned}
I_{add.}^{nclass.}[\chi_A^{+}, \chi_B^{+}] &= I_{nadd.}^{nclass.}[\chi_A^{+}, \chi_B^{+}] - I_{total}^{nclass.}[\chi_A^{+}, \chi_B^{+}] \\
&= 4[P_A^{0}\int p_A^{+}(\nabla \chi_B^{+})^2\, dr + P_B^{0}\int p_B^{+}(\nabla \chi_A^{+})^2\, dr] \\
&\quad + 8\int (P_A^{0}p_A^{+} + P_B^{0}p_B^{+})(\nabla \chi_A^{+})\cdot(\nabla \chi_B^{+})\, dr. \tag{108}
\end{aligned}$$

One also observes that for the internal equilibria in polarized subsystems $\alpha^+ = A^+, B^+$,

$$\psi_\alpha^{\;+} = \psi_\alpha \exp(i\chi_\alpha^{\;+}), \qquad \chi_\alpha^{\;+} = -(1/2)\ln p_\alpha^{\;+}, \tag{109}$$

the magnitudes of the entropy classical and nonclassical contributions do not equalize:

$$S^{class.}[\psi_R^{\;+}] = -\int p_R^{\;+}\ln p_R^{\;+}\, dr = -\int p_R^{\;+}\ln(P_A^{\;0}p_A^{\;+} + P_B^{\;0}p_B^{\;+})\, dr$$

$$\neq 2\int p_R^{\;+}\chi_R^{\;+}\, dr = -S^{nclass.}[\psi_R^{\;+}] = -\int p_R^{\;+}\ln(p_A^{\;+}p_B^{\;+})\, dr, \tag{110}$$

$$I^{class.}[\psi_R^{\;+}] = \int(\nabla p_R^{\;+})^2/p_R^{\;+}\, dr \neq 4\int p_R^{\;+}(\nabla\chi_R^{\;+})^2\, dr = I^{nclass.}[\psi_R^{\;+}]$$

$$= \int p_R^{\;+}\{[(\nabla p_A^{\;+})/p_A^{\;+}]^2 + [(\nabla p_B^{\;+})/p_B^{\;+}]^2$$

$$+ 2(\nabla p_A^{\;+}\cdot\nabla p_B^{\;'})/(p_A^{\;+}p_B^{\;+})\}dr. \tag{111}$$

However, the internal *phase*-equilibria in subsystems do imply

$$-S[\chi_\alpha^{\;'}] = S[p_\alpha^{\;+}], \; -\tilde{I}[\chi_\alpha^{\;+}] = I[\chi_\alpha^{\;+}] = I[p_\alpha^{\;+}], \text{ or}$$

$$S[\psi_\alpha^{\;+}] = \tilde{I}[\psi_\alpha^{\;+}] = 0 \text{ and } I[\psi_\alpha^{\;+}] = 2I[p_\alpha^{\;+}]. \tag{112}$$

This generates the vanishing internal populational derivatives of entropies in both reactants,

$$\partial S[\psi_\alpha^{\;+}]/\partial N_\alpha = \partial\tilde{I}[\psi_\alpha^{\;+}]/\partial N_\alpha = 0, \tag{113}$$

and hence a vanishing *in situ* CT contribution originating from the *additive* component of the resultant global and gradient entropies:

$$\partial S^{add.}[\psi_R^{\;+}]/\partial N_{CT} = \partial\tilde{I}^{add.}[\psi_R^{\;+}]/\partial N_{CT} = 0. \tag{114}$$

Hence, the nonvanishing CT derivatives of the resultant entropies in the polarized reactive system are solely due to their *nonadditive* components:

$$\partial S[\psi_R^+]/\partial N_{CT} = \partial S^{nadd.}[\psi_R^+]/\partial N_{CT} \quad \text{and}$$

$$\partial \tilde{I}[\psi_R^+]/\partial N_{CT} = \partial \tilde{I}^{nadd.}[\psi_R^+]/\partial N_{CT}. \tag{115}$$

Thus, contrary to the global equilibrium, the populational derivatives of resultant measures of the entropy/information content of equilibrium states of the polarized reactants do not vanish. Such *in situ* indices can be used to describe information in terms of the CT processes in the acid-base reactive systems.

8.7 CONCLUSION

In this work, we examined the information description of reactive systems at typical hypothetical stages of a bimolecular chemical reaction involving the mutually closed (nonbonded) or open (bonded) donor/acceptor reactants. We have emphasized the phase aspect of the subsystem equilibria at the polarization and charge transfer stages of this process. Implications of the *intra-* and *inter-*fragment equilibria have been discussed and the current promotion of polarized reactants has been explored. The additive and nonadditive components of the classical (probability) and nonclassical (phase/current) contributions to resultant measures of the entropy/information content of quantum electronic states have been determined and their *in situ* CT derivatives have been examined.

"Thermodynamic" phases of reactants and the reactive system as a whole offer a new class of reactivity descriptors through which the progress of chemical reactions can be monitored and indexed [10, 74]. Indeed, by linking the equilibrium phases to the distribution of electrons, one can also follow the *phase*-aspect of a chemical reaction, a reflection of changing reactant densities. The reactivity preferences can now be related to a possible *phase*-matching between the two subsystems, the current-promotion due to a presence of the reaction partner can be addressed, and the dynamical aspect of molecular relaxations, e.g., the promolecule-to-molecule transition, can be tackled [10, 44]. Additional insight into the bond breaking/forming processes comes from OCT, with the internal communications in subsystems shaping the *bond*-promotions in polarized reactants and the *inter*-reactant propagations reflecting the subsystem *external*-bonding status in chemical reactions [10, 74, 76].

KEYWORDS

- **acid-base systems**
- **charge transfer processes**
- **continuity relations**
- ***in situ* description**
- **phase-equilibria**
- **quantum entropy/information**

REFERENCES

1. Fisher, R. A., (2004). *Proc. Cambridge Phil. Soc., 22*, 700 (1925); see also: B. R. Frieden, *Physics from the Fisher Information – A Unification*, 2nd Ed. (Cambridge University Press, Cambridge).

2. Shannon, C. E., (1949). *Bell System Tech. J., 27*, 379, 623 (1948); Shannon, C. E., Weaver, W., *The Mathematical Theory of Communication* (University of Illinois, Urbana).

3. Kullback, S., & Leibler, R. A., (1951). *Ann. Math. Stat. 22*, 79.

4. Kullback, S., (1959). *Information Theory and Statistics* (Wiley, New York).

5. Abramson, N., (1963). *Information Theory and Coding* (McGraw-Hill, New York).

6. Pfeifer, P. E., (1978). *Concepts of Probability Theory*, 2nd Ed. (Dover, New York).

7. Nalewajski, R. F., (2006). *Information Theory of Molecular Systems*, (Elsevier, Amsterdam).

8. Nalewajski, R. F., (2010). *Information Origins of the Chemical Bond*, (Nova, New York).

9. Nalewajski, R. F., (2012). *Perspectives in Electronic Structure Theory*, (Springer, Heidelberg).

10. Nalewajski, R. F., (2016). *Quantum Information Theory of Molecular States*, (Nova Science Publishers, New York).

11. Nalewajski, R. F., Köster, A. M., Escalante, S., (2005). *J. Phys. Chem. A, 109*, 10038.

12. Luken, W. L., & Beratan, D. N., (1984). *Theoret. Chim. Acta (Berl.), 61*, 265.

13. Luken, W. L., & Culberson, J. C., (1984). *Theoret. Chim. Acta (Berl.), 66*, 279.

14. Becke, A. D., & Edgecombe, K. E., (1990*). J. Chem. Phys., 92*, 5397.

15. Silvi, B., & Savin, A., (1994). *Nature, 371,* 683.
16. Savin, A., Nesper, R., Wengert, S., & Fässler, T. F., (1997). *Angew. Chem. Int. Ed. Engl., 36,* 1808.
17. Hohenberg, P., & Kohn, W., (1964). *Phys. Rev., 136B,* 864.
18. Kohn, W., & Sham, L. J., (1965). *Phys. Rev., 140A,* 1133.
19. Levy, M., (1979). *Proc. Natl. Acad. Sci. USA, 76,* 6062.
20. Nalewajski, R. F, (2008). *Int. J. Quantum Chem., 108,* 2230, Nalewajski, R. F., de Silva, P., Mrozek, J., (2010). *J. Mol. Struct: THEOCHEM, 954,* 57.
21. Nalewajski, R. F., (2014). in *Advances in Quantum Systems Research,* ed. by Z. Ezziane (Nova Science Publishers, New York), pp. 119.
22. Nalewajski, R. F., (2005). *Theoret. Chem. Acc., 114,* 4.
23. Nalewajski, R. F., (2005). *J. Math, Chem., 38,* 43.
24. Nalewajski, R. F., (2005). *Mol. Phys., 103,* 451.
25. Nalewajski, R. F., (2006). *Mol. Phys., 104,* 365, 493, 2533, 3339.
26. Nalewajski, R. F., (2008). *J. Math. Chem., 43,* 265.
27. Nalewajski, R. F., (2007). *J. Phys. Chem. A., 111,* 4855.
28. Nalewajski, R. F., (2009). *Int. J. Quantum Chem. 109,* 425, 2495.
29. Nalewajski, R. F., (2009). *Adv. Quant. Chem., 56,* 217.
30. Nalewajski, R. F., Szczepanik, D., & Mrozek, J., (2011). *Adv. Quant. Chem. 61,* 1.
31. Nalewajski, R. F., (2010). *J. Math. Chem. 47,* 709, 808, *Ibid.,* 49, 592 (2011).
32. Nalewajski, R. F., Szczepanik, D., & Mrozek, J., (2012). *J. Math. Chem., 50,* 1437.
33. Nalewajski, R. F., (2011). *J. Math. Chem., 49,* 371.
34. Nalewajski, R. F., (2011). *J. Math. Chem., 49,* 546.
35. Nalewajski, R. F., (2011). *J. Math. Chem., 49,* 806.
36. Nalewajski, R. F., & Gurdek, P., (2011). *J. Math. Chem., 49,* 1226.
37. Nalewajski, R. F., (2012). *Int. J. Quantum Chem., 112,* 2355.
38. Nalewajski, R. F., & Gurdek, P., (2012). *Struct. Chem., 23,* 1383.
39. Nalewajski, R. F., (2011). J. Math. Chem., 49, 2308.
40. Nalewajski, R. F., (2013*). Ann. Phys. (Leipzig) 525,* 256.
41. Nalewajski, R. F., (2014). *J. Math. Chem., 52,* 588, 1292, 1921.
42. Nalewajski, R. F., (2014). *Mol. Phys., 112,* 2587.
43. Nalewajski, R. F., (2015). *Int. J. Quantum Chem., 115,* 1274.
44. Nalewajski, R. F., (2015). *J. Math. Chem., 53,* 1126.
45. Nalewajski, R. F., (2016). *J. Math. Chem., 54,* 1777.
46. Nalewajski, R. F., (2013*). J. Math. Chem., 51,* 369.
47. Harriman, J. E., (1981). *Phys. Rev. A, 24,* 680.
48. Zumbach, G., & Maschke, K., (1984). *Phys. Rev. A., 28,* 544 (1983); Erratum: *Phys. Rev., 29,* 1585.
49. Nalewajski, R. F., & Korchowiec, J., (1997). *Charge Sensitivity Approach to Electronic Structure and Chemical Reactivity,* (World Scientific, Singapore).
50. Nalewajski, R. F., & Korchowiec, J., (1996). A. Michalak, *Topics in Current Chemistry 183,* 25.
51. Geerlings, P., de Proft, F., & Langenaeker, W., (2003). *Chem. Rev. A., 103,* 1793.
52. Mulliken, R. S., (1934). *J. Chem. Phys., 2,* 782, Iczkowski, R. P., & Margrave, J. L., (1961). *J. Am. Chem. Soc., 83,* 3547.
53. Sanderson, R. T., (1952). *J. Am. Chem. Soc., 74,* 272.

54. Gyftopoulos, E. P., & Hatsopoulos, G. N., (1965). *Proc. Natl. Acad. Sci., USA 60,* 786.
55. Parr, R. G., Donnelly, R. A., Levy, M., & Palke, W. E., (1978). *J. Chem. Phys., 69,* 4431. Perdew, J. P., Parr, R. G., Levy, M., & Balduz, J. L., (1982). *Phys. Rev. Lett., 49,* 1691.
56. Pearson, R. G., (1973). *Hard and Soft Acids and Bases* Dowden, Hatchinson, Ross, Stroudsburg.
57. Parr, R. G., & Pearson, R. G., (1983). *J. Am. Chem. Soc., 105,* 7512.
58. Parr, R. G., & Yang, W., (1984). *J. Am. Chem. Soc., 106,* 4049.
59. Chattaraj, P. K., ed., (2009). *Chemical Reactivity Theory: A Density Functional View* (CRC/ Taylor & Francis, Boca Raton).
60. Liu, S., (2009). in *Chemical Reactivity Theory: A Density Functional View,* ed. by Chattaraj, P. K., (CRC/Taylor & Francis, Boca Raton), pp. 179.
61. Nalewajski, R. F., (2009). in *Chemical Reactivity Theory: A Density Functional View,* ed. by P. K. Chattaraj (CRC/Taylor & Francis, Boca Raton), pp. 453.
62. Nalewajski, R. F., (1995). in *Proceedings of the Nato ASI on Density Functional Theory,* ed. by Dreizler, R. M., & Gross, E. K. U., (Plenum, New York), pp. 339.
63. Nalewajski, R. F., Błażewicz, D., & Mrozek, J., (2008). *J. Math. Chem., 44,* 325.
64. Baekelandt, B. G., Janssens, G. O. A., Toufar, H., Mortier, W. J., Schoonheydt, R. A., & Nalewajski, R. F., (1995). *J. Phys. Chem., 99,* 9784.
65. Nalewajski, R. F., (2000). *Computers Chem., 24,* 243.
66. Nalewajski, R. F., (2006). *Adv. Quant. Chem., 51,* 235.
67. López-Rosa, S, Esquivel, R. O., Angulo, J. C., Antolín, J., Dehesa, J. S., & Flores-Gallegos, N., (2010). *J. Chem. Theory Comput., 6,* 145.
68. López-Rosa, S., (2010). *Information-Theoretic Measures of Atomic and Molecular Systems,* PhD Thesis, University of Granada.
69. Toro-Labbé, A., Gutiérez-Oliva, S., Politzer, P., & Murray, J. S., (2009). in *Chemical Reactivity Theory: A Density Functional View,* ed. by Chattaraj, P. K., (CRC/Taylor & Francis, Boca Raton), pp. 293.
70. Dirac, P. A. M., (1958). *The Principles of Quantum Mechanics,* 4th Ed. (Clarendon, Oxford).
71. Nalewajski, R. F., (2015). *J. Math. Chem., 53,* 1.
72. Nalewajski, R. F., (1994). *Int. J. Quantum Chem., 49,* 675.
73. Weizsäcker, C. F. von, (1935). *Z. Phys., 96,* 431.
74. Nalewajski, R. F., (2016). *J. Math. Chem.,* submitted.
75. Primas, H., (1981). *Chemistry, Quantum Mechanics and Reductionism: Perspectives in Theoretical Chemistry* (Springer, Berlin).
76. Nalewajski, R. F., (2014). *Indian J. Chem., 53 A,* 1010.

CHAPTER 9

FAILURES OF EMBEDDED CLUSTER MODELS FOR pK_a SHIFTS DOMINATED BY ELECTROSTATIC EFFECTS

AHMED A. K. MOHAMMED, STEVEN K. BURGER, and PAUL W. AYERS

Department of Chemistry and Chemical Biology, McMaster University, Hamilton, Ontario L8S 4M1, Canada

CONTENTS

ABSTRACT

The protonation state of amino acid residues in proteins depends on their respective pK_a values. Computational methods are particularly important for estimating the pK_a values of buried and active site residues, where experimental data are scarce. In this work, we used the cluster model approach to predict the pK_a of some challenging protein residues and for which methods based on the numerical solution of the Poisson–Boltzmann equation and empirical approaches fail. The ionizable residue and its close environment were treated by quantum mechanics, while the rest of the protein was replaced by a uniform dielectric continuum. The approach was found to overestimate the electrostatic interaction, leading to predicting lower pK_a values.

9.1   INTRODUCTION

Accurate prediction of protein residues' pK_a values, which determine the protonation state of ionizable residues at a given pH, is very important because these residues are involved in intraprotein, protein-solvent, and protein-ligand interactions. Consequently, the protonation state of a protein makes significant contributions to protein stability, solubility, folding, binding ability, and catalytic activity [1, 2]. Therefore, it is impossible to build a computational model for a protein or even to obtain a qualitative understanding of protein structure and function, without first determining the protonation state of the protein. While the protonation state can be determined empirically, in principle, experimental measurements (mostly using NMR) are difficult for large proteins, buried residues, and active site residues. This has motivated the development of computational methods to predict pK_a values for protein residues [1, 3–16]. Unfortunately, these methods are often unreliable for buried residues.

The most commonly used methods for protein pK_a prediction are based either on the numerical solution of Linearized Poisson–Boltzmann equation (LPBE) or on fast empirical methods built from quantitative structure-property relationships (QSPR). (The main challenge for developing empirical methods is that the specific properties that determine protein pK_a's are still not clear [17, 18].) For computational expediency, methods based on both these approaches usually model the protein with a molecular mechanics force field, immersed in a dielectric continuum. The dielectric constant of the solvent is set to 80, and a value between 4 and 20 is used for the protein. The shift in pK_a for the residue is calculated by comparing the electrostatic energy of its protonated and non-protonated forms. Then, the shift is added to the pK_a value of the model. The structure of the protein is assumed to not undergo any change upon (de)protonation, which is often a poor assumption, especially for buried residues [19, 20]. Methods based on these traditional approaches often fail to predict the pK_a of buried residues and those that have large pK_a shift.

Typical molecular mechanics models for electrostatic energies are limited in accuracy by the assumption that atoms of a certain type always have the same charge [1, 21–23]. This unrealistic simplification of electrostatic interactions can be overcome using flexible-charge molecular mechanics models [24–30] or, ideally, by quantum mechanics (QM) [31]. However, due to its computational cost, QM methods cannot be applied to the whole protein. The cluster approach—wherein the ionizable residue and its immediate environment are taken out of the protein and treated quantum mechanically—overcomes this obstacle. Since it was introduced two decades ago, the cluster approach has been successfully used to model enzyme reactions and the number of its applications is growing [32–34].

The cluster approach was used by Li and coworkers to determine the pK_a of protein residues [35]. In their approach, the part of the protein that includes the ionizable residue and its immediate environment is treated by QM, while the rest of the protein is neglected. The solvent is described by the polarized continuum solvation model. The method successfully predicted the pK_a of five residues in the turkey ovomucoid third domain (OMTKY3). The approach was also applied successfully to predict pK_a values for small organic molecules [36–40] and protein residues near metal atoms. However, all the residues were on surface of the enzyme, where the interactions are dominated by hydrogen bonding.

We wondered whether the cluster method would work when electrostatic and empirical pK_a prediction methods fail. To this end, we have tested the cluster method for two aspartate residues—one of which is buried in the protein interior, and one of which is solvent-exposed—in which the pK_a of the carboxyl group is strongly affected by nearby charged residue(s).

9.2 COMPUTATIONAL DETAILS

9.2.1 PK_A CALCULATIONS

We followed the method developed by Li and coworkers to calculate the pK_a for a protein residue [24]. Using the reaction,

$$\text{HA}(aq) + \text{C}_2\text{H}_5\text{COO}^-(aq) \underset{}{\overset{\Delta G}{\rightleftharpoons}} \text{A}^-(aq) + \text{C}_2\text{H}_5\text{COOH}(aq) \qquad (1)$$

the pK$_a$ value of the carboxyl group in the protein, HA, can be determined from the equation

$$\text{pK}_a = 4.87 + \Delta G/1.36 \, , \qquad (2)$$

where 4.87 is the experimental pK$_a$ value for propionic acid at 298 K and 1.36 is $2.303RT$ at $T = 298$ K. ΔG is the change in the standard free energy of reaction (1) in kcal/mol. To compare the accuracy of the quantum mechanical cluster model to more conventional approaches, we used two continuum electrostatic methods, namely the web-based version of KARLSBERG+ [41] and the MCCE code [42–44]. To represent the empirical methods, we used the popular PROPKA program [45].

9.2.2   FREE ENERGY CALCULATIONS

All QM calculations were performed with the Gaussian 09 program [46]. Solvation free energy calculations were performed with the Gaussian 03 program [47].

The free energy of a molecule is given by

$$G = E_{\text{ele}} + G_{\text{sol}} \qquad (3)$$

The electronic energies, E_{ele}, were computed at the MP2/6-31+G(2d,p) level by using structures that were optimized at the RHF/6-31+G(d) level of theory. The solvation free energies, G$_{\text{sol}}$, were calculated by the polarizable continuum model at the IEF-PCM/RHF/6-31+G(d) level of theory for the same geometries using the keywords RADII=UAHF, RET=100, TSNUM=240, and SCFVAC. To account for conformational flexibility in the cluster, all possible conformers of the acid were considered, and the final free energy was computed by Boltzmann weighting all conformers within 1 kcal/mol of the most stable conformer, i.e.,

$$G = -RT \ln \left[\sum_{\text{conformers}} \exp(-G_i/RT) \right] \qquad (4)$$

9.2.3 PROTEIN MODEL CONSTRUCTION

For each model, the coordinates for the atoms were taken from the PDB files downloaded from the PDB database. The files were stripped of metal atoms, cofactors, inhibitors, and solvent molecules. Atomic coordinates for the model for Asp75 were taken from the PDB code: 1a2p bacterial barnase [48]. Atomic coordinates for the model for Asp21 were taken from the PDB code: 1beo, which is fungal beta-cryptogein [49]. Hydrogen atoms were added using the web-based program molprobity software [50]. Where the residue was truncated, additional hydrogen atoms were added manually to satisfy the valence. The protein backbone was fixed, and all the hydrogen atoms were optimized at the HF/3-21G(d) level of theory; then, the side chain (CH_2COO^- or CH_2COOH) was optimized. One oxygen atom of the carboxyl group was fixed. The positions of the neighboring protons of OH groups were also minimized at the same level of theory.

9.3 RESULTS

9.3.1 ASPARTATE 75

Structure 9.1 shows the cluster model for Asp75. The model includes the immediate chemical environment of the ionizable residue. Three conformers were found for the protonated form. There are two hydrogen bonds between two hydrogens of Arg83 and two oxygens of the carboxyl group of Asp75. A pK_a of –4.63 pH units was obtained by the method developed by Li and coworkers, which is much lower than the experimental value (error = –7.4.) Negative results were also obtained by PROPKA and KARLSBERG+, see Table 9.1. Only the MCCE code gave a reasonable prediction of 3.6. One possible explanation is that these methods overestimate the stabilization from the charge-charge interaction between the carboxylate group of Asp75 and two nearby positively charged amides: Arg83 and Arg87. These interactions favor the ionized (non-protonated) form, leading to a lower pK_a value. Another possible explanation, suggested by Li and coworkers, is that the position of Arg83 in the crystal structure is different from its position in solution. To address this possibility, the geometries of Arg83 were optimized at the same level, and the

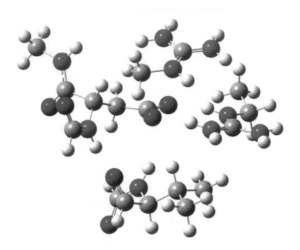

STRUCTURE 9.1   Model compound for Asp75 of bacterial barnase. The model contains the ionizable group and its immediate chemical environment.

TABLE 9.1   Comparison Between Experimental and Several Pk$_a$ Prediction Methods

PDB Code	Residue	Exp. pK$_a$	Current study	PROPKA[a]	MCCE[a]	KARL-SBERG+
1a2p	Asp75	3.1	−4.6	−1.3	3.6	−8.2
1beo	Asp21	2.5	−3.7	1.4	3.6	0.3

[a] Results were obtained from Ref. [64].

pK$_a$ was recalculated. This refined calculation did not improve the results. The solvation free energy was calculated for different values of dielectric constants, 4 and 20, but this did not improve the results either.

9.3.2 ASPARTATE 21

The model for Asp21 is represented in Figure 9.1. The pK$_a$ shift is primarily due to electrostatic interactions between the carboxyl group and the positively charged amide group of Lys62. A pK$_a$ of −3.72 pH units was obtained by the method developed by Li and coworkers, see Table 9.1. Lower pK$_a$ values were obtained by the rest of the methods (except for MCCE), which suggests that the explanation for these poor results is simi-

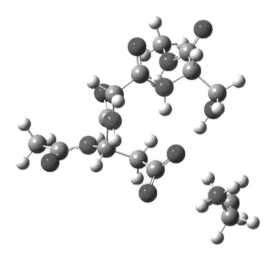

FIGURE 9.1   Model compound for Asp21 of fungal beta-cryptogein. The model contains the ionizable group and its immediate chemical environment.

lar to the explanation we deduced for Asp75: nearby positively charged residues cause these methods to overstabilize the ionized form of Asp21.

9.4   DISCUSSION

The calculated pK_a values for both aspartate residues we considered were far lower than the experimentally measured values for most of the methods. We attribute this to overestimation of Coulombic interactions between nearby positively charged residues and the carboxyl group, causing the models to overestimate the stabilization of the carboxylate anion. The trends are even not right: the experimental pK_a value of Asp75 is higher than that of Asp21, but the methods predict the opposite order. This is consistent with our hypothesis: Asp75 has two nearby positively charged Arg residues, while Asp21 has only one nearby positively charged Lys residue.

Asp75 is a buried residue. The prediction of pK_a for buried residues is extremely challenging, partly because desolvation is a key determinant of pK_a. The hydrophobic environment stabilizes the neutral form of Asp75, which increases the pK_a of the residue. Desolvation effects were found to raise the pK_a shifts by 4-6 pH units [51–54]. A study by Laurents et al. reported that the current continuum models underestimate the contribution of desolvation effect [55]. The inability of the continuum solvation models to adequately capture desolvation effects helps explain why Asp75 is erroneously predicted to be more acidic than Asp21 by the computational approaches we consider.

Although Asp21 is a surface residue, the cluster method still underestimates its pK_a value. Again, the likely culprit is the implicit solvation model, the polarized continuum model, which probably fails to effectively screen the charges, giving too large a shift in pK_a values. This observation is consistent with the well-known, but typically ignored, observation that the dielectric screening—a concept from macroscopic electrostatics—is an inappropriate model for short-range electrostatic interactions in the protein environment [56]–[61]. It is also known that pK_a shifts are affected by slight changes in protein structure, including those induced by crystal-packing forces [60, 62, 63]. Especially for buried residues, the minimization of the protein using a method like QM/MM is necessary to obtain the right structure.

Our results indicate that one needs to develop better continuum models for describing electrostatic interactions in proteins. In addition, explicit water molecules need to be included, as also suggested by a benchmark study [64]. The response of the neighboring protein residues, due to protein flexibility, to (de)protonation also needs to be included. There may also be other important effects. Indeed, the failure of most of these methods can be attributed to our lack of a deep understanding of what properties of proteins affect the pK_a and the absence of quantitative models for how these properties affect the pK_a.

9.5 CONCLUSIONS

The goal of this work was to use a cluster model, similar to those that are widely used to model enzymatic chemical reactions, to predict the pK_a values of amino acid residues in proteins. Results from the cluster model were compared to those from more traditional (and less computationally demanding approaches) including those based on the numerical solution of the Poisson–Boltzmann equation (MCCE and KARLSBERG+) and those based on empirical fitting (PROPKA). We tested the approaches on two aspartate residues whose pK_a shift is induced by nearby positively charged residues. One of the residues, Asp75, was buried. The other, Asp21, was solvent-exposed.

The cluster approach predicted pK_a values that were far too low. This contradicts earlier findings from Li and coworkers, where the cluster model gave excellent results. We attribute the inadequacy of the cluster model for our cases to the overestimation of the stabilization effect due to nearby positively charged residues and the underestimation of the hydrophobic effect by the continuum solvation model. It is also possible that (presumably slight) differences in the structure of the protein in solution, compared to the crystal structure, account for part of the discrepancy. Finally, the change in protein geometry upon (de)protonation may be important.

One obvious solution is to increase the size of the cluster model and (ideally) introduce explicit solvent molecules. This is not practical because the number of conformers grows exponentially as the size of the cluster model increases, causing a proportionate increase in the computational cost. Approaches based on the QM/MM method, where the protein environment is modeled with molecular mechanics force field, are more promising. Such models would provide a better description of the hydrophobic environment of the ionizable residue, but the computational cost is much higher. QM/MM approaches are not yet routinely practical, but they could be used for difficult residues in important proteins.

KEYWORDS

- **buried residues**
- **cluster model**
- **Poisson Boltzmann**
- **protein pK_a prediction**

REFERENCES

1. Warshel, A., Sharma, P. K., Kato, M., & Parson, W. W., (2006). Modeling electrostatic effects in proteins. Biochimica Et Biophysica Acta. *Proteins and Proteomics, 1764,* 1647–1676.

2. Schlick, T., (2002). *Molecular Modeling and Simulation*: an interdisciplinary guide: an interdisciplinary guide. Springer, New York,.

3. Burger, S. K., & Ayers, P. W., (2011). Empirical prediction of protein pKa values with residue mutation. *J. Comput. Chem., 32,* 2140–2148.

4. Burger, S. K., & Ayers, P. W., (2011). A parameterized, continuum electrostatic model for predicting protein pKa values. *Proteins: Structure, Function, and Bioinformatics, 79,* 2044–2052.

5. Fogolari, F., Brigo, A., & Molinari, H., (2002). The Poisson–Boltzmann equation for biomolecular electrostatics: a tool for structural biology. *J. Mol. Recognit., 15,* 377–392.

6. Bashford, D., & Karplus, M., (1990). pKa's of ionizable groups in proteins: atomic detail from a continuum electrostatic model. *Biochemistry., 29,* 10219–10225.

7. Yang, A. S., Gunner, M. R., Sampogna, R., Sharp, K., & Honig, B., (1993). On the calculation of pKas in proteins. *Proteins: Structure, Function, and Bioinformatics., 15,* 252–265.

8. Antosiewicz, J., McCammon, J. A., & Gilson, M. K., (1994). Prediction of pH-dependent properties of proteins. *J. Mol. Biol., 238,* 415–436.

9. Antosiewicz, J., Briggs, J. M., Elcock, A. H., Gilson, M. K., & McCammon, J. A., (1996). Computing ionization states of proteins with a detailed charge model. *J. Comput. Chem., 17,* 1633–1644.

10. Sham, Y. Y., Chu, Z. T., & Warshel, A., (1997). Computing ionization states of proteins with a detailed charge model. *J. Phys. Chem. B., 101,* 4458–4472.

11. Delbuono, G. S., Figueirido, F. E., & Levy, R. M., (1994). Intrinsic pKas of ionizable residues in proteins: an explicit solvent calculation for lysozyme. *Proteins: Structure, Function, and Bioinformatics., 20,* 85–97.

12. Warshel, A., Sussman, F., & King, G., (1986). Free energy of charges in solvated proteins: microscopic calculations using a reversible charging process. *Biochemistry, 25,* 8368–8372.

13. Kollman, P., (1993). Free energy calculations: applications to chemical and biochemical phenomena. *Chem. Rev., 93,* 2395–2417.

14. Mehler, E. L., & Guarnieri, F., (1999). A self-consistent, microenvironment modulated screened coulomb potential approximation to calculate pH-dependent electrostatic effects in proteins. *Biophysical Journal., 77,* 3–22.

15. Sandberg, L., & Edholm, O., (1999). A fast and simple method to calculate protonation states in proteins. *Proteins: Structure, Function, and Bioinformatics., 36,* 474–483.

16. Matthew J. B., & Gurd, F. R. N., (1986). Calculation of electrostatic interactions in proteins. *Methods Enzymol., 130,* 413–436.

17. Forsyth W. R., Antosiewiez, J. M., & Robertson, A. D., (2002). Empirical relationships between protein structure and carboxyl pKa values in proteins. *Proteins: Structure, Function, and Bioinformatics., 48,* 388–403.

18. Edgcomb, S. P., & Murphy, K. P., (2002). Variability in the pKa of histidine side-chains correlates with burial within proteins. *Proteins: Structure, Function, and Bioinformatics., 49*, 1–6.
19. Swails, J. M., & Roitberg, A. E., (2012). Enhancing conformation and protonation state sampling of hen egg white lysozyme using pH replica exchange molecular dynamics. *J. Chem. Theory Comp., 8*, 4393–4404.
20. Meng, Y. L., & Roitberg, A. E., (2010). Constant pH replica exchange molecular dynamics in biomolecules using a discrete protonation model. *J. Chem. Theory Comp., 6*, 1401–1412.
21. Ji, C. G., Mei, Y., & Zhang, J. Z. H., (2008). Developing polarized protein-specific charges for protein dynamics: MD free energy calculation of pK a shifts for Asp 26/Asp 20 in Thioredoxin. *Biophys. J., 95*, 1080–1088.
22. Tong, Y., Ji, C. G., Mei, Y., & Zhang, J. Z., (2009). Simulation of NMR data reveals that proteins' local structures are stabilized by electronic polarization. *J. Am. Chem. Soc., 131*, 8636–8641.
23. Lee, L. P., Cole, D. J., Skylaris, C. K., Jorgensen, W. L., & Payne, M. C., (2013). Polarized protein-specific charges from atoms-in-molecule electron density partitioning. *J. Chem. Theory Comp., 9*, 2981–2991.
24. Mortier, W. J., & Leuven, K. U., (1987). Electronegativity equalization and its applications. *Electronegativity, 66*, 125–143.
25. Mortier, W. J., Ghosh, S. K., & Shankar, S., (1986). Electronegativity-equalization method for the calculation of atomic charges in molecules. *J. Am. Chem. Soc., 108*, 4315–4320.
26. Mortier, W. J., Vangenechten, K., & Gasteiger, J., (1985). Electronegativity equalization: application and parametrization. *J. Am. Chem. Soc., 107*, 829–835.
27. Chelli, R., Procacci, P., Righini, R., & Califano, S., (1999). Electrical response in chemical potential equalization schemes. *J. Chem. Phys., 111*, 8569–8575.
28. Nistor, R. A., Polihronov, J. G., Muser, M. H., & Mosey, N. J., (2006). A generalization of the charge equilibration method for nonmetallic materials. *J. Chem. Phys., 125*, 094108.
29. Verstraelen, T., Ayers, P. W., Van Speybroeck, V., & Waroquier, M., (2013). ACKS2: Atom-condensed Kohn-Sham DFT approximated to second order. *J. Chem. Phys., 138*, 074108.
30. Van Duin, A. C. T., Dasgupta, S., Lorant, F., & Goddard, W. A., (2001). ReaxFF: a reactive force field for hydrocarbons. *J. Phys. Chem. A, 105*, 9396–9409.
31. Helgaker, T., Jørgensen, P., & Olsen, J., (2000). *Modern Electronic Structure Theory*. Wiley, Chichester.
32. Siegbahn, P. E. M., & Himo, F., (2009). Recent developments of the quantum chemical cluster approach for modeling enzyme reactions. *J. Biol. Inorg. Chem., 14*, 643–651.
33. Siegbahn, P. E. M., & Borowski, T., (2006). Modeling enzymatic reactions involving transition metals. *Acc. Chem. Res., 39*, 729–738.
34. Siegbahn, P. E. M., & Blomberg, M. R. A., (2000). Transition-metal systems in biochemistry studied by high-accuracy quantum chemical methods. *Chem. Rev., 100*, 421–437.

35. Li, H., Robertson, A. D., & Jensen, J. H., (2004). The determinants of carboxyl pKa values in turkey ovomucoid third domain. *Proteins: Structure, Function, and Bioinformatics., 55,* 689–704.

36. Lim, C., Bashford, D., & Karplus, M., (1991). Absolute pKa calculations with continuum dielectric methods. *J. Phys. Chem., 95,* 5610–5620.

37. Chen, J. L., Noodleman, L., Case, D. A., & Bashford, D., (1994). Incorporating solvation effects into density functional electronic structure calculations. *J. Phys. Chem., 98,* 11059–11068.

38. Richardson, W. H., Peng, C., Bashford, D., Noodleman, L., & Case, D. A., (1997). Incorporating solvation effects into density functional theory: calculation of absolute acidities. *Int. J. Quantum Chem., 61,* 207–217.

39. Topol, I. A., Tawa, G. J., Burt, S. K., & Rashin, A. A., (1997). Calculation of absolute and relative acidities of substituted imidazoles in aqueous solvent. *J. Phys. Chem. A., 101,* 10075–10081.

40. Li. H., Hains, A. W., Everts, J. E., Robertson, A. D., & Jensen, J. H., (2002). The prediction of protein p K a's using QM/MM: the pKa of lysine 55 in turkey ovomucoid third domain. *J. Phys. Chem. B., 106,* 3486–3494.

41. Kieseritzky, G., & Knapp, E. W., (2008). Optimizing pKa computation in proteins with pH adapted conformations. *Proteins: Structure, Function, and Bioinformatics., 71,* 1335–1348.

42. Song, Y., Mao, J., & Gunner, M. R., (2009). MCCE2: improving protein pKa calculations with extensive side chain rotamer sampling. *J. Comp. Chem., 30,* 2231–2247.

43. Georgescu, R. E., Alexov, E. G., & Gunner, M. R., (2002). Combining conformational flexibility and continuum electrostatics for calculating pKa s in proteins. *Biophys. J., 83,* 1731–1748.

44. Alexov, E., & Gunner, M. R., (1997). Incorporating protein conformational flexibility into the calculation of pH-dependent protein properties. *Biophys. J., 74,* 2075–2093.

45. Li, H., Robertson, A. D., & Jensen, J. H., (2005). Very fast empirical prediction and rationalization of protein pKa values., *61,* 704–721.

46. Frisch, M. J., Trucks, G. W., Schlegel, H. B., et al., (2009). Gaussian 09, Revision C.01, Gaussian, *Inc., Wallingford CT.*

47. Frisch, M. J., Trucks, G. W., Schlegel, H. B., et al., (2004). Gaussian 03, Revision D.01, Gaussian, Inc., Wallingford CT.

48. Oliveberg, M., Arcus, V. L., & Fersht, A. R., (1995). pKa values of carboxyl groups in the native and denatured states of barnase: the pKa values of the denatured state are on average 0.4 units lower than those of model compounds. *Biochemistry, 34,* 9424–9433.

49. Gooley, P. R., Keniry, M. A., Dimitrov, R. A., Marsh, D. E., Gayler, K. R., & Grant, B. R., (1998). The NMR solution structure and characterization of pH dependent chemical shifts of the β-elicitin, cryptogein. *J. Biol. NMR., 12,* 523–534.

50. Davis, I. W., et al., (2007). *Nucleic Acids Research, 35,* 375–383.

51. Mehler, E. L., Fuxreiter, M., & Garcia-Moreno, B., (2002). *Biophys., J., 82,* 1748.

52. Mehler, E. L., Fuxreiter, M., Simon, I., & Garcia-Moreno, E. B., (2002). The role of hydrophobic microenvironments in modulating pKa shifts in proteins. *Proteins: Structure, Function, and Bioinformatics., 48,* 283–292.

53. Dwyer, J. J., Gittis, A. G., Karp, D. A., Lattman, E. E., Spencer, D. S., Stites, W. E., & Garcia-Moreno. B., (2000). *Biophys. J., 79,* 1610–1620.
54. Lambeir, A. M., Backmann, J., Ruiz-Sanz, J., Filimonov, V., Nielsen, J. E., Kursula, I., Norledge, B. V., & Wierenga, R. K., (2000). The ionization of a buried glutamic acid is thermodynamically linked to the stability of Leishmania mexicana triose phosphate isomerase. *Eur. J. Biochem., 267,* 2516–2524.
55. Laurents. D. V., Huyghues-Despointes, B. M. P., Bruix, M., Thurlkill, R. L., Schell, D., Newsom, S., Grimsley, G. R., Shaw, K. L., Trevino, S., Rico, M., Briggs, J. M., Antosiewicz, J. M., Scholtz, J. M., & Pace, C. N., (2003). *J. Mol. Biol., 325,* 1077–1092.
56. Berkowitz, M. L., Bostick, D. L., & Pandit, S., (2006). Aqueous solutions next to phospholipid membrane surfaces: insights from simulations. *Chem. Rev., 106,* 1527–1539.
57. Mehler, E. L., & Guarnieri, F., (1999). A self-consistent, microenvironment modulated screened coulomb potential approximation to calculate pH-dependent electrostatic effects in proteins. *Biophys. J., 77,* 3–22.
58. Sandberg, L., & Edholm, O. A., (1999). A fast and simple method to calculate protonation states in proteins. *Proteins: Structure, Function, and Bioinformatics., 36,* 474–483.
59. Matthew, J. B., & Gurd, F. R. N., (1986). Calculation of electrostatic interactions in proteins. *Methods Enzymol., 130,* 413–436.
60. Nielsen, J. E., & Vriend, G., (2001). Optimizing the hydrogen-bond network in Poisson-Boltzmann equation-based pKa calculations. *Proteins: Structure, Function, and Bioinformatics., 43,* 403–412.
61. Schutz, C. N., & Warshel, A., (2001). What are the dielectric "constants" of proteins and how to validate electrostatic models? *Proteins: Structure, Function, and Bioinformatics., 44,* 400–417.
62. Nielsen J. E., & McCammon, J. A., (2003). On the evaluation and optimization of protein X-ray structures for pKa calculations. *Protein Sci., 12,* 313–326.
63. Georgescu, R. E., Alexov, E. G., & Gunner, M. R., (2002). Combining conformational flexibility and continuum electrostatics for calculating pKa s in proteins. *Biophys. J., 83,* 1731–1748.
64. Stanton, C. L., & Houk, K. N., (2008). Benchmarking pKa prediction methods for residues in proteins. *J. Chem. Theory Comp., 4,* 951–966.

CHAPTER 10

A STATISTICAL PERSPECTIVE ON MOLECULAR SIMILARITY

FARNAZ HEIDAR-ZADEH,[1-3] PAUL W. AYERS,[1]
and RAMON CARBÓ-DORCA[4]

[1]*Department of Chemistry and Chemical Biology, McMaster University, Hamilton, Ontario, Canada*

[2]*Department of Inorganic and Physical Chemistry, Ghent University, Krijgslaan 281 (S3), 9000 Gent, Belgium*

[3]*Center for Molecular Modeling, Ghent University, Technologiepark 903, 9052 Zwijnaarde, Belgium*

[4]*Institut de Química Computacional i Catàlisi, Universitat de Girona, Girona, Spain*

CONTENTS

10.1   INTRODUCTION

Although chemists have been using arguments based on molecular similarity to explain and explore chemistry since the days of medieval alchemy, quantitative approaches based on molecular similarity emerged only in the latter half of the twentieth century [1–9]. The use of quantum mechanical properties and descriptors, typically the electron density, to describe molecular similarity emerged even later, in the 1980s [10–22]. In these more modern approaches, the property of a substance indexed with i, π_i, is expressed as a reference value plus a correction [23],

$$\pi_i = \pi_{\text{ref}} + \sum_{j=1}^{n} z_{ij}\omega_j \qquad (1)$$

where z_{ij} measures the similarity between molecules i and j and $\mathbf{W} = \begin{bmatrix} \omega_1 & \omega_2 & \cdots & \omega_n \end{bmatrix}^T$ is a vector of fit coefficients that can be determined, for example, by linear regression. In matrix-vector notation, this can be expressed as

$$\pi = \pi_{\text{ref}}\mathbf{I} + \mathbf{Zw} \qquad (2)$$

where $\pi = \begin{bmatrix} \pi_1 & \pi_2 & \cdots & \pi_n \end{bmatrix}^T$ is the vector of property values and $\mathbf{Z}$ is a similarity matrix, each element of which, z_{ij}, represents the similarity between two molecules in the dataset. The goal of this chapter is to reinterpret and revise this framework based on statistical principles. However, we wish to preserve the key reasoning on which similarity-based arguments are based: the properties of an unknown substance are estimated based on the properties of n other known substances by using the intuitively appealing idea that the property of the unknown substance will resemble that of substances similar to it, but may be distinct from substances that are very different. Denoting the index of the unknown substance as "?," this suggests that the unknown property can be computed from an equation of the form,

$$\pi_? = \sum_{j=1}^{n} \omega_j \pi_j \qquad (3)$$

Note that the converse is not true: sometimes very different substances have similar properties (both arsenic and chlorine gases are deadly). Likewise, substances that are similar in one context (water and methanol have similar acidity) may be dissimilar in another context (one should not drink methanol!).

Equation (3) can be viewed as an interpolation between the properties of the known substances. Perhaps, the simplest method is (generalized) Shepard interpolation, where the interpolation weights have a form like

$$\omega_j = \frac{\gamma_{?j}^{-p}}{\sum_{j=1}^{n} \gamma_{?j}^{-p}} \tag{4}$$

where the similarity analogue of the semivariogram is defined as

$$\gamma_{ij} = \frac{1}{2}\left(z_{ii} + z_{jj} - 2z_{ij}\right) \tag{5}$$

Note that $\gamma_{ij} \geq 0$, with equality only if substances i and j are identical. In this expression, the exponent $p > 0$ is chosen so that as p becomes larger, properties of dissimilar substances contribute less and less to the estimate. In the limit as $p \to \infty$, the property of the unknown substance is assigned the property value of the most similar molecule in the training set. At the other extreme, if $p = 0$, then all substances in the training set contribute equally, and the property value of the unknown substances is the unweighted arithmetic mean of the property values of the training set.

Shepard interpolation is used quite frequently for interpolating points on potential energy surface in chemistry [24–30], but it does not seem popular in the molecular similarity field. This might be because Shepard interpolation is usually defined using the distance between data points, and there is no unambiguous way to define the distance between molecules. We decide instead on the formula in Eq. (5): if the similarity matrix were a covariance matrix, then this would be the statistical semivariogram. It does have several of the same appealing properties of the semivariogram, however, that $\gamma_{ij} \geq 0$ with equality only if $i = j$ and the fact that γ_{ij} is bounded from above. The former property allows us to use γ_{ij} as a replacement for the distance between i and j in conventional Shepard interpolation. The boundedness from above ensures that if one considers a case where the unknown molecule is very dissimilar to all the molecules in the training set, the property value predicted by Shepard interpolation approaches the arithmetic mean of the property values of the training data. This makes the semivariogram preferable to the distance, where this latter property is not true.

10.2   RELATIONSHIP BETWEEN MOLECULAR SIMILARITY AND GAUSSIAN PROCESSES (KRIGING)

The strength of Shepard interpolation is its simplicity; its weakness is its sensitivity to redundancy and/or clustering in the training data. For example, by including several extremely similar molecular structures (or, in the extreme case, repeating the same structure multiple times), one can increase the weight of this (family of) structure(s) arbitrarily. Avoiding this problem requires data-driven weights. That is, the weights in Eq. (3) must depend not only on the similarity of the target substance to the training dataset but also to the similarity between different substances in the training set. This motivates the following statistically motivated approach to property estimation based on molecular similarity. To achieve this, we could assume that the quantity in Eq. (5) is actually a semivariogram or that the similarity matrix is actually a covariance matrix. With these assumptions, we could derive a model based on kriging (also known as Gaussian processes) [31–34]. While the model we derive will be identical to kriging, we will derive the equations in the context of chemical similarity. The result is a new way to use molecular similarity measures for property prediction.

First, we assume that the unknown property value is a weighted average of the properties of the known substances, which means that the coefficients in Eq. (3) are constrained to satisfy

$$1 = \sum_{j=1}^{n} \omega_j \qquad (6)$$
$$0 \le \omega_j$$

In the context of kriging, this condition is related to statistical "unbiasedness" condition. Note that the Shepard interpolation weights in Eq. (4) are, in this sense, also unbiased.

Second, we would like for the error in the predicted property value to be as small as possible, i.e., the expectation value of the squared residual, $\left\langle \left(\pi_? - \pi_?^{\text{exact}} \right)^2 \right\rangle$, should be as small as possible. (Of course, we do not know the exact property value, $\pi_?^{\text{exact}}$.) This gives a variational principle for the weighting coefficients,

$$\min_{\left\{ \omega_i \left| \begin{array}{l} 0 \leq \omega_j \\ 1 = \sum_{j=1}^n \omega_j \end{array} \right. \right\}} \frac{1}{2} \left\langle \left(\pi_? - \pi_?^{\text{exact}} \right)^2 \right\rangle \tag{7}$$

with the Lagrangian

$$\mathcal{L}\left(\{\omega_i\}, \mu \right) = \frac{1}{2} \left\langle \left(\sum_{j=1}^n \omega_j \pi_j - \pi_?^{\text{exact}} \right)^2 \right\rangle - \mu \left(1 - \sum_{j=1}^n \omega_j \right) \tag{8}$$

The weighting factors can then be obtained by solving the equation

$$0 = \frac{\partial \mathcal{L}}{\partial \omega_i} = \left\langle \pi_i \left(\sum_{j=1}^n \omega_j \pi_j - \pi_?^{\text{exact}} \right) \right\rangle + \mu$$

$$0 = \frac{\partial \mathcal{L}}{\partial \mu} = \sum_{j=1}^n \omega_j - 1 \tag{9}$$

This is a system of $n + 1$ linear equations with $n + 1$ unknowns. We can simplify the equations to

$$\left\langle \pi_i \pi_?^{\text{exact}} \right\rangle = \sum_{j=1}^n \left\langle \pi_i \pi_j \right\rangle \omega_j + \mu$$

$$1 = \sum_{j=1}^n \omega_j \tag{10}$$

or, in matrix notation,

$$
\begin{bmatrix}
\langle \pi_1^2 \rangle & \langle \pi_1\pi_2 \rangle & \cdots & \langle \pi_1\pi_n \rangle & 1 \\
\langle \pi_1\pi_2 \rangle & \langle \pi_2^2 \rangle & \ddots & \vdots & \vdots \\
\vdots & \ddots & \ddots & \langle \pi_{n-1}\pi_n \rangle & 1 \\
\langle \pi_1\pi_n \rangle & \cdots & \langle \pi_{n-1}\pi_n \rangle & \langle \pi_n^2 \rangle & 1 \\
1 & \cdots & 1 & 1 & 0
\end{bmatrix}
\begin{bmatrix}
\omega_1 \\ \omega_2 \\ \vdots \\ \omega_n \\ \mu
\end{bmatrix}
=
\begin{bmatrix}
\langle \pi_1\pi_?^{\text{exact}} \rangle \\
\langle \pi_2\pi_?^{\text{exact}} \rangle \\
\vdots \\
\langle \pi_n\pi_?^{\text{exact}} \rangle \\
1
\end{bmatrix}
\tag{11}
$$

How should we interpret $\langle \pi_i\pi_?^{\text{exact}} \rangle$? This expectation value is related to how the property value for the ith molecule are related to the value of the unknown property value, and it requires a statistical model for the property values, which are regarded as random variables. However, we do not have a statistical model available. We do not know how these properties will be correlated in general; therefore, we estimate them by studying properties we do know for both substances, selecting those properties based on their perceived relevance to the property of interest. For example, if we know the properties $\{\sigma^{(m)}\}_{m=1}^{M}$ for both molecules, we could consider

$$
\langle \pi_i\pi_?^{\text{exact}} \rangle = \frac{1}{M}\sum_{m=1}^{M}\sigma_i^{(m)}\sigma_?^{(m)}
\tag{12}
$$

Alternatively, if we have a real-space descriptor of both molecules, $\sigma(\mathbf{r})$, we could consider

$$
\langle \pi_i\pi_?^{\text{exact}} \rangle = \int \sigma_i(\mathbf{r})\sigma_?(\mathbf{r})d\mathbf{r}
\tag{13}
$$

Equations (12) and (13) are, of course, exactly the forms that one commonly uses to measure chemical similarity [9, 10, 35]. This motivates us to use elements of the similarity matrix to estimate the unknown quantities in Eq. (11), $Z_{ij} \approx \langle \pi_i\pi j \rangle$ and $Z_{i?} \approx \langle \pi_i\pi_?^{\text{exact}} \rangle$, giving the equations

$$
\begin{bmatrix}
z_{11} & z_{12} & \cdots & z_{1n} & 1 \\
z_{12} & z_{22} & \ddots & \vdots & \vdots \\
\vdots & \ddots & \ddots & z_{n-1,n} & 1 \\
z_{1n} & \cdots & z_{n-1,n} & z_{nn} & 1 \\
1 & \cdots & 1 & 1 & 0
\end{bmatrix}
\begin{bmatrix}
\omega_1 \\ \omega_2 \\ \vdots \\ \omega_n \\ \mu
\end{bmatrix}
=
\begin{bmatrix}
z_{1?} \\ z_{2?} \\ \vdots \\ z_{n?} \\ 1
\end{bmatrix}
\tag{14}
$$

or in more compact notation,

$$\begin{bmatrix} \mathbf{Z} & \mathbf{1} \\ \mathbf{1}^T & 0 \end{bmatrix} \begin{bmatrix} \mathbf{w} \\ \mu \end{bmatrix} = \begin{bmatrix} \mathbf{z}_? \\ 1 \end{bmatrix} \tag{15}$$

Using the formula for the inverse of a block matrix, Eq. (15) can be rewritten as

$$\begin{bmatrix} \mathbf{w} \\ \mu \end{bmatrix} = \begin{bmatrix} \mathbf{Z}^{-1} - \left(\mathbf{d}^{\text{partial}}\right)\left(\mathbf{d}^{\text{partial}}\right)^T / d^{\text{total}} & \mathbf{d}^{\text{partial}} / d_{\text{total}} \\ \left(\mathbf{d}^{\text{partial}}\right)^T / d_{\text{total}} & -1/d_{\text{total}} \end{bmatrix} \begin{bmatrix} \mathbf{z}_? \\ 1 \end{bmatrix}$$

$$= \left(\begin{bmatrix} \mathbf{Z}^{-1} & \mathbf{0} \\ \mathbf{0}^T & 0 \end{bmatrix} - \frac{1}{d^{\text{total}}} \begin{bmatrix} \mathbf{d}^{\text{partial}} \\ -1 \end{bmatrix} \begin{bmatrix} \mathbf{d}^{\text{partial}} & -1 \end{bmatrix} \right) \begin{bmatrix} \mathbf{z}_? \\ 1 \end{bmatrix} \tag{16}$$

where

$$d_i^{\text{partial}} = \sum_{j=1}^{n} z_{ij}^{-1} \tag{17}$$

$$d^{\text{total}} = \sum_{i=1}^{n}\sum_{j=1}^{n} z_{ij}^{-1} \tag{18}$$

and Z_{ij}^{-1} are the elements of the inverse of the similarity matrix, $\mathbf{Z}^{-1}$. (If the similarity matrix is not positive definite, the (generalized) inverse of its positive-definite subspace should be used.) In Eq. (16), the product of a vector and a transposed vector defines a rank one matrix, *i.e.*, the elements of $\mathbf{VW}^T$ are $v_i w_j$. The predicted properties are obtained by multiplying Eq. (16) on the left by the vector of known properties, $\begin{bmatrix} \pi_1 & \pi_2 & \cdots & \pi_n & 0 \end{bmatrix}$, thus obtaining

$$\pi_? = \pi \cdot \left(\mathbf{Z}^{-1} \cdot \mathbf{z}_? - \mathbf{d}^{\text{partial}} \left(\frac{\left(\mathbf{d}^{\text{partial}} \cdot \mathbf{z}_? - 1\right)}{d^{\text{total}}} \right) \right) \tag{19}$$

or in more explicit notation,

$$\pi_? = \sum_{j=1}^{n} \pi_j \left(\sum_{k=1}^{n} z_{jk}^{-1} z_{k?} - \frac{\left(\sum_{m=1}^{n} z_{jm}^{-1} \right) \left(\sum_{k=1}^{n} \sum_{l=1}^{n} z_{k?} z_{kl}^{-1} - 1 \right)}{\sum_{h=1}^{n} \sum_{i=1}^{n} z_{hi}^{-1}} \right) \tag{20}$$

The second term in this expression arises only because of the constraint that the weights add up to one. It is reasonable to expect that this constraint is often nearly satisfied whenever the dataset is well curated; therefore, we expect the second term in this expression to be reasonably small. In such cases, the simple formula $\pi_? = \pi \mathbf{z}^{-1} z_?$ can be used. (This formula corresponds to simple kriging.) In either case, after the property has been estimated, the error in the prediction can be estimated using

$$\left\langle \left(\pi_? - \pi_?^{\text{exact}} \right)^2 \right\rangle = \sum_{i=1}^{n} \sum_{j=1}^{n} w_i z_{ij} w_j - 2 \sum_{i=1}^{n} w_i z_{i?} + z_{??}$$
$$= \mathbf{w}^T \mathbf{Z} \mathbf{w} - 2 \left(\mathbf{z}_? \cdot \mathbf{w} \right) + z_{??} \tag{21}$$

By itself, this estimate is not very accurate, because simply changing the scale of the similarity measure (by multiplying it by a constant) changes this quantity. This expression, however, can be useful for comparing the relative accuracy of different predictions. Alternatively, one can (approximately) determine the appropriate scale factor for the similarity matrix by cross-validation.

10.3 SUMMARY

To summarize our findings: if one is given a dataset of molecules with known properties and an appropriate measure of the chemical similarity between these molecules and one wishes to know the property of another molecule, then one (1) computes the similarity of the target molecule to the molecules in the database and (2) uses the similarity information and the known properties of the database to estimate the value of the property for the target molecule as a weighted average of the properties of the molecules in the dataset (cf. Eq. (19)). The relative accuracy of different property predictions from this formulation can be assessed with Eq. (21); thus, with this framework, one not only estimates the properties of targeted molecules but also learns about the reliability of those estimates. The approach in Eq. (19) uses data-driven weights to compensate for unbalanced, clustered training data set. Moreover, there is significant freedom in how one chooses to define the molecular similarity; this freedom allows one to customize the similarity measure so that it is appropriate to the property of interest. As discussed in the opening paragraphs, this is important because molecules that are similar in one chemical context may be dissimilar in another setting. Therefore, in our view, this similarity-based kriging method for property prediction is more appealing than previous approaches.

ACKNOWLEDGMENTS

The authors acknowledge support from NSERC and Compute Canada. FHZ acknowledges support from a Vanier-CGS fellowship from NSERC.

KEYWORDS

- covariance
- Gaussian processes
- generalized linear model
- Kriging
- machine learning
- molecular similarity
- quantitative structure property relationships
- Shepard interpolation

REFERENCES

1. Ballester, P. J., (2011). Ultrafast shape recognition: method and applications. *Future Medicinal Chemistry., 3,* 65–78.
2. Sukumar, N., & Das, S., (2011). Current Trends in Virtual High Throughput Screening Using Ligand-Based and Structure-Based Methods. *Combinatorial Chemistry & High Throughput Screening., 14,* 872–888.
3. Rupp, M., & Schneider, G., (2010). Graph Kernels for Molecular Similarity. *Molecular Informatics., 29,* 266–273.
4. Sukumar, N., Krein, M., & Breneman, C. M., (2008). Bioinformatics and cheminformatics: Where do the twain meet? *Current Opinion in Drug Discovery & Development., 11,* 311–319.
5. Ballester, P. J., & Richards, W. G., (2007). Ultrafast shape recognition to search compound databases for similar molecular shapes. *J. Comput. Chem., 28,* 1711–1723.
6. Ralaivola, L., Swamidass, S. J., Saigo, H., & Baldi, P., (2005). Graph kernels for chemical informatics. *Neural Networks., 18,* 1093–1110.
7. Bender, A., & Glen, R. C., (2004). Molecular similarity: a key technique in molecular informatics. *Organic & Biomolecular Chemistry., 2,* 3204–3218.
8. Klebe, G., (2000). Recent developments in structure-based drug design. *Journal of Molecular Medicine., 78,* 269–281.
9. Maggiora, G. M., & Shanmugasundaram, V., (2010). Molecular similarity measures. *Methods in Molecular Biology., 272,* 39–100.
10. Bultinck, P., Girones, X., & Carbó-Dorca, R., (2005). Molecular quantum similarity: Theory and applications. *Rev. Comput. Chem., 21,* 127–207.
11. Bultinck, P., & Carbó-Dorca, R., (2005). Molecular quantum similarity using conceptual DFT descriptors. *J. Chem. Sci., 117,* 425–435.
12. Geerlings, P., Boon, G., Van Alsenoy, C., & De Proft, F., (2005). Density functional theory and quantum similarity. *Int. J. Quantum Chem., 101,* 722–732.
13. Besalú, E., Girones, X., Amat, L., & Carbó-Dorca, R., (2002). Molecular quantum similarity and the fundamentals of QSAR. *Acc. Chem. Res., 35,* 289–295.
14. Carbó-Dorca, R., Amat, L., Besalú, E., Girones, X., & Robert, D., (2000). Quantum Mechanical origin of QSAR: theory and applications. *J. Mol. Struct.: THEOCHEM., 504,* 181–228.
15. Carbó-Dorca, R., Amat, L., Besalú, E., & Lobato, M., (1998). Quantum Similarity. *In: Advances in Molecular Similarity,* CarboDorca, R., Mezey, P. G. Eds., vol. *2,* pp. 1–42.
16. Carbo-Dorca, R., & Besalu, E., (1998). A General Survey of Molecular Quantum Similarity. *THEOCHEM-Journal of Molecular Structure., 451,* 11–23.
17. Carbó, R., Besalú, E., Amat, L., & Fradera, X., (1996). On Quantum Molecular Similarity Measures (QMSM) and Indices (QMSI). *J. Math. Chem., 19,* 47–56.
18. Carbó, R., & Calabuig, B., (1992). Molecular quantum similarity measures and N-dimensional representation of quantum objects. 1. Theoretical foundations. *Int. J. Quantum Chem., 42,* 1681–1693.

19. Carbó, R., & Calabuig, B., (1992). Molecular quantum similarity measures and N-dimensional representation of quantum objects. 2. Practical applications. *Int. J. Quantum Chem., 42,* 1695–1709.
20. Carbó, R., Leyda, L., & Arnau, M., (1980). How Similar Is A Molecule to Another - An Electron-Density Measure of Similarity Between 2 Molecular-Structures. *Int. J. Quantum Chem., 17,* 1185–1189.
21. Miranda-Quintana, R. A., Cruz-Rodes, R., Codorniu-Hernandez, E., & Batista-Leyva, A. J., (2010). Formal theory of the comparative relations: its application to the study of quantum similarity and dissimilarity measures and indices. *J. Math. Chem., 47,* 1344–1365.
22. Zadeh, F. H., & Ayers, P. W., (2013). Molecular alignment as a penalized permutation procrustes problem. *J. Math. Chem., 51,* 927–936.
23. Carbo-Dorca, R., (2016). Aromaticity, Quantum Multimolecular Polyhedra, and Quantum QSPR Fundamental Equation. *J. Comput. Chem., 37,* 78–82.
24. Crittenden, D. L., & Jordan, M. J. T., (2005). Interpolated potential energy surfaces: How accurate do the second derivatives have to be? J. *Chem. Phys., 122,* 044102.
25. Ishida, T., & Schatz, G. C., (2003). A local interpolation scheme using no derivatives in potential sampling: application to O(D-1)+H-2 system. *J. Comput. Chem., 24,* 1077–1086.
26. Bettens, R. P. A., & Collins, M. A., (1999). Learning to interpolate molecular potential energy surfaces with confidence: A Bayesian Approach. *J. Chem. Phys., 111,* 816–826.
27. Ishida, T., & Schatz, G. C., (1997). Automatic potential energy surface generation directly from ab initio calculations using Shepard interpolation: A test calculation for the H-2+H system. *J. Chem. Phys., 107,* 3558–3568.
28. Burger, S. K., & Ayers, P. W., (2010). Methods for finding transition states on reduced potential energy surfaces. *J. Chem. Phys., 132,* 234110.
29. Burger, S. K., & Ayers, P. W., (2010). Dual grid methods for finding the reaction path on reduced potential energy surfaces. *J. Chem. Theory Comp., 6,* 1490–1497.
30. Burger, S. K., Liu, Y. L., Sarkar, U., & Ayers, P. W., (2009). Moving least-squares enhanced Shepard interpolation for the fast marching and string methods. *J. Chem. Phys., 130,* 024103.
31. Wackernagel, H., (2003). *Multivariate Geostatistics: An Introduction With Applications.* New York, Springer, Verlag.
32. Kitanidis, P. K., (1993). Generalized covariance functions in estimation. *Mathematical Geology., 25,* 525–540.
33. Isaaks, E. H., & Srivastava, R. M., (1989). *An Introduction to Applied Geostatistics.* New York, Oxford University Press,.
34. Clark, I., (1979). *Practical Geostatistics.* London, Applied Science Publishers.
35. Carbó-Dorca, R., & Besalú, E., (1998). A general survey of molecular quantum similarity. *J. Mol. Struct.: THEOCHEM., 451,* 11–23.

CHAPTER 11

MODELING CHEMICAL REACTIONS WITH COMPUTERS

YULI LIU and PAUL W. AYERS

Department of Chemistry and Chemical Biology, McMaster University, Hamilton, Ontario L8S 4M1, Canada

CONTENTS

11.1   INTRODUCTION

The ultimate goal of computational chemistry is to model chemical reactions using computers. Suppose that we are given a set of molecules: instead of mixing them in beakers, we could load the information into a computer and, through computer simulation, predict what would happen and explain how it would happen. While computational chemists have made great progress in this quest, it is still far from being fully realized.

This chapter discusses theoretical developments and computational studies that advance our ability to simulate chemical reactions.

There are many ways to predict what happens in a chemical reaction. Some methods focus on a certain property of the reactants or a certain type of reaction: we call these "specific methods." For example, reactivity indicators provide a straightforward way to predict which site of a molecule will be attacked by a specific type of reagent [1, 2] or predict a molecule's susceptibility to a specific type of reaction (e.g., by predicting the quality of a leaving group) [3, 33]. General purpose methods are designed to work for all possible types of chemical processes and typically use the potential energy surface or free energy surface of the molecular system. For gas phase reactions, the potential energy surface (PES) provides important information about the reaction: the energy minima represent the reactant, the product, and potential reactive intermediates; the first-order saddle points represent the transition states linking these stable structures. Finding the stationary points[4] on the PES gives us detailed information about the chemical reaction mechanism(s). In most chemical systems, one reaction mechanism is dominant, and knowing the unique minimum energy reaction path linking the reactant(s) and product(s) [5] provides sufficient information to characterize the thermodynamics and kinetics of the chemical system. For gas-phase reactions at sufficiently high temperatures, condensed-phase reactions, and the reactions of complex biological systems, the potential energy surface cannot represent the system's behavior because the molecule fluctuates and statistically samples a range of different structures. Statistical sampling is required to achieve the thermodynamic properties of a macroscopic system, i.e., the free energy difference between two states. The same principles apply here, but the mechanism should be characterized using free energy surfaces instead of PESs.

This chapter discusses quantum mechanics (QM) and molecular mechanics (MM) tools and concepts used for finding minimum energy reaction paths on PES for gas phase reactions and determining free energy differences in complex biological systems using molecular dynamics (MD).

11.2   THE POTENTIAL ENERGY SURFACE AND THE BORN-OPPENHEIMER APPROXIMATION

The molecular PES is fundamental to reaction mechanism studies. It represents the electronic energy of a molecular system as a function of all the relevant atomic positions. The Born-Oppenheimer approximation, or another similar adiabatic approximation, is required to define the PES [6].

The Schrödinger equation for a molecule with n electrons, and N nuclei, is

$$\left[-\sum_{\alpha=1}^{N} \frac{\hbar^2}{2M_\alpha} \nabla_\alpha^2 - \sum_{i=1}^{n} \frac{\hbar^2}{2m_e} \nabla_i^2 + \frac{e^2}{4\pi\varepsilon_0} \left(-\sum_{\text{all } i,\alpha} \frac{Z_\alpha}{r_{i\alpha}} + \sum_{i<j} \frac{1}{r_{ij}} + \sum_{\alpha<\beta} \frac{Z_\alpha Z_\beta}{r_{\alpha\beta}} \right) \right] \quad (1)$$

$$\psi(\mathbf{x}, \mathbf{X}) = E_{total} \psi(\mathbf{x}, \mathbf{X})$$

Here, $\mathbf{x}$ and $\mathbf{X}$ represent the electronic and nuclear coordinates; M_α and Z_α are the mass and the atomic number (nuclear charge) of the αth nucleus; m_e is the mass of an electron; e is the charge on a proton and the magnitude of the charge on an electron; $r_{i\alpha}$, r_{ij}, and $r_{\alpha\beta}$ represent electron-nucleus, electron-electron, and nucleus-nucleus distances, respectively.

Because nuclei are thousands of times heavier than electrons (and the resting mass of an electron is approximately 1836 times smaller than that of the proton), they move much more slowly than electrons. Born and Oppenheimer proposed that electrons can be pictured as moving in the field of fixed nuclei. When the nuclei move, the electron density should adjust almost instantaneously. Thus, the electronic motion (described by the electron wave function $\psi_e(\mathbf{x};\mathbf{X})$) can be separated from the nuclear motion (described by the nuclear wave function $\psi_n(\mathbf{X})$),

$$\psi(\mathbf{x}, \mathbf{X}) = \psi_e(\mathbf{x}; \mathbf{X}) \psi_n(\mathbf{X}) \quad (2)$$

where $\psi_e(\mathbf{x};\mathbf{X})$ is a solution of the Schrödinger equation involving the electronic Hamiltonian or Hamiltonian describing the motion of n electrons in the field of N fixed nuclei (N point charges),

$$\hat{H}_e \psi_e(\mathbf{x}; \mathbf{X}) = V_e(\mathbf{X}) \psi_e(\mathbf{x}; \mathbf{X}) \quad (3)$$

where $\psi_e(\mathbf{x};\mathbf{X})$ is a function of the electronic coordinates $\mathbf{x}$ and depends only on the nuclear coordinates $\mathbf{X}$ parametrically, because it is solved for a particular choice of nuclear positions.

Because nuclei move much slower than electrons, the nuclear kinetic energy term

$$\left(-\sum_{\alpha=1}^{N} \frac{\hbar^2}{2M_\alpha} \nabla_\alpha^2 \right)$$

in Eq. (1) is often neglected. The nuclear potential energy term

$$\left(\frac{e^2}{4\pi\varepsilon_0} \sum_{\alpha<\beta} \frac{Z_\alpha Z_\beta}{r_{\alpha\beta}} \right)$$

does not depend on the electronic positions. Therefore, the electronic Hamiltonian is defined as,

$$\hat{\mathrm{H}}_e = -\sum_{i=1}^{n} \frac{\hbar^2}{2M_e} \nabla_i^2 + \frac{e^2}{4\pi\varepsilon_0} \left(-\sum_{\text{all } i,\alpha} \frac{Z_\alpha}{r_{i\alpha}} + \sum_{i<j} \frac{1}{r_{ij}} \right) \qquad (4)$$

The potential energy $V_e(\mathbf{X})$ of a particular nuclear configuration $\mathbf{X}$ is determined by the total electronic energy associated with that nuclear configuration and can be obtained by solving the Schrödinger Eq. (3) involving the electronic Hamiltonian. Adding the nuclear-nuclear repulsion term to $V_e(\mathbf{X})$ defines the PES, which is usually denoted as $V(\mathbf{X})$.

The PES $V(\mathbf{X})$ of a molecular system contains important information on its geometries and the relative energies of its locally stable structures as well as the most favorable reaction pathways between these structures. Figure 11.1 shows the 3-D surface plot and contour plot of a 4-well 2-D potential energy surface. The bottoms (green, yellow to orange) of the 4 wells represent 4 energy minima (reactant, product, or intermediates), and the first-order saddle points represent the transition states between each pair of minima. This surface is commonly used as a test case for algorithms.

11.3   MINIMUM ENERGY PATH

The most widely accepted definition of minimum energy path (MEP) is the intrinsic reaction coordinate (IRC) proposed by Fukui [7]; the IRC is

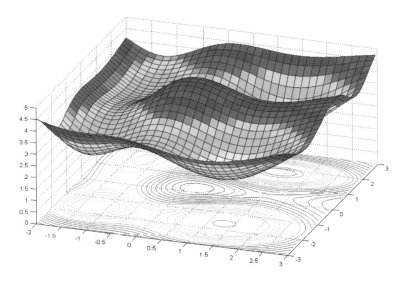

FIGURE 11.1   The surface plot and contour plot of a 4-well 2-D potential energy surface.

the steepest descent path (SDP) from the first-order saddle point down to the adjacent minima on the PES. In this thesis, the MEP is defined as SDP. For multistep reactions, the overall minimum energy path would be composed by the linking the MEPs of the individual steps in sequence. When multiple mechanisms (alternative reaction paths) exist, the MEP with the lowest overall energy barrier is the global MEP, while the others are local MEPs. The MEP provides critical information about chemical reactions, including information about the mechanism, the reaction rate, etc. Thermodynamic properties like the heat of reaction and the equilibrium constant[8] can also be derived from the MEP. Unsurprisingly, theoretical chemists have exerted great effort toward finding MEPs.

There are two families of algorithms for finding the MEP: the surface walking [9–11] algorithms (an "initial value" formulation) and the two end algorithms [12–17] (a "boundary value" formulation). The two end methods usually require a good guess for the path linking the reactant and product. Only the local MEP can be found using the two end methods. By contrast, surface walking methods only need the reactant configuration and then explore the PES to predict the products and the mechanism along the way. Unfortunately, surface walking algorithms usually are either very expensive or, if a heuristic is used to simplify the computation, they tend to be unreliable [18] for complicated systems.

The present authors developed a new surface walking algorithm, namely the fast marching method (FMM), for finding the global minimum energy reaction path. We also developed a new two-end algorithm, the QSM-NT method, that locates all the stationary points on a PES, which allows us to find several alternative minimum energy paths (and helps reduce the need for good guesses of the reaction path).

11.3.1 THE FAST MARCHING METHOD

Fast marching methods are numerical schemes that solve the eikonal equation. The FMM for determining MEP transforms a multiwell PES ($V(\mathbf{R})$) to a single-well energy cost surface ($U(\mathbf{R})$) by solving the eikonal equation that defines the cost of traveling from the initial configuration ($\mathbf{R}_0$, the reactant) to another ($\mathbf{R}$) on the PES, [5, 9, 11, 32]

$$\left|\nabla U_n(\mathbf{R})\right| = \left\{\sqrt{2(E - V(\mathbf{R}))}\right\}^n \tag{5}$$

with the boundary condition of $U(\mathbf{R}_0)=0$.

This eikonal equation describes wavefront propagation with the local speed function $\dfrac{1}{\left\{\sqrt{2(E-V(\mathbf{R}))}\right\}^n}$. Pictorially, we imagine flooding the PES, starting from the reactant "valley." The "water" level rises until it breaches the lowest-energy "mountain pass" (the transition state) and then races to the bottom of the next "valley" (the intermediate) along the SDP. The "valley flooding" process continues until the product is found. The contour lines that show what portion of the surface was underwater at a given point in time define a new single-well energy cost surface. The SDP from the product to the reactant is the MEP; it is constructed by a process called backtracing.

Because the "water" level will always go to the next "valley" through the lowest energy "mountain pass," the FMM can assure that the minima (bottom of valleys) are linked by the lowest energy transition states. Therefore, the global minimum energy path is found.

The FMM is a very general and reliable method. Without prior knowledge of the PES, it can always find the global MEP. But it is an expensive method. Moving least square-enhanced Shepard interpolation has been applied to reduce the computational cost [32]. The FMM has been successfully applied to analytical PES and small gas-phase chemical reactions. To apply FMM to larger systems, we can use the parallel FMM and compute many points on the surface at once. The current FMM method still has an exponential dependence on the dimensionality of the PES, however; therefore, parallelizing the program will not make it possible to look at systems with very many reactive degrees of freedom-like proteins with 10 or more ionizable residues. At this point, FMM is restricted to small systems. Medium-sized systems could be accessed if a better interpolation method could be designed, so that fewer ab initio calculations were required. For complex biological systems with only a few reactive bonds (e.g., enzyme reactions), FMM can be interfaced with QM/MM program packages such as Sigma to explore the reaction path.

11.3.2 TWO END METHODS

The string method is a very popular path-finding method [19]. It is a two end method. The string method divides the initial path into several nodes, which are connected by strings to define the path. The nodes are driven to the SDP by a normal force orthogonal to the tangent of the path. The tangent to the path is updated at each iteration and the nodes are redistributed (to maintain equal spacing between nodes) until a good approximation to the SDP is found.

The string method algorithm can be conceptually described as dropping an elastic pearl necklace on the PES, with the two ends of the necklace fixed on the reactant and the product. The pearls roll down from their initial position until the necklace settles into a local MEP.

The quadratic string method (QSM) [20] uses the same algorithmic structure as the string method, except that a local quadratic approximation of the PES is used; this reduces the number of energy and gradient calculation.

The Newton trajectory (NT) is an alternative reaction path that has been proposed by Quapp and his coworkers [21, 22]. A Newton trajectory is a curve on which all gradients are pointing in the same (or opposite) direction, called the searching direction of the NT. Because the magnitude of the gradient at the stationary point is zero, its direction is arbitrary. Therefore, a Newton trajectory passes all stationary points on the PES. If carefully chosen, a continuous NT without any turning points or higher-order saddle points can be found; Quapp proposes that this is a good model for the reaction path. The problem is that an NT can contain spurious turning points (nonstationary point), second-order or higher-order saddle points (stationary points), energy maxima (stationary points). All these points could be maxima on the energy profile of the NT and appear to be transition states, which might give a misleading reaction path. Without prior knowledge of the PES, it seems difficult to find a searching direction that defines a NT in which all turning points are minima or first-order saddle points.

To avoid the "turning point problem" associated with using a single NT as the reaction path, we developed a new method for finding the stationary points on the PES by locating the intersections of two or more NTs [4]. Because an NT passes all stationary points on the PES, the NTs intersect at stationary points; after finding the stationary points, we can determine the possible reaction pathways. We adapted the QSM algorithm to find NTs. This new method is called QSM-NT [4].

The QSM-NT method can find all stationary points on the PES, and accordingly all alternative reaction paths. The pitfalls of this method include: 1) discontinuous Newton trajectories might impede locating all stationary points and 2) multiple minima on the hyperplane might lead to the wrong path. The first problem is inherent to the characteristic of Newton trajectory. The workaround is to try more searching directions and to locate more Newton trajectories and their intersections. The second problem is associated with the path-finding algorithm (QSM). The use of a growing string algorithm (GSM) can solve this problem. GSM is more expensive than QSM; therefore, the next step of this program would be to design a method that automatically switches from QSM to GSM when the QSM calculations is failing to converge to the NT.

11.4   MOLECULAR MECHANICS (MM)

Macroscopic systems contain an enormous number of interacting particles. For gas phase reactions, molecules are so dilute and far apart that computational chemists usually consider only one set of reactant molecules for an ab initio potential energy calculation. For condensed phase systems, such as reaction in solution or the reactions of macromolecules, the interactions between molecules is too strong to be ignored, and thousands, or even millions, of atoms must be modelled. It is difficult to describe the evolution of such large and complex system in a deterministic way. Instead, statistical sampling is used to study the systems' average behavior. Just as in pathfinding methods for gas-phase reactions, the potential energy is the basic input, but it is now used for statistical sampling. In molecular mechanics (MM), the potential energy of the molecular system is modeled by classical mechanics, wherein the atoms and bonds are considered as charged (and perhaps polarizable) balls and springs, respectively. The energy function depends on force constants (to describe the springs' strength) and the displacement from the equilibrium.

11.4.1   POTENTIAL ENERGY FUNCTIONALS AND FORCE FIELDS

In MM, the potential energy of a system is calculated using force fields, which include the form of the potential energy function and the values of its associated parameters. The forms of the potential energy functions differ between different force fields, but the general form can be described using the following formulae [23–25],

$$V = V_{bonded} + V_{non-bonded}$$
$$V_{bonded} = V_{bond} + V_{angle} + V_{dihedral} \tag{6}$$
$$V_{non-bonded} = V_{van\ der\ Waals} + V_{electrostatic}$$

The bond and angle terms are usually modeled as harmonic oscillators. Because the torsion is periodic, the dihedral or torsional terms are modeled by periodic functions, e.g., a Fourier series. The van der Waals terms are typically modeled using a 6-12 Lennard-Jones potential. The electrostatic terms are modeled using the Coulomb interaction, with set atomic charges for different types of atoms in the molecule. For example, the potential energy functional for the AMBER force field [26] is in the following form,

$$V = V_{bond} + V_{angle} + V_{dihedral} + V_{van\ der\ Waals} + V_{electrostatic}$$

$$= \sum_{bonds} \frac{1}{2} k_b \left(b - b^0 \right)^2 + \sum_{angles} \frac{1}{2} k_\theta \left(\theta - \theta^0 \right)^2 + \sum_{torsions} \frac{1}{2} V_n \left[1 + \cos \left(n\omega - \gamma \right) \right]$$

$$+ \sum_{j=1}^{N-1} \sum_{i=j+1}^{N} \left\{ \varepsilon_{i,j} \left[\left(\frac{r_{ij}^0}{r_{ij}} \right)^{12} - 2 \left(\frac{r_{ij}^0}{r_{ij}} \right)^6 \right] + \frac{q_i q_j}{4\pi\varepsilon_0 r_{ij}} \right\} \qquad (7)$$

Here, k_b is the stretching force constant and b^0 is the equilibrium bond length; k_θ is the bending force constant and θ^0 is the equilibrium bond angle; V_n is the height of the torsional barrier, and n is its periodicity (the number of maxima per full revolution); ω is the torsional angle value, and γ is the phase angle (which is usually 0 or π depending on the periodicity: $\gamma = 0$ if n is odd, and $\gamma = \pi$ if n is even); $\varepsilon_{i,j}$ is the depth of the Lennard-Jones potential well and r_{ij} is the distance between atoms i and j when the potential reaches its minimum $-\varepsilon_{i,j}$; q_i and q_j are partial charges assigned to atoms i and j, respectively; and ε_0 is the electric constant. These parameters are defined for each type of atoms, bonded or non-bonded atom pairs, and bonded triplets (angles) or quadruplets (torsions). For macromolecules, their parameters are usually chosen to reproduce experimental measurements and/or reproduce quantum-mechanical calculations on small molecules.

11.4.2 FROM MICROSCOPIC TO MACROSCOPIC

MM is around a million times faster than QM at computing PESs. This means that MM can be applied to much larger systems than that for which QM can be applied. Although the minimum energy structure on the PES represents the most stable structure of the molecule, molecules are not static; they do not stay at the minimum energy structure, but fluctuate around it. According to Einstein and Stern's expression for the zero point energy in 1913, a molecule preserves a residual vibrational energy of $\frac{1}{2}h\nu$ at absolute zero temperature ($T = 0°K$) [27]. All quantum mechanical systems undergo structural fluctuations, even in their ground state. In real life, the molecule statistically samples a range of different structures. If the molecule is quite small and rigid in structure, and if the temperature is low, then the fluctuations are usually tightly clustered around the minimum energy structure. In this circumstance, it is reasonable to use the minimum energy structure on the PES to model the molecule's structure. For molecules that are large and/or floppy, the idea of a unique molecular structure is inadequate, and the molecule should be modeled as a statistical distribution of the structures on the PES. The properties of such systems are no longer determined by a single state, but by averaging over all possible microstates that satisfy the given constraints that define the thermodynamic system; the molecule is represented by the statistical ensemble comprising these microstates. The macroscopic "thermodynamic" properties of the system are the average properties of the ensemble [28].

The ensemble average can be obtained by statistical sampling methods such as Monte Carlo simulation and MD simulation. In this chapter, we discuss only MD simulation.

11.4.3 MOLECULAR DYNAMICS (MD) SIMULATION

According to the ergodic hypothesis, all accessible microstates are equiprobable over a long period of time. Therefore, we can assume that the averaging over the statistical ensemble is equivalent to averaging over the time evolution of the system. MD uses a force field (either from molecular mechanics (MM/MD) or quantum mechanics (QM/MD)) to determine the physical movements of the atoms in a chemical system in time. The basic idea of MD simulation is to predict the evolution of a system over a long period of time and then use the time average to calculate the ensemble average.

Some experimental techniques can measure macromolecular systems at atomic resolution. For example, a scanning electron microscope (SEM) can probe a molecular surface and reveal details to less than 1 nm; the most advanced transmission electron microscope (TEM) can even achieve resolution below 0.5 Å; X-ray crystallography can take "snapshots" of crystal structures; nuclear magnetic resonance (NMR) spectroscopy can probe certain features of molecular motions. But the applications of existing experimental techniques are greatly restrained by sample preparation and sample strength. Moreover, it is still challenging to access both the static and dynamic structures with atomic resolution in the laboratory. MD simulation provides us the opportunity to peer at the motion of the individual atoms in a complex chemical system in a way that is not yet experimentally feasible.

Although MD simulation provides us a way to "visualize" atomic motions, there are some inherent problems and errors associated with MD simulation.

The first problem is related to the ergodic hypothesis. Although it is very hard to prove ergodicity, it is believed that almost all many-body systems are ergodic. However, complex chemical and biological systems might show "nonergodic behavior" during MD simulations, meaning that the systems do not properly explore the phase space. The causes of "nonergodic behavior" include: 1) Large systems diffuse so slowly that the volume in the phase space explored during the computer simulation is insufficient to estimate the ensemble average by the time average. In other words, the simulation time is not long enough to apply the ergodic hypothesis. 2) Different volumes of the phase space are separated by such high-energy barriers that the transitions between these volumes become rare events that occur so infrequently that proper sampling of the phase space cannot be achieved. 3) Different volumes of the phase space are connected by very narrow regions (so-called "entropy bottlenecks"); hence, the transitions between them are rarely sampled [29]. When a system appears "nonergodic" during MD simulation, more advanced techniques like stratification (also called multistage sampling) or importance sampling are required to ensure better exploration of the phase space.

The errors of a MD simulation usually come from the following sources: 1) Discretization of time. In MD simulation of a complex system, the classical equation of motion for the atomic nuclei will be solved numerically. This is performed by discretizing the time with a finite timestep Δt, which is the time length between evaluations of the potential. The force on each atom is held constant during the time span Δt. The timestep has to be smaller than the fastest vibrational frequency in the system. Usually, it is set to be $1fs$ ($10^{-15}s$) or less. By using algorithms such as SHAKE, which constrain the vibrations of the fastest atoms, the timestep can be increased. 2) Errors in force fields. In classical MD (or MM/MD), force fields are used to calculate the potential energy of the system. Force fields are usually derived from experimental studies and quantum mechanical computations on small molecules; therefore, they are not exact. They can be very inaccurate if the chemical characteristic (e.g., bonding pattern) of the molecule changes. 3) Nonbonded cutoffs. In an MD simulation, the most time-consuming part is the evaluation of the potential energy as a function of the atomic position. The nonbonded terms are the most expensive part of an MM force field because there is an energy term from every pair of nonbonded atoms. Therefore, in most MD simulations, non-bonded cutoffs are applied to reduce the computational cost. Neglecting the Coulomb and Lennard–Jones terms between atoms separated by more than the cutoff distance not only increases the speed of the calculation but also introduces errors.

11.5   FREE ENERGY CALCULATION

As we discussed in Section 11.4.2, the thermodynamic properties of a macroscopic system cannot be represented by a single state and require, instead, approximating the ensemble average. For small-molecule gas phase reactions, the minimum energy reaction path on the PES can pro-vide a good approximation to the reaction mechanism. But for condensed phase reactions and large complex systems, there is an entire family of reaction pathways, determined by the free energy behavior of the system. Because it is often impossible to select a single reaction coordinate that captures the full range of possible reaction mechanisms, in these cases, we usually focus not on reaction pathways; instead, we settle for computing free energy differences between key chemical species.

The free energy is usually expressed as the Helmhotz free energy, A, or the Gibbs free energy, G. The Helmholtz free energy is the thermody-namic potential of the canonical (NVT: concentration, volume, temper-ature) ensemble. The Gibbs free energy is the thermodynamic potential of the isothermal-isobaric (NPT: concentration, pressure, temperature) ensemble. They can be expressed in terms of the partition function. For example, the Helmholtz free energy can be expressed as,

$$A(N,V,T) = -k_B T \ln Z(N,V,T) \tag{8}$$

where the canonical partition function can be written as,

$$Z(N,V,T) = \sum_{\upsilon} e^{-\frac{E_\upsilon}{k_B T}} \tag{9}$$

if all points in the phase space are visited; or

$$Z(N,V,T) = \iint d\mathbf{p}^N d\mathbf{r}^N \exp\left(\frac{-E(\mathbf{p}^N, \mathbf{r}^N)}{k_B T}\right) \tag{10}$$

if integrating over the classical phase space. The Helmhotz free energy can be expressed as,

$$A = k_B T \ln \frac{1}{Z(N,V,T)} = k_B T \ln\left(\iint d\mathbf{p}^N d\mathbf{r}^N \exp\left(\frac{E(\mathbf{p}^N, \mathbf{r}^N)}{k_B T}\right) \rho\left(\mathbf{p}^N, \mathbf{r}^N\right)\right) \tag{11}$$

where $\rho\left(\mathbf{p}^N, \mathbf{r}^N\right) = \dfrac{\exp\left(\dfrac{-E(\mathbf{p}^N, \mathbf{r}^N)}{k_B T}\right)}{Z(N,V,T)}$ is the probability of the state with energy $E(\mathbf{p}^N, \mathbf{r}^N)$.

If it were possible to visit all points in the phase space, then the partition function could be calculated using Eq. (9). In general, accurate estimation of the partition function is impossible due to inadequate sampling during the finite simulation time [30]. Calculating the free energy from direct MD or Monte Carlo simulation is very difficult because MD or Monte Carlo simulation preferentially generate states of low energy, and according to Eq. (11), high energy states can also make significant contribution to the free energy. Fortunately, in most cases, we are only interested in the free energy differences between two systems or two states of a system. Because free energy is a state function and it does not depend on the path, but only depends on the initial and final state, the calculation of free energy differences can be carried out through a series of mixed states using the free energy perturbation theory or thermodynamic integration.

11.5.1 THERMODYNAMIC INTEGRATION

In thermodynamic integration, the order parameter, λ, is defined so that the free energy is a continuous function of λ[31],

$$\Delta A = \int_0^1 \frac{\partial A(\lambda)}{\partial \lambda} d\lambda = \int_0^1 \left\langle \frac{\partial H(\mathbf{p}^N, \mathbf{r}^N, \lambda)}{\partial \lambda} \right\rangle_\lambda d\lambda \qquad (12)$$

When using empirical MM force fields, the kinetic energy portion of the Hamiltonian does not depend on λ; thus, the only contribution is from the potential function $V(\mathbf{r}^N, \lambda)$. In practice, numerical integration such as a Gaussian quadrature formula, $\Delta A = \sum_i \omega_i \left\langle \frac{\partial V(\lambda)}{\partial \lambda} \right\rangle_i$, is used to approximate the integral in Eq. (12). For each discrete value, λ_i, MD or Monte Carlo simulation is performed to estimate the value of $\left\langle \frac{\partial V(\lambda)}{\partial \lambda} \right\rangle_i$. The potential function for each λ value can be expressed as a weighted average of the initial ($\lambda = 0$) and final perturbed state ($\lambda = 1$),

$$V(\mathbf{r}^N, \lambda) = (1-\lambda)^k V(\mathbf{r}^N, 0) + \left[1 - (1-\lambda)^k\right] V(\mathbf{r}^N, 1) \qquad (13)$$

where $k = 1$ represents linear mixing when no atom appears or disappears in the perturbed state, and $k > 1$ is typically used when dummy atoms are used in the perturbed state.

11.5.2 pK$_a$ CALCULATION

One of the most important examples of a free energy difference in biochemistry is the pK$_a$ value of amino acid residues in a protein. Recall that the pK$_a$ is proportional to the free energy of deprotonating an ionizable group,

$$pK_a = \frac{\Delta A}{2.3026RT} \qquad (14)$$

Therefore, the essence of pK_a calculation is directly related to the calculation of free energy difference. Usually, we are interested in the pK_a difference of the free amino acid in solvent and in protein environment, in which case, the pK_a shift (ΔpK_a) is calculated through the difference of the deprotonation free energy between the model compound (e.g., the amino acid in a dipeptide chain) and the amino acid in the protein.

11.5.3 MM/PBSA AND MM/GBSA BINDING ENERGY CALCULATION

The molecular mechanics/Poisson–Boltzmann surface area (MM/PBSA) and its complementary molecular mechanics/General Born surface area (MM/GBSA) approach are postprocessing methods to calculate the binding free energy based on the sets of structures from an MD trajectory or a Monte Carlo simulation. The MM/PBSA method combines the molecular mechanical energies with continuum solvent so that the binding free energy between two species (ligand-protein, protein-protein, etc.) can be expressed as,

$$\Delta G_{bind} = \left\langle G_{complex} \right\rangle - \left(\left\langle G_{protein} \right\rangle + \left\langle G_{ligand} \right\rangle \right)$$

$$= \Delta E_{MM} + \Delta G_{solv} - T\Delta S \qquad (15)$$

where ΔE_{MM} is the molecular mechanical energy difference between the bound state (complex) and unbound state (receptor and ligand), which can be evaluated based on the structures (called snapshots) taken from an MD trajectory. ΔG_{solv} is the solvation free energy difference between the bound and unbound state. The solvation free energy includes two components: the electrostatic energy for transferring the solute from the vacuum to the solvent and the non-electrostatic contribution that combines the free energy required to form the cavity and van der Waals terms. In MM/PBSA, the electrostatic component of the solvation free energy is calculated by solving the Poisson–Boltzmann equation in the vacuum and the solvent, $\Delta G_{PB} = \frac{1}{2}\sum_i q_i \left(\phi_i^{80} - \phi_i^1 \right)$. The nonpolar contribution is calculated using an empirical solvent accessible surface area formula, $\Delta G_{nonpolar} = \gamma SASA$, $\gamma = 0.0072 \text{kcal} \text{Å}^{-2}$. ΔS is the entropy change upon binding. The MM/GBSA method is the same as MM/PBSA, except that in MM/GBSA, the electrostatic component of solvation free energy is calculated using the generalized-Born continuum solvent model.

11.6 SUMMARY

This chapter overviews the methods to elucidate chemical reactions. Three different methods were discussed. For small molecules, it is computationally feasible to find minimum energy reaction path on the PES, with the PES computed from quantum mechanical models. This gives very detailed information about the mechanism of the chemical reaction. Biological systems are too large, and entropic effects are too important, for the full reaction path to be determined. Instead, free energy differences between key structures are computed by sampling the PES with MD; in these applications, the PES is modeled using the ball-and-spring-type models known as MM.

For small molecules, there are both one-end methods and two end methods for determining reaction pathways. As an example of the one-end methods, we discussed the fast marching method (FMM) for finding minimum energy path (MEP) in section 11.3.1. FMM is one of the most general and reliable surface walking algorithms for finding MEP. Without any prior knowledge about the PES, it can always find the global MEP. Unfortunately, FMM is an expensive method.

A hybrid between the one end and two end methods is the QSM-NT method for finding all stationary points on the PES presented in Section 11.3.2. Usually path-finding methods can find only one reaction path, and the reliability of the results depends on the initial guess. QSM-NT can find all stationary points, and accordingly all alternative reaction paths, which could be a great advantage while studying reactions with several alternative reaction mechanisms or when trying to analyze and compare different postulated mechanisms. QSM-NT was proven to be efficient and reliable through successful applications to analytical potentials and chemical reactions; however, it is not a "black box" method and, unlike FMM, sometimes fails [4].

For complex biological systems, the properties of the system can no longer be represented by a single state, but by averaging over all possible microstates consistent with given restraints instead. Statistical sampling methods such as MD simulation are used to calculate the ensemble average of the system. In these cases, the reaction path is not found; instead, only the free energy difference between key stationary points on free energy surface is determined.

Finding and characterizing the pathways between reactants and products is very important for studying the mechanisms and energetics of chemical reactions, not only in the gas phase but also in complex environments like enzymes. Existing path-finding methods either require a good initial guess in order to locate the desired path or are very expensive because a significant portion of the PES must be explored. Our research on these topics is only the start of the long journey; there are still lots of work to do.

ACKNOWLEDGMENTS

PWA and YL thank NSERC for financial support.

KEYWORDS

- chemical reaction mechanism
- fast-marching method
- free-energy calculations
- intrinsic reaction coordinate
- minimum energy path
- Newton trajectory
- potential energy surface
- string method

REFERENCES

1. Anderson, J. S. M., Melin, J., & Ayers, P. W., (2007). Conceptual density-functional theory for general chemical reactions, including those that are neither charge- nor frontier-orbital-controlled. 2. Application to molecules where frontier molecular orbital theory fails. *Journal of Chemical Theory and Computation, 3*, 375–389.

2. Anderson, J. S. M., Melin, J., & Ayers, P. W., (2007). Conceptual density-functional theory for general chemical reactions, including those that are neither charge- nor frontier-orbital-controlled. 1. Theory and derivation of a general-purpose reactivity indicator. *Journal of Chemical Theory and Computation., 3*, 358–374.

3. Anderson, J. S. M., Liu, Y. L., Thomson, J. W., & Ayers, P. W., (2010). Predicting the quality of leaving groups in organic chemistry: Tests against experimental data. *Journal of Molecular Structure-Theochem., 943*, 168–177.

4. Liu, Y., Burger, S. K., & Ayers, P. W., (2011). Newton trajectories for finding stationary points on molecular potential energy surfaces. *J. Math. Chem.,49*, 1915–1927.

5. Liu, Y., & Ayers, P. W., (2011). Finding minimum energy reaction paths on ab initio potential energy surfaces using the fast marching method. *Journal of Mathematical Chemistry, 49*, 1291–1301.

6. David, J. W., (2003). *The Born-Oppenheimer Approximation and Normal Modes*. In Energy Landscapes, Cambridge University Press, pp. 119–160.

7. Fukui, K., (1970). A Formulation of the Reaction Coordinate. *J. Phys. Chem., 74*, 4161.

8. Dey, B. K., Janicki, M. R., & Ayers, P. W., (2004). Hamilton-Jacobi equation for the least-action/least-time dynamical path based on fast marching method. *J. Chem. Phys., 121*, 6667–6679.

9. Dey, B. K., & Ayers, P. W., (2006). A Hamilton-Jacobi type equation for computing minimum potential energy paths. *Molecular Physics., 104*, 541–558.

10. Irikura, K. K., & Johnson, R. D., (2000). Predicting unexpected chemical reactions by isopotential searching. *Journal of Physical Chemistry A., 104*, 2191–2194.

11. Dey, B. K., & Ayers, P. W., (2006). A Hamilton-Jacobi type equation for computing minimum potential energy paths. *Molecular Physics., 104*, 541–558.

12. Chu, J. W., Trout, B. L., & Brooks, B. R., (2003). A super-linear minimization scheme for the nudged elastic band method. *119*, 12708–12717.

13. Weinan, E., & Weiqing, R., (2005). Finite temperature string method for the study of rare events. *J. Chem. Phys.,109*, 6688–6693.

14. Ren, E, W. N., W. Q., & Vanden-Eijnden, E., (2002). String method for the study of rare events. *Physical Review B., 66*(5), 052301.

15. Peters, B., Heyden, A., Bell, A. T., & Chakraborty, A., (2004). A growing string method for determining transition states: Comparison to the nudged elastic band and string methods. *J. Chem. Phys., 120*, 7877–7886.

16. Trygubenko, S. A., & Wales, D. J., (2004). A Doubly Nudged Elastic Band Method for Finding Transition States. *J. Chem. Phys., 120*, 2082–2094.

17. Xie, L., Liu, H. Y., & Yang, W. T., (2004). Adapting the nudged elastic band method for determining minimum-energy paths of chemical reactions in enzymes. *J. Chem. Phys., 120*, 8039–8052.

18. Peters, B., Heyden, A., Bell, A. T., & Chakraborty, A., (2004). A Growing string method for determining transition states: Comparison to the nudged elastic band and string methods. *J. Chem. Phys., 120*, 7877–7886.
19. E, W. N., Ren, W. Q., & Vanden-Eijnden, E., (2002). String method for the study of rare events. *Physical Review B., 66*(5), 052301.
20. Burger, S. K., & Yang, W. T., (2006). Quadratic string method for determining the minimum-energy path based on multi objective optimization. *J. Chem. Phys., 124*, 054109.
21. Quapp, W., (2004). Newton trajectories in the curvilinear metric of internal coordinates. *Journal of Mathematical Chemistry, 36*, 365–379.
22. Quapp, W., Hirsch, M., & Heidrich, D., (2004). An approach to reaction path branching using valley-ridge inflection points of potential-energy surfaces. *Theoretical Chemistry Accounts., 112*, 40–51.
23. Andrew R. L., (1999). Empirical Force Field Models: Molecular Mechanics. *In Molecular Modelling: Principles and Applications,* Longman, pp 131–210.
24. Christopher J., (2004). Cramer Molecular Mechanics. In Essentials of Computational Chemistry: *Theories and Models*, Wiley, pp 17–68.
25. Tamar Schlick, (2002). Theoretical and Computational Approaches to Biomolecular Structure. In *Molecular Modeling and Simulation: An Interdisciplinary Guide*, Springer: New York, pp. 199–224.
26. Case, D. A., Darden, T. A., Cheatham, T. E., Simmerling, C. L., Wang, J., Duke, R. E., Luo, R., Crowley, M., Walker, R. C., Zhang, W., Merz, K. M., Wang, B., Hayik, S., Roitberg, A., Seabra, G., Kolossvary, I., Wong, K. F., Paesani, F., Vanicek, J., Wu, X., Brozell, S. R., Steinbrecher, T., Gohlke, H., Yang, L., Tan, C., Mongan, J., Hornak, V., Cui, G., Mathews, D. H., Seetin, M. G., Sagui, C., Babin, V., & Kollman, P. A., (2008). *AMBER 10*, University of California, San Francisco.
27. Einstein, A., & Stern, O., (1913). Einige Argumente fur die Annahmeeinermolekularen Agitation beimabsoluten Null punkt. *Ann. Phys.,345*, 551–560.
28. Christopher, J. C., (2004). Simulation of Molecular Ensembles. In *Essentials of Computational Chemistry: Theories and Models,* Wiley, pp. 69–104.
29. Christohe Chipot, M., & Scott Shell, (2007). Andrew Pohorille Introduction. *In Free Energy Calculations*, Springer, pp. 1–32.
30. Andrew R., (1999). Leach Computer Simulation Methods. *In Molecular Modelling: Principles and Applications*, Longman. pp. 261–312.
31. Andrew R., (1996). Leach Three Challenges in Molecular Modelling: Free Energies, Solvation and Simulating Reactions. *In Molecular Modelling: Principles and Applications,* Longman, pp. 481–542.
32. Burger, S. K., Liu, Y., Sarkar, U., & Ayers, P. W., (2009). Moving Least-Squares Enhanced Shepard Interpolation for the Fast Marching and String Methods. *J. Chem. Phys. (accepted)., 130*, 024103.
33. Anderson, J. S. M., Liu, Y. L., Thomson, J. W., & Ayers, P. W., (2010). Predicting the quality of leaving groups in organic chemistry: Tests against experimental data. *Journal of Molecular Structure-Theochem, 943*, 168–177.
34. Thompson, D. C., Liu, Y. L., & Ayers, P. W., (2006). A confined noninteracting many electron system: Accurate corrections to a statistical model. *Phys. Lett. A., 351*, 439–445.

CHAPTER 12

CALCULATION OF PROTON AFFINITY, GAS-PHASE BASICITY, AND ENTHALPY OF DEPROTONATION OF POLYFUNCTIONAL COMPOUNDS BASED ON HIGH LEVEL DENSITY FUNCTIONAL THEORY

ZAKI S. SAFI and WALAA FARES

Department of Chemistry, Al Azhar University-Gaza, Gaza City, Palestine

CONTENTS

12.1   INTRODUCTION

Protonation [Eq. (1)] and deprotonation [Eq. (2)] processes of poly-functional polar molecules play very crucial role in organic chemistry, biochemistry, medicine, and pharmacy and are the first steps in many fundamental chemical rearrangements studies [1, 2].

$$B + H^+ \rightarrow BH^+ \tag{1}$$

$$AH \rightarrow A^+ + H^+ \tag{2}$$

where B and AH are the base and the acid, respectively. The capability of an atom or molecule in the gas phase to accept a proton can be characterized by calculating, from the above reaction, the proton affinity (PA), the molecular basicity in the gas phase basicity (GB), and the enthalpy of deprotonation, (ΔH_{dep}).

Using standard conditions, the PA is defined as the negative of the enthalpy change, ΔH, for the gas phase reaction (Eq. (1)) and the GB is defined as the negative of the free energy change, $\Delta_{base} G$, associated with the same reaction in **(Eq. (1))**, whereas the enthalpy of deprotonation is the energy associated with the deprotonation energy (Eq. (2)). The PA and GB quantities are site-specific values and therefore must be calculated at each chemically different binding site in the molecule. This means that some molecules will have multiple PAs and/or multiple GB values. Protonation and deprotonation give a deep understanding of the correlations between molecular structures, molecular stability, reactivity, prediction of the preferred site for nucleophilic and electrophilic attachment in the organic compounds, reaction mechanisms occurring in several biological systems and of the organic molecules [3], and the possible role played by the protonation processes in decreasing the activation barrier for the 1,3-hydrogen transfer process [4–7]. In addition, PA can be applied to different aspects such as determination of the stability of cationic species [8, 9], the hydrogen bonding ability, acidity of solid catalysts [10], and relation between PA and pKa [11, 12].

Recent years have witnessed a great progress in experimental and theoretical investigations of the protonation of polyfunctional compounds [3, 13–16]. Unfortunately, measuring of the proton affinities of polar polyfunctional molecules, an attribute common to most biologically interesting molecules, is an experimental challenge. It was pointed out that direct measurements of the PA are not easy, and are possible for only a few molecules, mainly olefins and carbonyl. The basis for this scale is described by Hunter and Lias [17]. On the other hand, based on the phenomenal growth in computational power in recent years, much attention has been given to the possibility of calculating these parameters by quantum methods. Ab initio approaches are very successful in providing reliable values of PA and GB for small molecules even at lower levels of theory [18, 19]. In the last two decades, the progress in the density functional theory (DFT) approaches makes this method another candidate for reliable calculation of proton affinities. It has been shown that the PA values computed using the B3LYP method are as effective as high level ab initio results, and the results were comparable up to ~1–2 kcal/mol in comparison with the measured values [20–29]. In addition to the direct calculation of the proton affinity and gas phase basicity, several descriptors were used to predict the proton affinity and gas phase basicities such as the molecular electrostatic potential (MEP) on the nuclei of the basic centers and the valence natural atomic orbital energies (NAO) of the basic centers [28, 30]. Yuang et al. [28] suggested that the more negative the MEP value, the stronger is the molecular basicity, and the larger is the proton affinity and gas phase basicity. They also proposed that for systems with multiple sites of the same basic element type, the most basic site (largest energy decrease) has the most negative MEP value.

Additionally, the ^1H NMR chemical shift (δ^1H) of the incoming proton [31] and the topological properties at the bond critical point (BCP) of the B-H$^+$ bond (including electron density, ρ_{BCP}, and its Laplacian, $\nabla^2\rho$) calculated by means of the Bader theory of atoms in molecules [32]. Strong linear correlations were found between these descriptors and the values of PA or GB in the gas phase. These results suggested that these quantum descriptors (MEP, NAO, δ^1H, and ρ_{BCP}) can be easily used to approximately estimate the PAs or GBs in the gas phase.

Understanding of the influence of the substituents on proton affinities and gas phase basicities has been studied experimentally and theoretically (see Ref. [22] and Refs. therein). Extensive works have been conducted to study the presence of the substituent on the aromatic ring, not necessarily a phenyl ring [33–35]. Kukol et al. [33] summarized the effect of the *ortho-*, *meta-*, and *para-*substituents on aromatic carbonyl compounds on the proton affinity and gas phase basicity. Substituent effects were subdivided into polar and steric effects. In general, polar effects were rather in terms of strong resonance interaction with electron donating/withdrawing characteristic of the substituent.

A lot of heterocycles have many acidic and basic centers in their structures; therefore, the study of the protonation and deprotonation of these systems will be a subject of interest, particularly, for the biologically important heterocycles. This chapter deals mainly to test the validity of the DFT method to calculate the intrinsic properties of polyfunctional compounds. This aim is achieved by firstly, calculating the proton affinity, gas phase basicity, and the enthalpy of deprotonation of some polyfunctional compounds in gas phase and in solution; secondly, comparing our calculated results with the available experimental data; and finally, studying the influence of the substituent effect on the PA and GB of the compounds under probe proton affinities. The selected compounds are hydantoin, isorhodanine, rhodanine and their thio derivatives, and benzamide and its m- and p- derivatives. All calculations were performed using the famous DFT at the B3LYP and BP86 6–311++G(2df,2p)//6–311+G(d,p) level of theory in the gas phase and in solution (water). Another important aspects that will be presented in this chapter is calculation of some quantum descriptors, namely, molecular electrostatic potential (MEP) on the nuclei of the base, natural atomic orbital energy (NAO) of the base, proton ^{1}H NMR chemical shifts (δ^1H) of the incoming proton, and the electron density at the O-H$^+$ bond critical point (ρ_{bcp}) and examine the possibility of using them to estimate/predict the intrinsic basicity of the polyfunctional compounds. The systems under consideration are hydantoin and its thio derivatives; mono and di-substituted hydantoin derivatives; and rhodamine, isorhodanine, and benzamide and their derivatives.

12.2   COMPUTATIONAL DETAILS

The standard hybrid DFT in the framework of B3LYP [36, 37] functional at the 6–311+G(d,p) [38, 39] basis function was used for geometry optimization. The harmonic vibrational frequencies of the different stationary points of the potential energy surface (PES) were calculated at the same level of theory that was used to check that all the structures are minima with no imaginary frequencies as well as to estimate the corresponding zero point energy corrections (ZPE) that were scaled by the empirical factor 0.9806 proposed by Scott and Radom [40]. In order to obtain more reliable energies for the local minima, final energies were evaluated by using the same functional combined with the 6–311++G(2d,2p) basis set. It has been shown that this approach is well suited for the study of this kind of systems, yielding PAs, GBs, and ΔH_{dep} in good agreement with the experimental values. A single point calculation was used to compute the molecular electrostatic potential (MEP) on each of the nuclei followed by a full natural bonding orbital (NBO) calculations [41] to obtain the natural atomic orbital energies. The molecular electrostatic potential (MEP) of a molecule is a real physical property, and it can be determined experimentally by X-ray diffraction techniques [42]. The MEP on the nuclei of a molecule is originally invoked to study electrophilic reactivity. $V(\vec{r})$ (MEP) that is created at a point $\vec{r}$ by electrons and nuclei of a molecule is given as (Eq. (3)):

$$MEP = V(\vec{r}) = \sum_{A} \frac{Z_A}{|R_A - \vec{r}|} - \int \frac{\rho(\vec{r}')}{|\vec{r}' - \vec{r}|} d\vec{r} \qquad (3)$$

where Z_A is the charge on nucleus A, located at R_A, and $\rho(\vec{r})$ is the molecule's electron density [43].

In order to compute the ^{1}H NMR absolute shielding constants (δ^1H values) of the associated proton, another single point calculation at the B3LYP/6–311+G(d,p) level with the gauge-independent atomic orbital (GIAO) method [44] was performed. Furthermore, the calculated chemical shielding was converted into chemical shift by subtracting 32.5976, the 1H shielding of tetramethylsilane computed at the same level of theory and basis set.

Analysis of the electron densities at bond critical points (bcps) of the optimized structures was performed by generating the wave functions through a single point calculation on the geometrized structures and analyzing these wave functions by means of the atoms-in-molecules (AIM) theory proposed by Bader et al. [32] as implemented in AIM2000 program package [45]. We have additionally carried out calculations in solution using the integral equation formalism polarizable continuum model

(IEF-PCM) in which the solvent is represented by an infinite dielectric medium at the same level of theory. All calculations were performed with the Gaussian 09 program [46] with tight self-consistent field convergence, and in addition, ultrafine integration grids were used for MEP calculations.

The gas phase PA of a molecule B was estimated from the reaction involving the addition of a proton to the neutral species (protonation reaction) (Eq. (1)); the equation for the PA is written as (Eq. (4)):

$$PA = -\Delta H^{298} = H^{298}(BH^+_{(g)}) - \left[H^{298}(B_{(g)}) + H^{298}(H^+_{(g)})\right], \qquad (4)$$

where $H^{298}(BH^+_{(g)})$ is the sum of electronic and thermal enthalpies for the protonated species and $H^{298}(B_{(g)})$ is the sum of the electronic and thermal enthalpies for the neutral species under investigation. Under the standard state conditions, the values of the enthalpy and entropy of the gas-phase proton are $\Delta^0_g H(H+)=2.5RT=1.48$ kcal/mol. This value corresponds to the translational proton energy (a loss of three degrees of freedom is 3/2(RT) plus the PV work-term (=RT for ideal gas)).

Thermodynamically, GB is related to PA and the change of entropy ΔS at constant temperature (298.15 K) by (Eq. (5)):

$$GB = PA - T\Delta S \qquad (5)$$

For proton, only the translational entropy $S_{translational}(H^+)$ is not equal to zero and is determined to be 26.04 cal/mol and the value of Gibbs free energy of the gas has been used as shown in (Eq. (6)) [47–49].

$$\Delta^0_g G(H^+) = \frac{5}{2}RT - T\Delta S^0 = 1.48\text{-}7.76 \ = \ \text{-}6.28 \text{ kcal/mol} \qquad (6)$$

To calculate the proton affinity and molecular basicities in the solvent (water in our case), the reaction to be considered is identical to one used in the gas phase (Eq. (1)), except that all constituents are solvated. PCM thermodynamic results, obtained from the harmonic frequency analysis, were used to evaluate the PA value according to the following equation (Eq. (7)):

$$PA = -\Delta protH^{298} = -\left[H^{298}(BH^{+}_{solv}) - H^{298}(B_{solv}) - H^{298}(H^{+}_{solv}) \right] \quad (7)$$

In this case, the proton enthalpy is determined from summing the solvation enthalpy of the proton $\Delta_{solv}H^{298}(H^{+})=-275.12$ kcal/mol [50], which is very close to the experimental value (-275.14 kcal/mol)[51], with proton gas-phase enthalpy $H^{298}(H^{+}(g))=1.48$ kcal/mol In order to calculate the molecular basicity in solution, Eq (the constant $pK_{b}(B)$ for a base B is given by the well-known thermodynamic relation (Eq. (8)):

$$pK_{b}(B) = \frac{\Delta_{prot}G(solv)}{2.303RT} \quad (8)$$

where R, T, and $\Delta_{prot}G(solv)$ (1 atm, standard state) are the gas phase constant, the temperature (298.15 K), and the standard free enthalpy of base protonation reaction in solution, respectively. $\Delta_{prot}G(solv)$ is determined from the following equation (Eq. (9)):

$$\Delta protG^{298}(solv) = G^{298}(BH_{+}(solv) - G^{298}(B(solv) - G^{298}(H^{+}(solv) \quad (9)$$

In water, the aqueous free enthalpy $G^{298}(H^{+}(aq))$ is determined by summing $\Delta_{aq}G^{298}(H^{+}) = -266.13$ kcal/mol[51, 52] and $G^{298}(H^{+}(g)) = -6.3$ kcal/mol [53].

To calculate the deprotonation enthalpy (DPE, ΔH^{298}) for removing H^{+} according to the deprotonation reaction (Eq. (2)), we follow (Eq. (10)):

$$\Delta H^{298}_{DPE} = H^{298}(AH_{(g)}) - \left[H^{298}(A^{-}_{(g)}) - H^{298}(H^{+}_{(g)}) \right] \quad (10)$$

where $H^{298}(AH_{(g)})$is the sum of electronic and thermal enthalpies for the isolated molecules under probe, $H^{298}(A^{-}_{(g)})$is the sum of the electronic and thermal enthalpies for the anion species for the corresponding molecules, and $H^{298}(H^{+}_{(g)})$is the translational energy term for the H^{+} (i.e., the thermal correction term), whose value is 0.889 kcal/mol.

12.3 RESULTS AND DISCUSSION

12.3.1 *HYDANTOIN AND ITS THIO DERIVATIVES*

12.3.1.1 Proton Affinity of Hydantoin and its Thio Derivatives

To our knowledge, there is no experimental data related to the proton affinity of hydantoin. The available data are from the study of Silva et al. [54] and Bouchoux [55, 56] who used the G3MP2B3 and G4MP2 level of theory, respectively. In Table 12.1, we list the DFT B3LYP results of PAs of hydantoin and its thio derivatives (Scheme 1), together with the avail-

TABLE 12.1 Proton Affinities of Different Basic Sites of Hydantoin Calculated Using Different Theoretical Methods

Species	Method	O2(S2)	O4(S4)	N1	N2
Hydantoin	B3LYP*	197.8	193.7	186.1	175.9
	G3MP2B3[a]	196.5	192.1	185.7	175.7
	G4MP2[b]	198.0	198.0		
2-thiohydantoin	B3LYP*	205.9	192.9	186.2	174.5
	G3MP2B3[a]	203.3	191.9	186.4	175.8
4-thiohydantoin	B3LYP*	197.8	202.3	187.2	176.4
2,4-dithiohydantoin	B3LYP*	206.0	202.0	186.2	174.5

[a] Values taken from Ref [54]; [b] Values taken from Ref[55]; *Results of this work

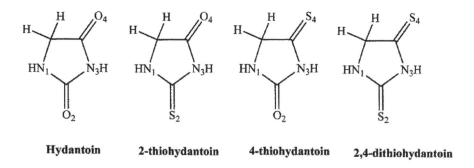

Hydantoin   **2-thiohydantoin**   **4-thiohydantoin**   **2,4-dithiohydantoin**

SCHEME 1 Schematic representation of hydantoin and its thio derivatives.

able theoretical results [54–56]. Results revealed that the calculated PAs using the B3LYP method are in good agreement with those of G3MP2B3 and G4MP2. An excellent linear relationship was observed between the B3LYP and G3MP2B3 results (see Figure 12.1).

The reported results show a very strong linear relationship with correlation coefficients $R^2 = 0.999$, reflecting the validity and effectiveness of the DFT B3LYP method in calculating the intrinsic basicity of the polyfunctional molecules. It is worth mentioning that DFT methods consume much lower computation time than the ab initio method. Analysis of Table 12.1 indicates that thiohydantoin derivatives are more basic than the hydantoin one and the sulfur atom at position 2 (S2), which is located between two amidic groups, is more basic than the one at position 4 (S4). The PA trend of the heteroatoms in hydantoin and its thio derivatives can be arranged from highest to lowest basicity as follows: S2 > S4 > O2 > O4 > N1 > N4. These results agree with those reported for analogue systems [3, 13].

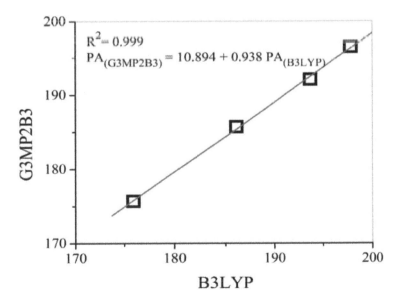

FIGURE 12.1 Linear relationship between the proton affinity calculated using B3LYP and the PAs calculated using the G3MP2B3 method of the basic centers of the hydantoin molecule. All values are in kcal/mol.

12.3.1.2 Substituent Effect on the Proton Affinity of Hydantoin

This subsection focuses on studying the influence of the mono- and di-substituent(s) on the proton affinity of hydantoin molecules. The species under probe are shown in Scheme 2. The results of this study are presented in Table 12.2. As can be observed in the table, ongoing from electron with drawing substituents to electron donating ones, the proton affinity of

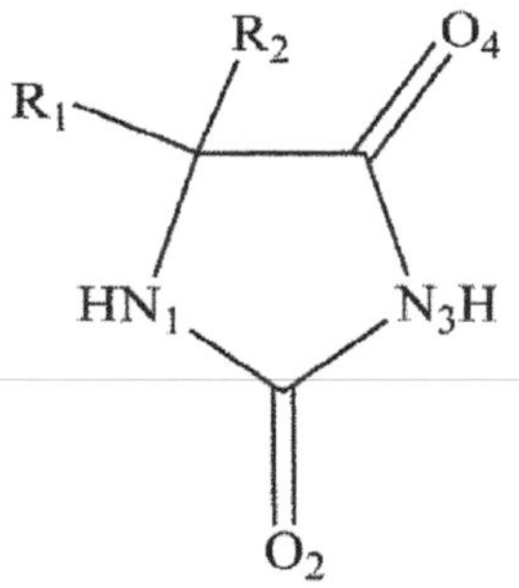

R1=R2=H	Hydantoin
R1=CH$_3$, R2=H	5-methylhydantoin
R1=R2=CH$_3$	5,5-dimethylhydantoin
R1=NH$_2$, R2=H	5-aminohydantoin
R1=R2=NH$_2$	5,5-diaminohydantoin
R1=ph, R2=H	5-phenylhydantoin
R1=R2=ph	5,5-diphenylhydantoin
R1=NO$_2$, R2=H	5-nitrohydantoin
R1=R2=NO$_2$	5,5-dimethylhydantoin
R1=Cl, R2=H	5-chlorohydantoin
R1=R2=Cl	5,5-cholrohydantoin

SCHEME 2 Schematic representation of mono and di-substituted hydantoins.

TABLE 12.2 Proton Affinities (kcal/mol) of Mono- and Di-Substituted Hydantoin Calculated Using B3LYP Method

Species	O2	O4
5-phenylhydantoin	200.5	199.8
5,5-diphenylhydantoin	206.2	203.8
5-aminohydantoin	197.7	197.3
5,5-diaminohydantoin	202.7	201.3
5-methylhydantoin	183.8	179.3
5,5-dimethylhydantoin	201.0	196.1
Hydantoin	197.8	193.7
5-chlorohydantoin	188.4	186.5
5,5-dichlorohydantoin	184.6	183.9
5-nitrohydantoin	182.9	183.9
5,5-dinitrohydantoin	173.5	177.6

hydantoin is increased. For example, the proton affinity of 5-aminohydantoin (197.7 kcal/mol) is higher than that of 5-nitrohydantoin (182.9 kcal/mol).

It is also obvious that proton affinities of the 5,5-di-substituted species are larger than those of the 5-monosubstituted ones. For example, the proton affinity of 5,5-diaminohydantoin is ~5 kcal/mol higher than that of 5-aminohydantoin. In contrast, the proton affinity of 5,5-dinitrohydantoin is found to be ~9.5 kcal/mol lower than that 5-nitrohydantoin. This finding indicates that the presence of two electron withdrawing groups largely decreases the intrinsic basicity of the basic center, whereas the reverse is true in the case of the electron donating groups.

12.3.1.3 5-Methylhydantoin and Its Thio Derivatives

In this part, we introduce the results obtained by Safi and Frenking [24] for 5-methylhydantoin and its thio derivatives shown in Scheme 3. The PAs were obtained by using the two level of theory, B3LYP and BP68, with 6–311+G(2fd,2p)//6–311+G(d,p) basis set.

Table 12.3 shows that the agreement between both two levels of theory is satisfactory. Our results in Table 12.3 deduced that the atom with the most intrinsic basic character is the heteroatoms of the carbonyl and/or thiocarbonyl group attached to position 2 and/or 4. The intrinsic basicity of these ranges between 193 and 205 kcal/mol. These results led us to conclude that the four hydantoin systems are bases with moderate strength in the gas phase. In addition, it was found that the strongest base among

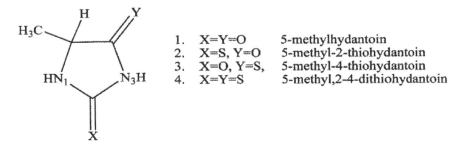

1. $X=Y=O$ 5-methylhydantoin
2. $X=S, Y=O$ 5-methyl-2-thiohydantoin
3. $X=O, Y=S,$ 5-methyl-4-thiohydantoin
4. $X=Y=S$ 5-methyl,2-4-dithiohydantoin

SCHEME 3   5-methylhydantoin and its thio derivatives.

TABLE 12.3 Calculated Gas Phase Proton Affinities of 5-Methylhydantoin and Its Thio Derivatives at B3LYP/6–311+G(3df,2p)//B3LYP/6–311+G(d,p) and BP686–311+G(3df,2p)//B3LYP/6–311+G(d,p) Levels of Theory

		N1	O2	N3	O4	S2	S4
1	B3LYP	187	197	175	193	--	--
	BP68	188	197	--	193	--	--
2	B3LYP	186	--	177	192	204	--
	BP68	188	--	176	193	204	--
3	B3LYP	187	197	174	--	--	200
	BP68	188	197	176	--	--	201
4	B3LYP	188	--	--	--	205	200
	BP68	189	--	--	--	204	200

Results are taken from Ref. 24.

the four is compound **4**, but the basicity gap is small when compared with the other molecules.

The PA values in Table 12.3 indicate that the basicities of the lone pairs at O4 and S4 are lower than that of the lone pairs of O2 and S2 by 4 and 5 kcal/mol, respectively. These results are in agreement with those obtained for the analogue systems, namely uracil [27], thiouracil derivatives [57], triazepinesthio derivatives [58], and aliphatic amides [59]. Furthermore, the difference between the PAs of the heteroatoms of the carbonyl and thiocarbonyl groups and the two nitrogen atoms, N1 and N3, is quite substantial and ranges between 10 to 20 kcal/mol. This is in agreement with ab initio calculations, which showed that amides derived from weak acids protonate preferentially at the acid residue rather than at the nitrogen with energy differences between 10 and 30 kcal/mol [27, 60]. In summary, for 5-methylhydantoin and its thio derivatives, the most basic site is always sulfur atom at position 2, which leads us to classify the order of basicity sites in these molecules as follows: S2 > S4 > O2 > O4 > N1 > N3. To compare the accuracy of the DFT results with highly accurate coupled cluster calculations, the proton affinities of water and hydrogen sulfide were calculated. Table 12.4 summarizes the results at B3LYP/6–311+(2df,2f), B3LYP/6–311+(d,f), CCSD(T)/6–311+(2df,2f), and CCSD(T)/6–311+(d,f), together with experimental data [61]. The data show that the B3LYP results deviate only slightly from the CCSD(T) val-

TABLE 12.4   Proton Affinity Values (kcal/mol) of H_2O and H_2S Obtained at DFT and CCSD(T) Levels of Theory, with Experimental Values Given for Calibration

Species	DFT[a]	CCSD(T)[b]	Expt[c]
H_2O	164.2	164.9	165.2
H_2S	169.8	169.7	168.5

[a] Calculated at B3LYP/6–311+G(2df,2p)// B3LYP/6–311+G(d,p) level of theory.
[b] Calculated at CCSD(T)/6–311+G(2df,2p)// B3LYP/6–311+G(d,p).
[c] Data taken from Ref. [61].

ues, which in turn agree very well with the experimental findings. The results suggest that our DFT values for the heterocyclic compounds should be highly accurate.

Table 12.5 displays the calculated acidities for neutral 5-methylhydantoin and its thio derivatives (Scheme 3) together with their radical cations. The deprotonation energies were obtained using two levels of theory, B3LYP and BP86, with 6–31G(d), 6–311G(d,p) and B3LYP/6–311+G(2df,2p) basis functions. For sake of comparison, we included the acidity of the protonated molecule obtained at the same basis sets. As it is found in literature [62–64], data reported in Table 12.5 indicate that

TABLE 12.5   Calculated 298 K Deprotonation Energies, ΔH^{298}, of the 5-Methylhydantoin and Its Thio Derivatives (Scheme 3)

Species*	Method	Neutral		Radical	
		N1	N3	N1	N3
1	B3LYP[a]	356	351	201	196
	B3LYP[b]	348	343	199	193
2	B3LYP[a]	341	341	216	214
	B3LYP[b]	338	337	221	219
3	B3LYP[a]	346	335	220	214
	B3LYP[b]	341	336	220	217
4	B3LYP[a]	334	334	218	217
	B3LYP[b]	333	332	226	222

[a]B3LYP/6-311G(d,p).
[b]B3LYP/6-311+G(2df,2p)//6-311G(d,p).
All values are in kcal/mol.
$\Delta H^{298} = \Delta E_{elec} + \Delta ZPVE + \Delta(H^{298}-H^0) + 6.2$ kJ; scaled ZPVE by 0.9806 empirical factor.
Results are taken from Ref. 23.

the most acidic site of 5-methylhydantoin and its thio derivatives is the N3-H group. For hydantoins, the imide proton N3-H is more acidic than N1-H, and hence, this position is more reactive toward electrophiles. It is noted that there is an inverse relationship between the magnitude of ΔH^{298} and the strength of the acid. The larger the value of ΔH^{298}, the weaker is the acid. The first conspicuous fact is that the agreement between both two levels of theory is satisfactory, and the energy gap between the two levels is 2–3 kcal/mol. Another conspicuous fact is that the acidity values obtained from the 6–31G(d) and 6–311G(d,p) basis are more than ~10 and 8 kcal/mol, respectively, which are higher than those obtained using 6–311+G(2fd,2p)//6–311+G(d,p) basis functions, in agreement with those results reported for uracil [65]. These findings show that diffuse functions are required to lower computed deprotonation energies of anions [66].

For compounds 1 and 3, it is found that the acidity difference between N3 and N1 atoms is quite substantial and amounts to about 5.0 kcal/mol in favor of the former. For compounds 2 and 4, this acidity difference decreases to about 1.0 kcal/mol. The reason for this behavior is probably due to the high polarizability and size of the sulfur atom over the oxygen. In summary, our results suggest that the gas phase acidity trend of the molecules under investigation is as follows: 4 > 2 > 3 > 1. These results agree with those reported experimentally [64], which showed that compound 2 (pKa=8.5) is slightly stronger acid than compound 1 (pKa=9.0). Again, these results argued the validity and efficiency of the DFT method to predict and estimate the gas phase acidity of the polyfunctional compounds.

12.3.2   THIAZOLIDINE–2,4-DIONE AND ITS 5-SUBSTITUTED DERIVATIVES

In this subsection, we introduce and discuss the DFT B3LYP results of the proton affinities (PAs), molecular electrostatic potential (MEP), and natural valence atomic orbital energies (NVAO) of the possible basic centers (O2, N3, and O4) existing for thiazolidine–2–4,dione (Scheme 4) [64].

Table 12.6 presents the PAs of the basic centers (O2, O4, and N3) existing for the system under probe. The PAs were calculated using the DFT method at the B3LYP/6–311++G(3df,2p)//B3LYP6–311+G(d,p) level of theory. An analysis of Table 12.6 indicates that the most basic center is O2.

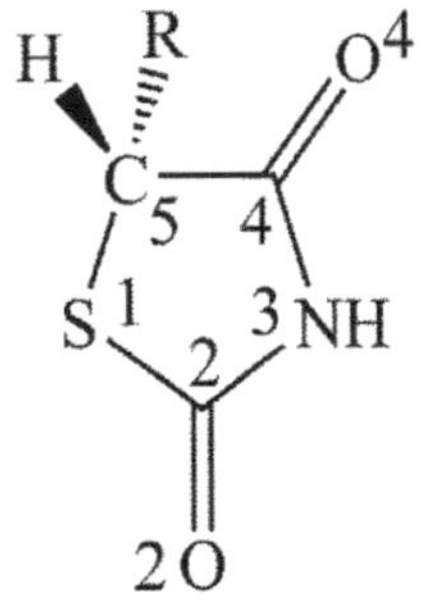

$R = -H, -CH_3, -Cl, -CN, -F, -NO_2$

SCHEME 4 Thiazolidine–2,4-dione and their 5-substituted derivatives.

TABLE 12.6 Proton Affinities (PAs) of the Basic Sites (in kcal mol⁻1), Molecular Electrostatic Potential on the Nuclei of the Basic Sites (MEP) (in a.u.) and the Valence Natural Atomic Orbital Energies (NNAO) (in a.u.) of the Thiazolidine–2,4-dione and Its Derivatives

	PA			MEP			NVAO		
	O2	O4	N3	O2	O4	N3	O2	O4	N3
H	192	190	168	−22.356	−22.358	−18.313	−1.842	−1.823	−1.525
CH3	195	192	171	−22.359	−22.36	−18.316	1.829	−1.814	−1.511
Cl	187	186	163	−22.343	−22.343	−18.301	−1.89	−1.873	−1.571
CN	181	179	156	−22.335	−22.334	−18.29	−1.92	−1.911	−1.611
F	185	183	161	−22.341	−22.341	−18.299	−1.898	−1.879	−1.577
NO2	181	184	158	−22.33	−22.33	−18.288	−1.935	−1.921	−1.615
NH2	199	197	-	−22.356	−22.355	−18.313	−1.841	−1.834	−1.525

Results are taken from Ref. 40.

The calculated PA of O2 of the parent thiazolidine–2,4-dione compound is ~192 kcal/mol and it is ~ 2 and 23 kcal/mol higher than that of O4 and N3, respectively. Furthermore, it is found that the PA is increased when changing from the electron withdrawing group to the electron releasing group. Our results show that the highest PA corresponds to 5-amino derivatives (~199 kcal/mol), whereas the smallest one corresponds to the 5-cyano derivatives (181 kcal/mol). The electron releasing group makes the electron pairs on the oxygen atom more available by donating the electron to the system. The reverse is true in the case of the electron withdrawing group. These results lead us to classify the compound under probe as a moderate basic compound, and the basic centers in this compound can be arranged in order of its strength as follows: O2 > O4 > N3. Based on our results of MEP and NVAO values from Table 12.6, it is found that the MEP trend order is $MEP(O_2) >$ $MEP(O_4) > MEP(N3)$. The absolute value of the $MEP(O_2)$ is higher than that of $MEP(N_2)$ by about 4.042–4.045 a.u. These results are in agreement with PA values, which suggest that O2 is the most basic center among all the centers existing for the thiazolidine–2,4-dione and its derivatives. The same conclusion can be also reached when the NVAO results are considered.

An adequate linear relationship between the proton affinity and MEP is shown in Figure 12.2. The correlation coefficient ($R^2 = -0.9419$) and the linear equation are shown in the figure. Based on these findings, one expects that the two oxygen atoms should be the site to preferably bond with the incoming proton. These numerical data led us to suggest, in accordance with our studies on benzamide derivatives, that the analysis of MEP on the nuclei of the basic centers can be used as a good descriptor to estimate the intrinsic basicities of the heterocyclic compounds containing different type of the basic centers.

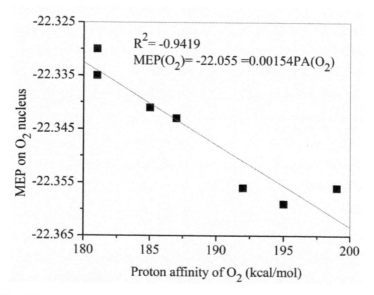

FIGURE 12.2 Linear relationships between the MEP on O2 nucleus (in a.u.) and the proton affinity of O2 (in kcal.mol) of the thiazolidine–2,4-dione derivatives.

12.3.3   ISORHODANINE

Table 12.7 displays the values of PAs, GB, and ΔH_{dep} of isorhodanine in gas phase and water (Scheme 5).

12.3.4   BENZAMIDE AND ITS DERIVATIVES

In this section, we introduce the results of the proton affinities and gas phase basicities of benzamide and its m- and p-(NH_2, $-CH_3$, OH, OCH_3, NO_2, CF_3, and Cl) derivatives in gas phase and in solution (Scheme 6). Table 12.8 summarizes the calculated PAs and GBs [67], together with the available experimental value for all investigated species [22, 68].

TABLE 12.7   Gas and Solvent Phase Proton Affinity, PA / kcal mol⁻¹, Gas-Phase Basicity, GB/kcal mol⁻¹, and the Enthalpy of Deprotonation, ΔH_{dep}/kcal mol⁻¹, of Isorhodanine Obtained at B3LYP/6–11++(3df,2p)// B3LYP/6–11+(d,p Level of Theory

	Gasphase			Water		
	PA	GB	DH$_{dep}$	PA	GB	DH$_{dep}$
S1	172.2	165.0		−43.3	−45.2	-
O6	190.1	184.6	-	−33.7	−32.5	-
N3	166.9	160.2	-	−53.3	−51.4	-
S7	194.4	187.9	-	−25.8	−22.7	-
N3-H8	-	-	324.2	-	-	554.8
C5-H2(3)	-	-	330.0	-	-	560.0

Results are taken from Ref. 29.

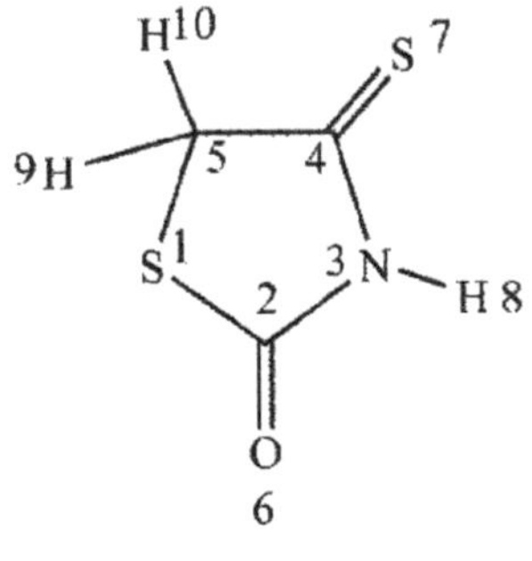

SCHEME 5   Isorhodanine.

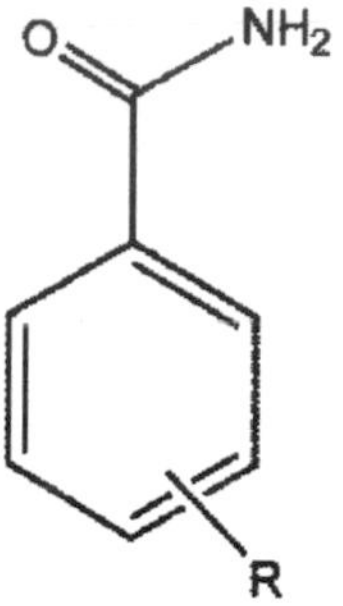

$$R1 = -H, -NH_2, -CH_3, -C_6H_5, -Cl, -F, -CF_3, -NO_2$$

SCHEME 6   Benzamide and its m- and p- derivatives.

A closer analysis of Table 12.8 shows interesting linear relationships between the calculated PAs and GBs of the investigated benzamides and the available experimental ones (Figure 12.3), i.e., the PA (ΔH) in Figure 12.3a and the GB (ΔG) in Figure 12.3b, with determination coefficients 0.9787 for PAs and 0.9782 for GBs, respectively. Therefore, linear correlations are derived between the calculated PA and GB and the corresponding experimental data as shown in Figure 12.3. These results indicate, in agreement with previous studies, that the B3LYP method may be considered as a useful method to calculate the proton affinities and gas-phase basicities of the polyfunctional compounds.

Another topic that should be considered here is the electron densities at the O-H$^+$ BCP, ρ_{BCP}, and their Laplacian, $\nabla^2{}_\rho$. These topological properties were evaluated at the B3LYP/6–311+G(d,p) level of theory by means of the AIM2000 approach. The values of ρ_{BCP} and $\nabla^2{}_\rho$ are included in Table 12.7. As can be seen, the positive values of ρ_{BCP} and the negative value of $\nabla^2{}_\rho$ suggest that the O-H$^+$ interaction have, in principle, a covalent character. Indeed, the ρ_{BCP}. Our results revealed an adequate relationship between the electron density at bcp of the O-H$^+$, which results due to the protonation of the basic center (O atom) by the proton (H$^+$) and the PA and/or GB (Figure 12.4). The correlation coefficient (R^2) is very close to 0.9. These results led us to conclude that the electron density at bcp can be used to estimate the PA and GB.

TABLE 12.8 Calculated Proton Affinities, PA and Gas-Phase Basicities, GB, in Gas Phase Calculated at B3LYP/6–311++G (2df,2p)//6–311++G(d,p) Level of Theory Together with the Available Experimental and Theoretical (ab initio and PM3 Semiempirical Level) of Benzamides and Its Derivatives

X	PA_{calc}	PA_{exp} [a]	GB_{calc}	GB_{exp} [a]	MEP(O)	MEP(N)	NPA(O)	NPA(N)	ρ_{bcp}	$\nabla^2\rho$	δ^1H
p-OCH$_3$	219.1	215.2	211.4	207.8	−22.409	−18.373	−1.634	−1.313	0.358	−2.56	7.1
m-OCH$_3$	214.9	215.3	207.2	207.9	−22.406	−18.37	−1.645	−1.323	0.357	−2.56	7.6
p-OH	216.7	-	209	-	−22.407	−18.371	−1.644	−1.322	0.358	−2.56	7.1
m-OH	213.3	-	205.5	-	−22.404	−18.368	−1.653	−1.331	0.356	−2.56	7.6
p-NH$_2$	223.6 214.2[a]	221.6	216.1	214.4	−22.414	−18.378	−1.617	−1.295	0.358	−2.55	6.7
m-NH$_2$	216.5	215.3	208.9	207.9	−22.405	−18.371	−1.64	−1.32	0.356	−2.54	7.5
p-CH$_3$	217	215.3	209.4	207.9	−22.405	−18.378	−1.643	−1.322	0.356	−2.54	7.4
m-CH$_3$	214.5	215.3	206	207.9	−22.404	−18.37	−1.647	−1.326	0.356	−2.54	7.5
H	208.4[a] 212.6 224.5[b]	213.2	204.9	205.8	−22.403	−18.367	−1.656	−1.334	0.355	−2.54	7.6
p-NO$_2$	202.1 197.2[b]	201.2	194.6	196.7	−22.381	−18.347	−1.732	−1.411	0.354	−2.54	8
m-NO$_2$	203.1	204.3	195.3	196.7	−22.384	−18.35	−1.723	−1.4	0.354	−2.54	7.9
p-Cl	210.7	209.7	203	202.3	−22.394	−18.362	−1.681	−1.358	0.355	−2.54	7.5
m-Cl	208.9	209.7	201.1	202.3	−22.393	−18.361	−1.685	−1.363	0.356	−2.54	7.7
p-CF$_3$	205.9	206.2	198.1	198.8	−22.39	−18.354	−1.706	−1.385	0.354	−2.54	7.8
m-CF$_3$	206.4	206.7	198.9	199.8	−22.391	−18.355	−1.702	−1.38	0.355	−2.55	7.8

[a]Values taken from Refs. [68, 69]. [b] Values taken from Ref. [69]
Results are taken from Ref. 22.

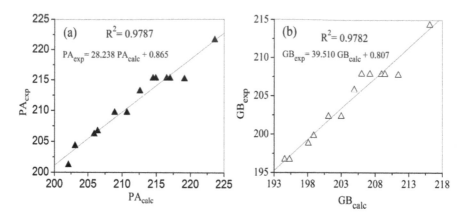

FIGURE 12.3   Linear relationships for (a) calculated proton affinity and the corresponding experimental data and (b) calculated gas-phase basicity and the corresponding experimental data.

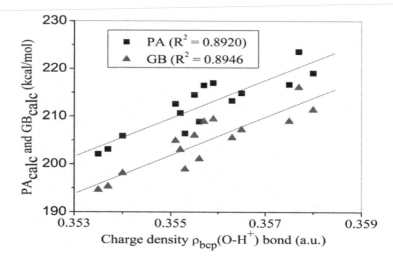

FIGURE 12.4   Linear correlation between the electron density at the bond critical point of the O-H$^+$ bond (in a.u.) and the proton affinity and gas phase basicity (in kcal/mol) of m- and p-substituted benzamides.

A good linear relationship between the ^{1}H NMR chemical shift (δ^1H) in (ppm) and the PA and GB of the benzamides is shown in Figure 12.5. Good linear relationships are obtained with correlation coefficients of 0.9173 *(PA vs. δ^1H)* and 0.9201 *(GB vs. δ^1H)*. In agreement with recent

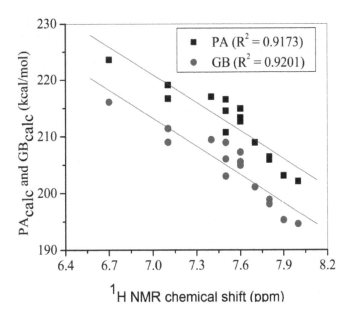

FIGURE 12.5 Linear correlation between the ^1H NMR of the incoming proton δ^1H (in ppm) and the proton affinity and gas phase basicity (in kcal/mol) of m- and p-substituted benzamides.

studies [62, 68, 70], these strong correlations permit us to conclude, among the abovementioned parameters, that the PA and GB of the species under probe can be estimated by other quantum descriptors such as MEP, NAVO, δ^1H, and ρ_{bcp}.

Table 12.9 presents the protonation energy, $\Delta_{prot}E$, proton affinities, PA, Gibbs free energy of deprotonation, $\Delta_{prot}G$, and the base equilibrium constant, pK_b of *para*- and *meta*-substituted benzamides calculated at the B3LYP/6–311++G(2df,2p)//6–311++G(d,p) level of theory in solution (water), together with the available experimental pK_b. A closer look at Table 12.8 indicates that quantitative agreement between theory and experiment is poor for water. These pK_b values were calculated by using different techniques (see Ref. [71] and the references. therein). Plot of pK_b (calc.) vs. pKb (exp) exhibits qualitative agreement, with a correlation coefficient of 0.9728 (see Figure 12.6). If one assumes linear behavior, the fitted equation should be given by the expression given in Figure 12.6. However, these

TABLE 12.9 $\Delta_{prot}E$, PA, $\Delta_{prot}G$ and pK(B)) of Para- and Meta-Substituted Benzamides Calculated at B3LYP/6–311++G(2df,2p)//6–311++G(d,p) Level of Theory in Solution (Water), Together with the Available Experimental pK$_b$ Values

X	PA	$\Delta_{prot}G$	pK$_b$	pK$_b$(experimental)		
				*	*	**
p-OCH$_3$	−15.1	−13.7	−10	-	-	-
m-OCH$_3$	−17.6	−16.2	−11.8	-	-	-
p-OH	−15.7	−14.7	−10.7	-	-	-
m-OH	−17.8	−16.4	−12	-	-	-
p-NH$_2$	−12.4	−10.8	−7.8	-	-	-
m-NH$_2$	−16.6	−15.4	−11.2	-	-	-
p-CH$_3$	−16.4	−15.4	−11.2	1.46	1.44	1.67[a]
m-CH$_3$	−17.3	−16.7	−12.2	1.33	1.37	1.76 [a]
H	−17.3	−15.9	−11.6	1.4	1.43	1.54[e,] 1.45[e], 1.54f, 1.38[c], 1.65[d], 1.74[a], 1.45[b],
p-NO$_2$	−21.5	−20.5	−14.9	2.36	2.28	2.70[a], 2.13[d]
m-NO$_2$	−21.1	−20.1	−14.6	2.04	2.01	2.42[a]
p-Cl	−18.3	−16.9	−12.3	1.6	1.66	1.97[a]
m-Cl	−19.3	−18.3	−13.3	1.63	1.65	2.09[a]
p-CF$_3$	−20	−18.9	−13.8	-	-	-
m- CF$_3$	−19.8	−19.1	−13.9	-	-	-

(*) Values are taken from Ref.[72]; (**) values are taken from Refs ((a):[73], (b):[74], (c):[75], (d):[76], (e):[77] , and (f): [78].

All values are in kcal/mol.

Results are taken from Ref. 22.

values of the correlation coefficients R are not representatives of a true linear relation (for which R ≥ 0.99), but they are indicative of a quasi linear correlation. According to our results, although the quantitative comparison with experimental methods is poor, one can conclude that the computation of the proton free enthalpy in solution with high accuracy is fundamental in solvation study if reliable experimental data are not available (Table 12.9).

12.4   CONCLUSION

In this chapter, we have introduced the calculation of the proton affinities (PA), gas phase basicities (GB), deprotonation energies (ΔH_{dep}), and

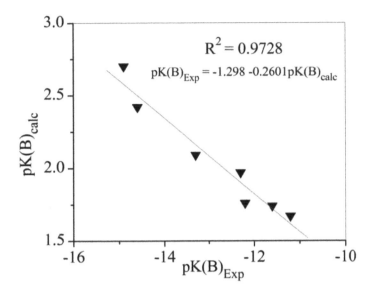

FIGURE 12.6   Linear correlation between the experimental PK_b (experimental) and theoretical PK_b values. (Reprinted with permission from Safi, Z., & Omar, S. Proton affinity and molecular basicity of m- and p-substituted benzamides in gas phase and in solution: A theoretical study. Chemical Physics Letters, 601–602, pp. 321–330. © 2014 Elsevier.)

some quantum descriptors such as molecular electrostatic potential on the nuclei, natural valence orbital energies, 1H NMR chemical shifts, and electron density at bond critical points of some polyfunctional compounds in the framework of density functional theory (DFT) using the famous B3LYP at different basis sets. The obtained results were compared with the available experimental data. A good agreement between the calculated and the experimental data was obtained. Finally, our findings led us to conclude that the DFT method is as effective as ab initio to calculate the PA, GB, and ΔH_{dep}.

ACKNOWLEDGMENTS

A generous allocation of computational time at the Scientific Computational Center (CCC) of the Universidad Autónoma de Madrid (Spain) acknowledged the University of Marburg (Germany).

KEYWORDS

- B3LYP
- benzamide
- density functional theory
- gas-phase basicity
- hydantoin
- isorhodanine
- proton affinity

REFERENCES

1. Enami, S., H. M., Hoffmann, M. R., & Colussi, A. J., (2012). Protonation and oligo-merization of gaseous isoprene on mildly acidic surfaces: Implications for atmospheric chemistry. *J. Phys. Chem. A., 116*(24), pp. 6027–6032.
2. Kennedy, R. A., C. A. M., Thomas, R., Watts, P., & Inter. P., (2003). *J. Mass Spectrom., 223*, pp. 627.
3. Abboud, J. L. M., O. M., de Paz, J. L. G., Yanez, M., Esseffar, M., Bouab, W., & Mouhtadi, El-M., Mokhlisse, R., Ballesteros, E., Herreros, M., Homan, H., Lopez-Mardomingo, C., Notario, R., (1993). Thiocarbonyl versus carbonyl compounds: A comparison of intrinsic reactivities. *J. Am. Chem. Soc., 115,* pp. 12468–12476.
4. Van Beelen, S. E., et al., (2004). *J. Phys. Chem. A,. 108*, pp. 2787.
5. Berthelot, M., M. D., Gal, C. J. F., Laurence, C., Le Questel, J. Y., Maria, P. C., & Tortajada, J., (1991). Gas-phase basicity and site of protonation of polyfunctional molecules of biological interest: FT-ICR experiments and AM1 calculations on nicotines, nicotinic acid derivatives, and related compounds. *J. Org. Chem., 56*, pp. 4490–4494.
6. Zulejko, J. E., & McMahon, T. B., (1993). Progress toward an absolute gas-phase proton affinity scale. *J. Am. Chem. Soc., 115*, pp. 7839–7848.
7. Wieczorek, R., & Dannenberg, J. J., (2004). *J. Am. Chem. Soc., 126.* pp. 12278.
8. Lee, I. C., & Masel, R. I., (2002). In *J. Phys. Chem. B.*, pp. 3902–3908.
9. Lee, I. C., & R. I. M., (2002). In *Catal. Lett.*, pp. 43.
10. Zheng, A., et al., (2007). *J. Phys. Chem. B.*, pp. 3085.
11. Pankratov, A. N., (2005). *Chem. Heterocycl. Comp.*, pp. 340.
12. Versees, W., et al., (2004). *J. Mol. Biol.*, pp. 1.
13. Lamsabhi, M., et al., (2000). *J. Phys. Chem. A.,106*, pp. 32.
14. Dang, P., & Madan, A. K., (1994). Structure-activity study on anticonvulsant (thio) hydantoins using molecular connectivity indices. *J. Phys. Chem. A., 34*(5), pp. 1162–1166.

15. Locock, R. A., & Coutts, R. T., (1970). The mass spectra of succinimides, hydanto-ins, oxazolidinediones and other medicinal anti-epileptic agents. *Org. Mass Spectrom, 3,* pp. 735–745.

16. Mo, O., De Paz, J. L. G., & Yanez, M., (1987). Protonation of three-membered ring heterocycles: an ab initio molecular orbital study. *J. Phys. Chem. A., 91,* pp. 6484–6490.

17. Hunter, E. P., & Lias, S. G., (1998). Evaluated Gas Phase Basicities and Proton Affinities of Molecules: An Update. *J. Phys. Chem. Ref. Data, 27*(3), pp. 413–656.

18. Hehre, W. J., et al., (1986). *Ab Initio Molecular Orbital Theory,* New York, John Willey & Sons.

19. Hehre, W. J., L. R., Schleyer, P. V. R., (1986). Ople, J. A., *Ab Initio Molecular Orbital Theory,* New York, John Willey & Sons.

20. Chandra, A. K., & Goursot, A., (1996). Calculation of proton affinities using density functional procedures: a critical study. *J. Phys. Chem., 100*(28), pp. 11599–11599.

21. Remko, M., Licdl, K. R., & Rode, B. M., (1996). Structure and gas-phase acidity of oxalic acid and its disila derivative. A theoretical study by means of the DFT quantum theoretical method. *J. Chem. Soc. Perkin Trans. 2, 50*(1), pp. 1743–1748.

22. Safi, Z., & Omar, S., (2014). Proton affinity and molecular basicity of m- and p-substituted benzamides in gas phase and in solution: A theoretical study. *Chemical Physics Letters, 601–602,* pp. 321–330.

23. Safi, Z. S., (2012). Theoretical density functional study of gas-phase tautomerization and acidity of 5-methylhydantoin and its thioderivatives. *European J. Chem., 3*(3), pp. 348–355

24. Safi, Z. S., & Frenking, G., (2013). *Protonation of 5-Methylhydantoin and Its Thio Derivatives in the Gas Phase: A Theoretical Study. Int. J. Quantum Chem.,* pp. 908.

25. Remko, M., & Rode, B. M., (1999). How Acidic Are the Selenocarboxylic Acids RCSeOH and RCOSeH (R = H, F, Cl, NH2, CH3)? *J. Phys. Chem. A., 103*(3), pp. 431–435.

26. Remko, M., K. R. L., Rode, B. M., (1997). *J. Mol. Struct. (THEOCHEM),* pp. 179.

27. Nguyen, T., A. K. C., Zeegers-Huyskens, T., (1998). Protonation and deprotonation energies of uracil Implications for the uracil–water complex. *J. Chem. Soc., Faraday Trans., 94*(9), pp. 1277–1280.

28. Yuang, H., L. L., Shubin, L., (2013). Towards understanding proton affinity and gas-phase basicity with density functional reactivity theory. *Chem. Phys. Lett.,* pp. 73–78.

29. Safi, Z. S., (2016). Tautomeric Study of Neutral, Protonated and Deprotonated Isorhodanine Based On High Level Density Functional Theory. *Oriental Journal of Chemistry, 2.*

30. Ebrahimi, A., Khorasani, S. M. H., & Jahantab, M., (2011). Additivity of substituent effects on the proton affinity and gas-phase basicity of pyridines. *Computational and Theoretical Chemistry, 966,* pp. 31–37.

31. Delian, Yi., Hailu, Z., & Zongwu, D., (2010). *J. Mol. Catal. A, Chem., 326,* pp. 88.

32. Bader, R. F. W., (1990). *Atoms in Molecules-A Quantum Theory.,* Oxford, UK, Oxford University Press.

33. Kukol, A., et al., (1993). Methyl group effect on the proton affinity of methylated acetophenones studied by two mass spectrometric techniques. *Org. Mass Spectrom*, pp. 1107–1110.

34. Böhm, S., J. F. G., Maria, P. C., Kulhánek, J., & Exner, J., (2005). Steric Effects in Isolated Molecules: Gas-Phase Basicity of Methyl-Substituted Acetophenones. *In Eur. J. Org. Chem.*, pp. 2580–2588.

35. Deakyne, C. A., (2003). Proton affinities and gas-phase basicities: theoretical methods and structural effects. *Int. J. Mass Spectrom.*, pp. 601–616.

36. Becke, A. D., (1988). Density-functional exchange-energy approximation with correct asymptotic behavior. *Phys. Rev. A.*, pp. 3098–3100.

37. Lee, C., W. Y., & Parr, R. G., (1988). *In. Phys. Rev. B.*, pp. 785.

38. Krishnan, R., J. S. B., Seeger, R., & Pople, J. A., (1980). *In. J. Chem. Phys.*, pp. 650.

39. Clark, T., J. C., Spitznagel, G. W., & Schleyer, P. V. R., (1983). *In J. Comput. Chem.*, pp. 294.

40. Scott, A. P., & L. R., (1996). *J. Phys. Chem.*, pp. 16502.

41. Reed, A. E., R. B.W., & Weinhold, F., (1985). *J. Phys. Chem*, pp. 735.

42. Suresh, C. H., (2006). Molecular electrostatic potential approach to determining the steric effect of phosphine ligands in organometallic chemistry. *Inorganic Chemistry. 45*(13), pp. 4982–4986.

43. Correa, J. V., P. J., Olah, J., Toro-Labbe, A., & Geerlings, P., (2009). *In. Chem. Phys. Lett.*, pp. 180.

44. Wolinski, K. J. F. H., & Pulay, P., (1990). *J. Am. Chem. Soc.*, pp. 8251.

45. Biegler-König, F., (2000). *A. I.M2000*, Bielefeld, Germany, University of Applied Sciences.

46. Frisch, M. J., et al., (2009). In *GAUSSIAN 09, Revision B.01, Gaussian, Inc.*, Wallingford, CT.

47. Martin, J. M. L., J. P. F., & Gijbels, R., (1989). *In:J. Comput. Chem.*, pp. 346.

48. Hwang, S., Jang, Y. H., & Chung, D. S., (2005). In *Bull. Korean Chem. Soc.*, pp. 585.

49. Topol, I. A., G. J. T., Burt, S. K., & Rashin, A. A., (1997). *In J. Phys. Chem. A.*, pp. 10075.

50. Mejías, J. A., & S. L., (2000). *J. Chem. Phys.*, pp. 7306.

51. Tissandier, M. D., K. A. C., Feng, W. Y., Gundlach, E., Cohen, M. J., Earhart, A. D., & Coe, J. V., (1998). *J. Phys. Chem. A.*, pp. 7787–7794.

52. Camaioni, D. M., & C. A. S., (2005). *J. Phys. Chem. A.*, pp. 10795.

53. Liptak, M. D., & G. C. S., (2001). *Int: J. Quantum Chem.*, pp. 727.

54. Silva, A. L. R., Cimas, Á., Vale, N., Gomes, P., Monte, M. J. S., & Ribeiro da Silva, M. D. M. C., (2013). Experimental and computational study of the energetics of hydantoin and 2-thiohydantoin. *J. Chem. Thermodynamics.*, pp. 158–165.

55. Bouchoux, G., (2015). Gas-phase basicities of polyfunctional molecules. Part 4: carbonyl groups as basic sites., *Mass Spectrometry Reviews.*, pp. 493–534.

56. Bouchoux, G., (2015). *In:Mass Spectrometry Reviews*, pp. 493–534.

57. Lamsabhi, M., M. E., Bouab, W., El Messaoudi, M., El Messaoudi, Abboud, J. L. M., Alcamı, M., & Yanez, M., (2000). *J. Phys. Chem. A.*, pp. 32.

58. Lamsabhi, M., M. E., Bouab, W., El Messaoudi, T., El Messaoudi, M., Abboud, J. L. M., Alcamı, M., & Yanez, M., (2002). *J. Phys. Chem. A.*, pp. 32.

59. Bouchoux, G., D. L., Bertrand, W., McMahon, T. B., Szulejko, J. E., Penaud, F. B., Mo, O., & Yáñez, M., (2005). *J. Phys. Chem. A.*, pp. 11851.
60. Bagno, A., & G. S., (1996). *In J. Phys. Chem.*, pp. 1536.
61. Hunter, E. P., & S. G. L., (1998). *J. Phys. Chem. Ref. Data*, pp. 413.
62. http://webbook.nist.gov/chemistry/.
63. Avendaño, C., & J. C. M., (2000). Hydantoin and Its Derivatives, in Kirk-Othmer Encyclopedia of Chemical Technology, *Kirk-Othmer Encyclopedia of Chemical Technology*, pp. 512–533.
64. Bene, J. E. D., (1993). *J. Phys. Chem.*, pp. 107.
65. Puszynska-Tuszkanow, M., M. D., Maciejewska, G., Adach, A., & Cieslak-Golonka, M., (2010). *Struct.Chem.*, pp. 315–332.
66. Bellanato, J., J. C. A., Ballesteros, P., & Martınez, M., (1979). *Spectrochim. Acta.*, pp. 807.
67. Safi, Z. S., (2016). A theoretical study on the structure of thiazolidine-2,4-dione and its 5-substituted derivatives in the gas phase. Implications for the thiazolidine-2,4-dione-water complex, *Arabian Journal of Chemistry*, pp. 616–625.
68. Grutzmacher, H. F., & A. C., (1994). *J. Am. Soc. Mass Spectrom*, pp. 826.
69. Grutzmacher, H. F., & A. C., (1998). *Eur. J. Mass Spectrom*, pp. 349.
70. Ebrahimi, A., S. M. H. K., & Jahant, M., (2011). *Comput. Theor. Chem.*, pp. 2995–3002.
71. Zheng, A., H. Z., Chen, L., Yue, Y., Yue, C., & Deng, F., (2007). *J. Phys. Chem. B.* 2007, pp. 3085.
72. Safi, Z. S., & S. O., (2014). *Chemical Physics Letters*, pp. 321–330.
73. Stojkovi, G., E. P., (2006). *J. Serb, Chem. Soc.*, pp. 1061.
74. Yates, K., & J. B. S., (1965). *In:Can. J. Chem.*, pp. 529.
75. Garcia, B., R. M. C., Castillo, J., Ibeas, S., Domingo, I., & Leal, J. M., (1993). *J. Phys. Org. Chem.*, pp. 101.
76. Haldna, U., (1990). *Prog. Phys. Org. Chem.*, pp. 65.
77. Edward, J. T., & S. C. W., (1977). *J. Am. Chem. Soc.*, pp. 4229.

CHAPTER 13

TAUTOMERISM AND DENSITY FUNCTIONAL THEORY

ZAKI S. SAFI

Department of Chemistry, Al Azhar University-Gaza, Gaza City, Palestine

CONTENTS

ABSTRACT

This chapter is devoted to investigate the tautomerization process using the density functional theory (DFT) in the neutral, protonated, and deprotonated forms. In addition, the effect of the interaction with the transition metal cations on the tautomerization process is also considered.

13.1   INTRODUCTION

To the best of our knowledge, according to Baker [1] in the first tautomeric book, the priority in defining tautomerism was given by Berzelius, who in 1832 used the term "metamerism" to explain reciprocal conversion of cyanic and cyanuric acid. In 2005, Raczynska et al. [2] pointed that the term "tautomerism" refers to a compound existing in equilibrium between two or more isomeric forms called tautomers. Along this line, it is known that if a molecule contains a proton donor (N-H or C-H group) and a proton acceptor (O, N, S, and/or Se atoms) in close proximity, an intramolecular 1,3-proton transfer (IPT) process (prototropic tautomerisation) takes place, which is of great interest to medicinal and biochemical applications. On similar lines, Lapworth and Hann [3] defined the prototropic tautomerism (PT) process as "the addition of a proton at one molecular site and its removal from another." Hence, IPT is clearly distinguished from ionization; it is one of the most important phenomena in organic chemistry despite the relatively small proportion of molecules in which it can occur. IPT is also defined as the movement of H^+ from a bonded position in one molecule to another position in the same molecule [4]. In the detailed description of proton transfer reactions, especially of rapid proton transfers between electronegative atoms, it should always be specified whether the term is used to refer to the overall process (including the more-or-less encounter-controlled formation of a hydrogen bonded complex and the separation of the products; see microscopic diffusion control) or just to the proton transfer event (including solvent rearrangement) by itself.

Enantiomers, or cis and trans isomers, possess a formulaic identity just as tautomers do but are difficult to interconvert and hence easy to isolate. Tautomers are different. Tautomers are the chameleons of chemistry, capable of changing by a simple change of phase from an apparently established structure to another (not perhaps until suspected) and then back again when the original conditions are restored, and of doing this in an instant: intriguing, disconcerting, and perhaps at times exasperating. A change in structure also means corresponding changes in properties.

Understanding of the relative stabilities of heterocyclic tautomers and any subsequent conversions between tautomeric forms is very vital for both structural chemists and biologists [5, 6]. Importantly, relative stabilities of various tautomeric structures of five-, six- and seven-membered rings (dioxo, oxo/thio, oxo/seleno, dithio, and diseleno combinations) were investigated using both theoretical and experimental tools in both aqueous and gas phases [7–18]. The last two decades have witnessed an extensive use of the computational methods to investigate the tautomeric equilibria in hetero-cyclic systems, which represent a central component of synthetic chemistry and biochemical processes. One of these methods is the density functional theory (DFT), which is considered as a powerful method to investigate this phenomenon.

Our main objective in the present chapter is to introduce some of the recent studies that used the DFT method to investigate the tautomerization of some heterocyclic compounds. In the first section, the tautomerization process of the neutral species is introduced. The second section investi-gates the tautomerization process of the protonated and deprotonated spe-cies. Finally, the effect of the interaction of the heterocyclic compounds with transition metal cations (Cu^+ and Zn^{2+}) on the tautomerization process is presented in the last section.

13.2 COMPUTATIONAL DETAILS

The geometries of the different species under consideration were fully optimized using the hybrid density functional B3LYP [19–22]. These approaches have been shown to yield reliable geometries for a wide variety of systems. All calculations were performed using the 6-31G(d), 6-31+G(d), 6-311+G(d,p), and 6-311++G(d,p) basis sets with the Gauss-ian 03 [23] and Gaussian 09 [24] series of programs. The harmonic vibra-tional frequencies of the different stationary points of the potential energy surface (PES) were calculated at the same level of theory used for their optimization in order to identify the local minima and the transition states (TS) as well as to estimate the corresponding zero point energy (ZPE) cor-rections.

In order to obtain more reliable energies for the local minima, final energies were evaluated by using the same functional combined with the 6-311+G(2df,2p) basis set for all atoms except for Cu^+ and Zn^{2+}, where the (14s9p5d/9s5p3d) basis set of Wachters [25] and Hay [26] was used, sup-plemented with a set of (1s2p1d) diffuse functions and with two sets of f functions and one set of g functions. It has been shown that this approach

is well suited for the study of this kind of systems, yielding binding energies in good agreement with the experimental values [27–29].

The corresponding Zn^{2+} and Cu^+ binding energies were evaluated
by subtracting, from the energy of the complex, the energy of the neutral and that of Zn^{2+} and Cu^+, after including the corresponding scaled
ZPE corrections by a factor of 0.9806 [30]. The basis set superposition
error (BSSE) has not been considered in the present study because, as
has been previously reported, this error is usually small for DFT and
DFT/HF hybrid methods, when the basis set expansion is sufficiently
flexible [31].

13.3   RESULTS AND DISCUSSION

13.3.1   TAUTOMERIZATION OF NEUTRAL HYDANTOIN

Safi and Fares [32] studied the relative stabilities of the different tautomers and the IPT process of hydantoin in the gas phase and in solution (see Scheme 1). Several hydantoin tautomers and their rotamors
and the possible transition states, which connect between the global
minimum and the most stable enol structures, were investigated. DFT
results (Table 13.1) showed that the most stable tautomers corresponds
to the diketo form (**T1**), followed by oxo-enol forms (**T2-T4**), while the
di-enol forms (**not shown**) are the least stable ones. The energy differences between the global minimum and the keto-enol forms ranged from
15.6 to 24.8 kcal/mol. The dienol forms were found to be less stable
by at least ~33 kJ/mol than the global minimum. The same conclusions
were also reached when the IPT processes are carried out in solution
(THF, methanol, DMSO, and water). These results are in agreement with
the previous semi-empirical methods (AM1 and PM3) [12, 33]. Results
revealed that the relative energy trend in the gas phase has the following order: Gas phase: T1 > T2 > T4 > T3. In solution, the trend is as
follows: T1 > T4 > T3 > T2. This change in the stability trend of the
oxo-enol forms may be referred to the polarity of solvents. Importantly,
in the gas phase, the results revealed that the tautomers T1 and T4 are
almost degenerate with only 0.2 difference in their relative energies in
favor of T2, which reflects the possible competitiveness in the intrinsic

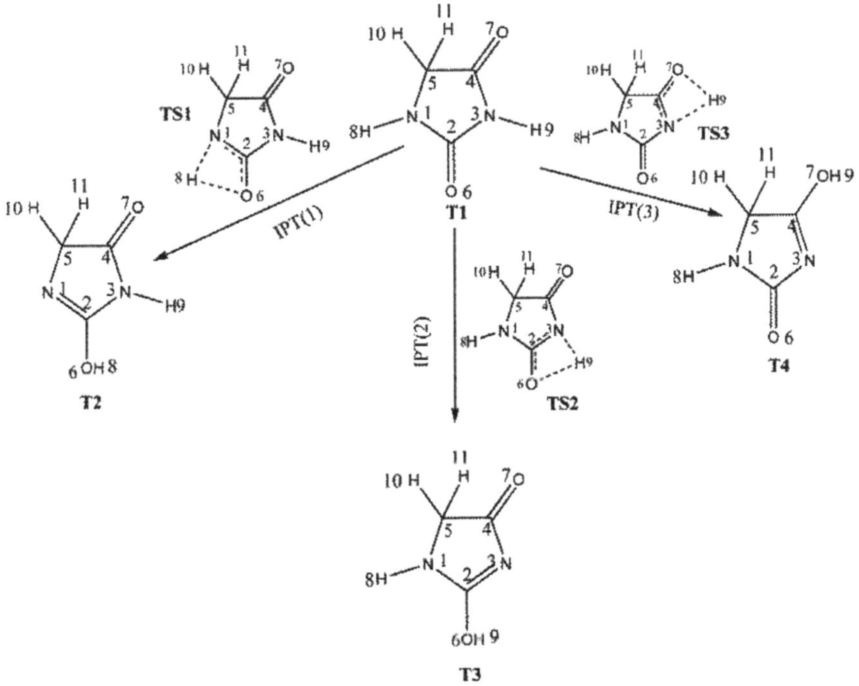

SCHEME 1   Schematic representation of hydantoin tautomers.

TABLE 13.1   Relative Energies (in kcal/mol) of the Different Tautomers of Hydantoin in Gas Phase and in Solution, Together With the Available Semiempircal Results [33, 34]

Species	Gas	THF	Methanol	DMSO	H₂O	AM1	PM3
T1	0.0	0.0	0.0	0.0	0.0	0.0	0.0
T2	17.0	18.5	18.8	18.9	21.8	15.6	10.6
T3	18.9	17.2	16.9	16.0	16.8	20.4	19.3
T4	17.2	16.0	15.7	15.6	15.6	20.2	-
TS1	52.0	48.9	48.9	48.9	48.9	-	-
TS2	54.3	47.3	51.3	51.3	54.2	-	-

basicities of the two carbonyl groups (C2=O6 and C4=O7). The results suggested that the C2=O6 group is slightly more basic than the C4=O7 group, which can be referred to the presence of two amino groups surrounded the C2=O6 group.

The competitiveness in the intrinsic basicity of the two carbonyl groups was confirmed by locating the transition states, which connect T1 with T2 and T4. Results showed that the relative energies of the two transition states are almost degenerate with only 2 kcal/mol difference in favor of TS1. The relative energies of the transition states were found to be very high, which indicated, in all cases that the T1 is the predominant tautomer and the IPT is thermodynamically impossible under the normal conditions.

13.3.2 TAUTOMERIZATION OF NEUTRAL 5-METHYLHYDANTION AND ITS THIO DERIVATIVES

In this subsection, the studies of tautomerization of neutral [34], protonated [10], and deprotonated [15] 5-methylhydantoin and its thio derivatives are presented. The main objectives in these studies was to predict which is more basic, i.e., the carbonyl group or the thiocarbonyl one, and to study the effect of protonation and deprotonation processes on the IPT process. Safi and Abu Awaad used hybrid density functional at the B3LYP level of theory using 6-31(d) and 6-311+(2df,2p) basis sets with full geometry optimization to study the IPT process of four compounds of 5-methylhydantoin and its thio derivatives (Scheme 2). They computed relative tautomerization energies, enthalpies, entropies, Gibbs free energies, and dipole moments for the compounds under probe in the gas phase. In the addition to the parent compound, four tautomers (**T1-T4**) and their rotamers were studied for each combination (see Scheme 2). The transition states that connect between the global minima and the most stable enolic/thiolic structures were also located and computed. The most important results are summarized in Table 13.2.

In agreement with hydantoin, the most stable tautomers correspond to diketo and the dithio forms. In summary, it was concluded that among all the investigated species, diketo/dithione forms are predominant in the gas phase and are thermodynamically more stable than enol/thiol forms, with a high energy barrier. On the other hand, contrary to previous findings for uracil, thiouracil, and oxo/thiotriazepines, the tautomerization process was always favored at the hetero atom (sulfur or oxygen) attached to position 4 within the five-membered ring, though a competitiveness is clearly observed

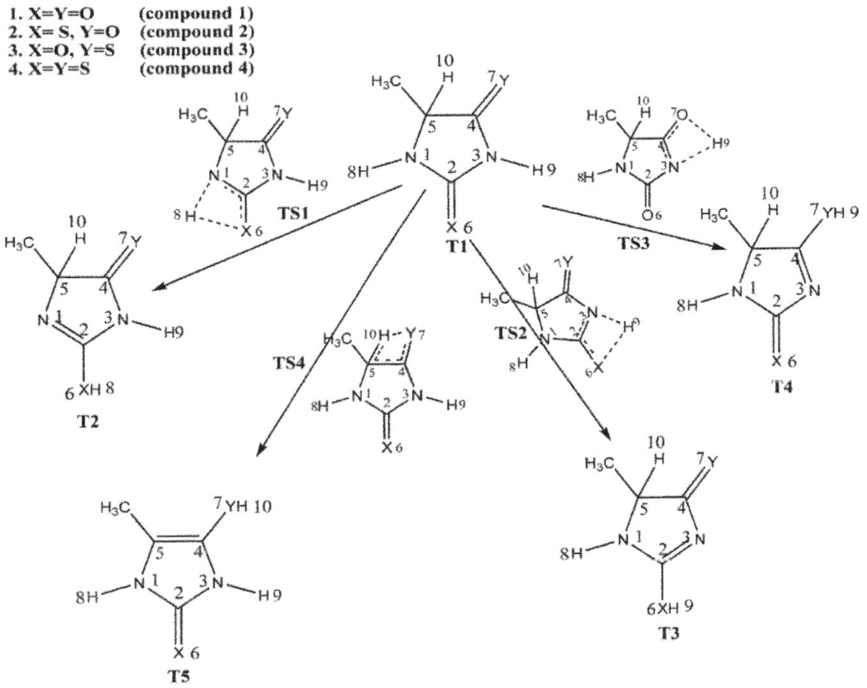

SCHEME 2 Schematic representation of 5-methylhydantion and its thio derivatives.

when compound (1) is the system of concern. Hence, the two oxygen atoms of the carbonyl groups attached to positions 2 and 4 have the same possibility to undergo the tautomerization process. Moreover, comparison of the relative energies of the transition states in Tables 13.1 and 13.2 indicated that the trend of relative stability of the tautomers is impacted by the electron donating group, CH_3, attached to position five in the hydantoin ring. For example, TS1 in the case of unsubstituted hydantoin was found to be 1.3 kcal/mol more stable than that in the case of 5-methylhydantoin.

13.3.3 TAUTOMERIZATION OF NEUTRAL ISORHODANINE AND ITS THIO DERIVATIVES

Very recently, high level DFT using B3LYP/6-311++G(2df,2p)//6-311+G(d,p) calculations were performed to compute the relative stabili-

TABLE 13.2 Relative Energies (in kcal/mol) for the Different Hydantoin and Its Thio Analogous in Gas Phase

Species	Compound (1)	Compound (2)	Compound (3)	Compound (4)
T1	0.0	0.0	0.0	0.0
T2	17.4	18.9	14.3	13.5
T3	19.0	16.8	19.7	17.1
T4	17.4	13.8	17.3	15.3
T5	21.5	17.5	6.0	2.6
TS1 (T1-T2)	53.5	55.3	42.6	43.2
TS2 (T1-T3)	51.4	40.2	52.1	40.2
TS3 (T1-T4)	54.3	41.0	55.2	42.0
TS4 (T1-T5)	78.0	75.6	44.7	44.6

ties and structures of the neutral isorhodanine (thiazolidin-2-one-4-thione) tautomeric forms in the gas phase and in solution [35]. For that, several tautomeric and rotameric forms of the species under investigation were investigated (see Scheme 3).

The relative energies, ΔE, single point energies, zero point energies, and thermal corrections to energy of all the species under probe in both gas phase and water are shown in Table 13.3. All structures that are listed are energy minima. The results revealed that the oxo/thione form (R1) is the most stable one among all the investigated species in both the gas phase and the solution. The calculations predict that the oxo/thione form (R1) is ~ 5.2 kcal/mol more stable than the oxo/thiol form R3. These results are in agreement with the previous results [13, 36–39]. Moreover, the results revealed

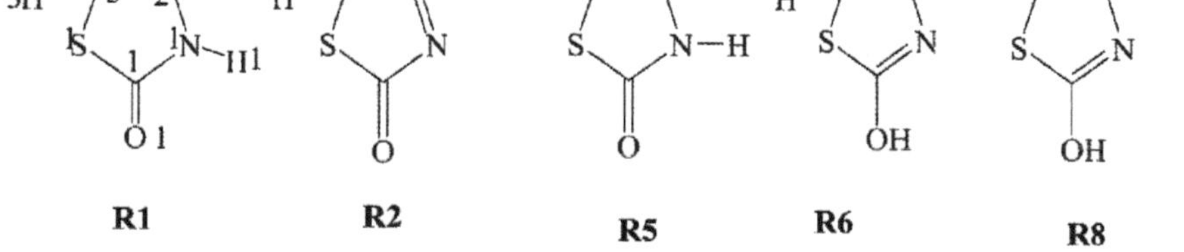

SCHEME 3   Tautomeric forms of isorhodanine.

TABLE 13.3 Gas and Solvent Phase Single Point Energy (E In Hartree), Zero Point Energy (ZPE in Hartree), Thermal Correction to Energy (TCE) in Hartree, and Relative Energies, (ΔE In Kcal Mol^{-1}) of Neutral, Protonated and Deprotonated Isorhodanine

Species	Gas				Water			
	E^a	ZPE^b	TCE^b	ΔE^c	E^a	ZPE^b	TCE^b	ΔE^c
R1	−1042.6537	0.0625	0.0688	0.0	−1042.6640	0.0624	0.0687	0.0
R2	−1042.6335	0.0586	0.0653	8.1	−1042.6459	0.0586	0.0652	6.8
R3	−1042.6385	0.0587	0.0656	5.2	−1042.6487	0.0586	0.0655	5.3
R4	−1042.6241	0.0619	0.0683	17.9	−1042.6359	0.0617	0.0680	16.8
R5	−1042.6223	0.0583	0.0653	15.0	−1042.6290	0.0581	0.0651	17.1

Values taken from Ref. 36.

that the neutral oxo/thiol forms (R2 and R3) are more stable than the enol/thio form (R4). Further, the enol/thio or oxo/thiol forms (R2-R4) were found to be more stable the enol/thiol form (R5). These results indicate that IPT (1,3-H transfer) is preferred at the sulfur atom (C2=S7) than at the oxygen one (C2=O6). These results can be explained in terms of the higher polarizability and consequently the size of the sulfur atom than of the oxygen atom. Moreover, from the results shown in Table 13.1, the stability trend of the different tautomers of isorhodanine is R1 > R3 > R2 > R4 > R5.

13.3.4 TAUTOMERIZATION OF NEUTRAL THIAZOLIDINE-2,4-DIONE AND ITS 5-SUBSTITUTED DERIVATIVES IN THE GAS PHASE

In this subsection, the hybrid density functional theory (B3LYP) was used to investigate five tautomeric structures of thiazolidine-2,4-dione and their 5-substituted derivatives (-CH$_3$, -NH$_2$, -Cl, -F, -CN-, and -NO$_2$) (Scheme 4) [15]. The relative energies of the species under consideration were calculated at the B3LYP6–311++G(3df,2p)//B3LYP/311+G(d,p) level of theory in the gas phase. Tautomeric forms of thiazolidine-2,4-dione are given in Scheme 3. Some important results are shown in Table 13.4.

Calculated thermodynamic parameters, such as relative energies, ΔE, enthalpies, ΔH, and Gibbs free energies, ΔG, of the various tautomers of thiazolidine-2,4-dione and its derivatives are presented in Table 13.4. ΔE, ΔH, and ΔH of each tautomer are defined as the difference between its

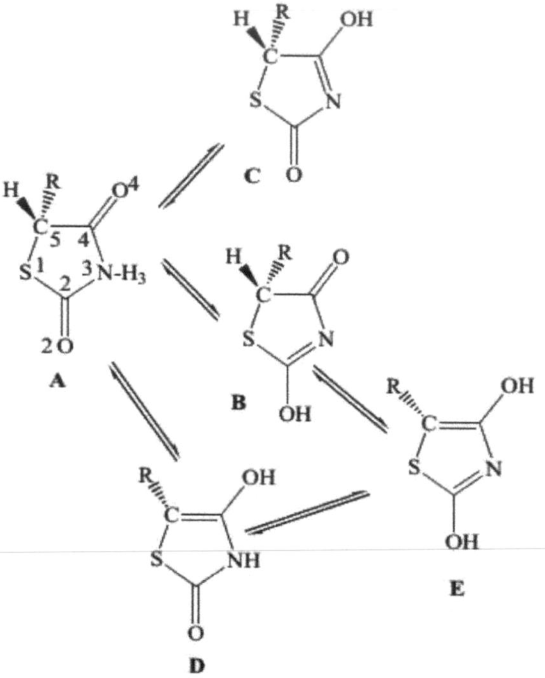

SCHEME 4   Tautomeric forms of thiazolidine-2,4-dione. (Reprinted with permission from the author from Safi, Z. S. (2016). A theoretical study on the structure of thiazolidine-2, 4-dione and its 5-substituted derivatives in the gas phase. Implications for the thiazolidine-2, 4-dione-water complex. Arabian Journal of Chemistry, 9(5), 616-625. © 2018 King Saud University.)

total energy or enthalpy or Gibbs free energy with respect to the most stable tautomer **A**. Results of Table 13.4 indicated that in all cases, the most stable tautomeric structure corresponds to the diketo structure, tautomer **A**. The second interesting finding was that, in contrast to the hydantoin system [40] and in agreement with the analog rhodamine [10], the next low lying tautomer correspond to the C tautomer, which is formed by a 1,3-hydrogen transfer from the hydrogen atom at position 3 to the carbonyl oxygen at position 4. The relative energy difference is ~ 14.5 kcal/ mol. The relative stability order (ΔE in kcal/mol) of the tautomeric forms of the compound under probe is as follows: A> C> B> D> E. Our results show that the **A** tautomer is about 60–106 kJ mol^{-1} more stable than the other tautomers (**B–E**); that is, the high positive ΔH demonstrates that the

TABLE 13.4 Relative energies, enthalpies and Gibbs free energies (in kcal/mol), of thiazolidine-2,4-dione derivatives calculated at B3LYP/6-311++G(3df,2p)/6-311+G(d,p).

	ΔE					ΔH					ΔG				
	A	B	C	D	E	A	B	C	D	E	A	B	C	D	E
H	0.0	17.1	14.1	18.9	25.6	0.0	17.3	14.2	19.5	26.1	0.0	17.3	14.3	18.8	25.7
CH$_3$	0.0	17.2	14.8	24.3	19.1	0.0	17.5	14.8	25.0	20.0	0.0	17.4	15.1	23.9	18.5
Cl	0.0	16.9	14.0	24.1	16.2	0.0	17.2	14.1	24.5	16.6	0.0	17.0	14.2	24.2	16.0
CN	0.0	15.7	14.0	11.5	14.3	0.0	16.0	14.1	11.9	14.4	0.0	15.9	14.3	11.3	14.7
F	0.0	17.3	14.2	21.9	29.3	0.0	17.5	14.3	22.6	30.0	0.0	17.4	14.5	21.5	29.0
NO$_2$	0.0	16.5	13.9	18.2	15.7	0.0	16.8	14.0	18.3	16.3	0.0	16.6	14.1	18.9	15.7
NH$_2$	0.0	17.1	17.0	14.6	24.7	0.0	17.3	17.3	14.7	25.4	0.0	17.0	17.2	14.9	24.7

(Data used with permission from the author from Safi, Z. S. (2016). A theoretical study on the structure of thiazolidine-2, 4-dione and its 5-substituted derivatives in the gas phase. Implications for the thiazolidine-2, 4-dione-water complex. Arabian Journal of Chemistry, 9(5), 616-625. © 2018 King Saud University.)

tautomerization process of the investigated system is highly endothermic. As it is shown in Figure 13.1, for the transformation from **A-B** through the transition state **TSAB,** the activation barrier is ~ 46 kcal/mol.

Highly positive ΔG values indicate that the prototropic isomerization process, which leads to the transition from keto to enol form is quite disfavored, i.e., it is nonspontaneous. It implies, in agreement with analog polyfunctional heterocyclic systems [10, 39], that this reaction is thermodynamically and kinetically unfavored in the gas phase, meaning that the reverse reaction is both thermodynamically and kinetically favored and the **A** tautomer is the dominant in the gas phase.

13.3.5   TAUTOMERIZATION OF (1:1) OXAZOLIDINE AND ITS THIO DERIVATIVES-ZN^{2+} COMPLEXES

This part of the chapter focuses on the effect of the association of the heterocyclic compounds with Zn^{2+} on the tautomerization process. Safi

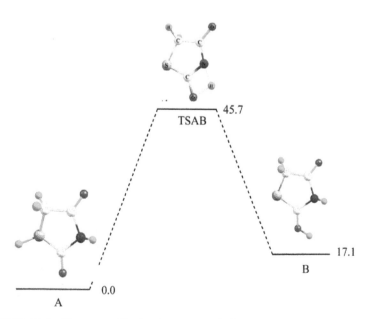

FIGURE 13.1   Energy profile for the prototropic isomerization process of thiazolidine-2,4-dione. All values are in kJ/mol.

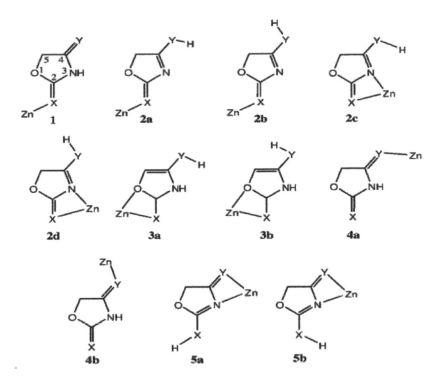

SCHEME 5 Schematic representation of different tautomers of oxazolidine–Zn^{2+} (X=Y=O). (Reprinted with permission from Safi, Z. S., & Lamsabhi, A. M. A theoretical density functional study of association of Zn2+ with oxazolidine and its thio derivatives in the gas phase. Journal of Physical Organic Chemistry, 23(8), 751-758. © 2010 John Wiley and Sons.)

and Lamsabhi [15] performed DFT calculations in order to study the gas-phase interaction of oxo- and thio-oxazolidine derivatives with Zn^{2+} (**Scheme 5**). The calculations were performed at the B3LYP/6-311G (2df,2p) level of theory. The results of this study are summarized in Table 13.5. Safi and Lamsabhi found that the alteration of the stability trends was observed for the isolated neutrals (Figure 13.2b). The stability order changes noticeably when the Zn^{2+} association occurs. While in the isolated neutrals, the dioxo form is the most stable, in their complexed homologs, the enol–thiol one (5a) becomes the most stable in the PES.

It was also noted that the interaction of Zn^{2+} with oxazolidine has a considerable catalytic effect on the IPT process. This finding was in con-

TABLE 13.5   Relative Energies (ΔE_{SCF} in kcal/mol) of [oxazolidine–Zn]$^{2+}$ Complexes Calculated at B3LYP/6-311þG(2df,2p)//B3LYP/6-311G(d,p) Level of Theory

Species	ΔE_{SCF}
1	11.6
2a	15.2
2b	19.9
2c	5.6
2d	3.8
3a	11.2
3b	8.0
4a	28.3
4b	25.8
5a	0.0
5b	5.7
TS (4a-4b)	29.3
TS (4a-5b)	68.0
TS (1-2a)	62.9
TS (2c-2d)	14.8
TS (2a-2c)	18.3
TS (5a-5b)	11.6

The ZPE corrections are included. Results are taken from Ref. 15

trast to what was found in the case of thiouracil-Cu^{2+} association (reference 1) and in agreement with the thiouracil-Ca^{2+} complexation (reference 2). In fact, the reduction of the energetic barrier was noticeable when comparing the energetic barriers of the processes 1→2 and 4→5 in the neutral case (Figure 13.2) and their homologs in Zn^{2+}-complexes. These results exhibited similar tendencies as those found in thiouracil-Cu$^+$ (reference 3), thiouacil-Ca^{2+} (reference 2), and thymine-M^{2+} (M=Ni, Cu, Zn) complexes (4).

13.4   CONCLUSION

In this chapter, we presented the tautomerization of some heterocyclic compounds by employing the hybrid density functional theory at the dif-

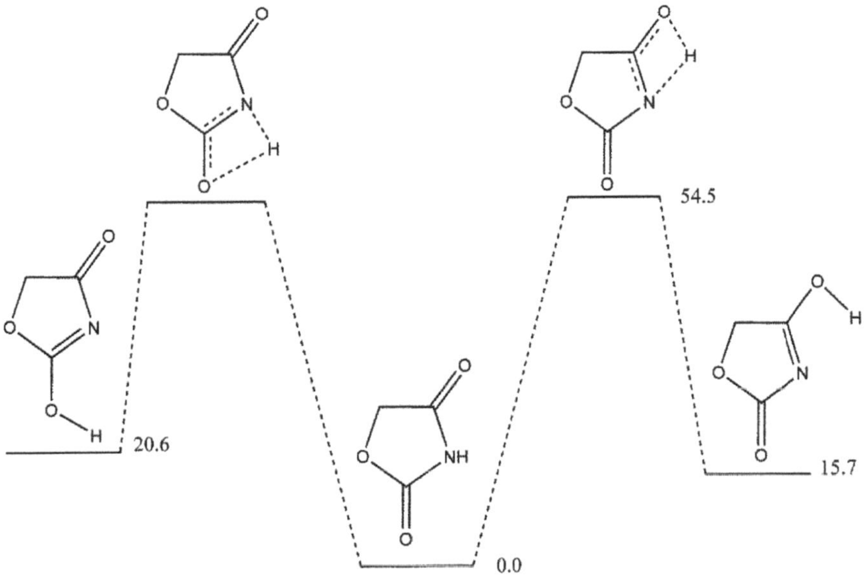

FIGURE 13.2 Energy profiles for oxazolidine tautomers. Relative energies are in kcal/mol. (Reprinted with permission from Safi, Z. S., & Lamsabhi, A. M. A theoretical density functional study of association of Zn2+ with oxazolidine and its thio derivatives in the gas phase. Journal of Physical Organic Chemistry, 23(8), 751-758. © 2010 John Wiley and Sons.)

ferent level theory. The results clearly indicated that DFT is very useful tool to study the tautomerization and the intramolecular proton transfer processes and compares well with other theoretical and experimental methods. Results revealed that the heterocyclic compounds preferred to exist as the diketo, oxo/thio, or dithio forms in all the considered species. The results also suggested that the oxo/thio and oxo/thio or thiol/oxo are more stable than the dienol or enol/thiol forms.

KEYWORDS

- **DFT B3LYP and BP86**
- **intramolecular proton transfer reaction**
- **tautomerism**

REFERENCES

1. Lamsabhi, M., Alcamí, M., Mó, O., Yáñez, M., (2006). *J. Phys. Chem. A, 110* (5), 1943–1950.
2. Trujillo, C., Lamsabhi, M., Mó, O., Yáñez, M., Salpin, J-Y., (2008). *Org. Biomol. Chem., 6,* 3695–3702.
3. Lamsabhi, M., Safi, Z., (2010). *J. Phys. Org. Chem., 23*(8), 751–758.
4. Rincon, E., Yáñez, M., Toro-Labbe, A., Mó, O., (2007). *Phys. Chem., 9,* 2531.
5. http://www.chemicool.com/definition/proton transfer reaction.html.
6. Elguero, J., A. R. K., & Denisko, O., (2000). *Adv. Heterocycl. Chem.*, pp. 1–84.
7. Minkin, V. I., A. D. G., Elguero, J., Katritzky, A. R., & Denisko, O., (2002). *Adv. Heterocycl. Chem.*, pp. 1110–1120.
8. Yekeler, H., (2005). *J. Mol. Struct. (Theochem),*. pp. 201.
9. Trifonov, R. E., I. A. L. K., Ostrovskii, V. A., & El guero, J., (2004). *J . Mol. Struct. (Theochem),* pp. 123.
10. Safi, Z. S., & F. A. A., (2005). *Electronic Journal of Chem.*, pp. 884.
11. Lamsabhi, M., M. A., Mo, O., Bouab, W., Esseffar, M., Abboud, J. L. M., & Yanez, M., (2000). *J. Phys. Chem. A.*, pp. 5122.
12. Fares, W. M., & Z. S. S., (2014). *Oriental Journal of Chemistry.* pp. 1045–1054.
13. Tahmasseb, D., (2003). *J. Mol. Struct., (Theochem),*. pp. 11.
14. Lamsabhi, M., M. E., Bouab, W. El Messaoudi, T., Abboud, J. L. M., Alcami, M., & Yanez, M., (2002). *J. Phys. Chem. A.*, pp. 7383.
15. Lamsabhi, M., Safi, Z., (2010). *J. Phys. Org. Chem., 23*(8), 751–758.
16. Agelova, S., V. E., Markova, N., Denkova, P., & Kostova, K., (2004). *J. Mol. Struct. (THEOCHEM).*, pp. 201–207.
17. Santos, M., M. J., Oliviera, S., da Silva, J., Lima, M., Galdino, S., & Pitta, I., (2005). *J. Mol. Struct., (THEOCHEM),* pp. 191–201.
18. Allegretti, P., G. L., Sierra, M., & Furing, J., (2000). Afinidad LVII, Enero-Febrero, 485, *J., Afinidad LVII, Enero-Febrero.*, pp. 41–49.
19. Shabanian, M., H. M., Hajibeygi, M., & Mohamadi, A., (2012). *Electronical Journal of Chemistry.* pp. 107–112.
20. Becke, A. D., (1993). *J. Chem. Phys.*, pp. 5648.
21. Lee, C., W. Y., & Parr, R. G., (1988). *Phys. ReV. B.*, pp. 785.
22. Becke, A. D., (1988). *Phys. Rev. A1, 38,* pp. 3098.
23. Perdew, J. P., (1986). *Phys. Rev. B.,* pp. 8822.
24. *Gaussian 03, Revision C.02,* 2004, Gaussian, Inc.,Wallingford, CT.
25. Frisch, M. J., et al., (2009). *Gaussian, Inc.,* Wallingford CT,.
26. Wachters, A. J. H., (1970). *J. Chem. Phys.*, pp. 1033.
27. Hay, P. J., (1977). *J. Chem. Phys.,*. pp. 4377.
28. Luna, A., B. A., Morizur, J. P., Tortajada, J., Mo, O., & Yanez, M., (1997). *J. Phys. Chem. A.*, pp. 5931.
29. Belcastro, M., T. M., Russo, N., & Toscano, M., (2005). *J. Mass Spectrom.*, pp. 300.
30. Rincon, E., M. Y., Toro-Labbe, A., & Mo, O., (2009). *Phys. Chem. Chem. Phys.,* pp. 2531.
31. Scott, A. P., & L. R., (1996). *J. Phys. Chem.*, pp. 16502.

32. Lamsabhi, A. M., M. A., Mo, O., Yanez, M., Tortajada, J., & Salpin, J. Y., (2007). *Chem Phys Chem.*, pp. 181.

33. Piasek, Z., & T. U., (1962). *Tetrahedron Lett.*, pp. 723–727.

34. Hawkes, G. E., E. W. R., & Hull, W. E., (1977). *Chem. Soc., Perkin,* pp. 1268–1275.

35. Safi, Z., (2012). *European J. of Chemistry,*. pp. 348–355.

36. Safi, Z. S., (2016). *Oriental Journal of Chemistry,*. pp. 2371–2381.

37. Enchev, V., S. C., & Jordanov, B., (2002). *Chemistry of Heterocyclic Compounds,*. pp. 1110–1120.

38. Andreocci, M. V., C. C., & Sestili, L., (1984). *Spectrochim. Acta A,*. pp. 1087–1096.

39. Al-Sehemi, A., & T. M. E. G., (2008). *J. Mol. Struct. (THEOCHEM),*. pp. 66–73.

40. Safi, Z. S., (2016). *Arabian Journal of Chemistry,*. pp. 616–625.

CHAPTER 14

IONIZATION ENERGIES OF ATOMS OF 103 ELEMENTS OF PERIODIC TABLE USING SEMIEMPRICAL AND DFT METHODS

NAZMUL ISLAM,[1] SAVAŞ KAYA,[2] and DULAL C. GHOSH[3]

[1]*Theoretical and Computational Chemistry Research Laboratory, Govt. Engineering College, Ramgarh, Jharkhand, PIN-825101, India*

[2]*Department of Chemistry, Faculty of Science, Cumhuriyet University, Sivas, 58140, Turkey*

[3]*Department of Chemistry, University of Kalyani, Kalyani, 741235, India*

CONTENTS

ABSTRACT

The ionization energies of the atoms are computed using the experimentally determined spectroscopic wave numbers invoking Bohr model and Rydberg formula. Because the basic input is the experimentally determined wave number, it is expected and found that all the factors that control the motion and energy of the electrons of the atoms–the electron correlation, orbital relaxation, and relativistic effects–are automatically subsumed in the experimental spectral data to make the computed ionization energy values corrected of such discrepancies prevalent in the Hartree–Fock self-consistent field (SCF) methods. Computed ionization energies have very good correlation with the experimental counter parts. In addition to the express periodicity of groups and periods, the inertness of Hg, the extreme reactivity of Cs atom, and the half-filled stability of N atom are also reproduced in the evaluated ionization energy data.

14.1   INTRODUCTION

The minimum energy necessary to remove an electron from a multielectron atom is the first ionization energy of the atoms. Although the periodic table does not follow from Schrödinger equation, the shell structure of atoms and the Pauli exclusion principle explain the chemical periodicity. The ionization energy is a periodic property of the elements. The first ionization energy of the atom reflects the energy with which the electron is bound to the system, and hence, it is an important descriptor of the atoms playing an important role in the determination of many physical and chemical properties of atoms. Efforts have been made to evaluate the ionization energy experimentally [1–4] as well as theoretically [5–14].

Experimentally, the ionization energy of a chemical element is usually measured in an electric discharge tube in which a fast-moving electron generated by an electric current collides with a gaseous atom of the element ejecting one of its electrons. Another important experimental technique is the photo electron spectroscopy (PES), which measures the ionization energy of molecules when electrons are ejected from different orbitals [1, 2].

Density functional theory (DFT) is one of the important tools considered to predict the chemical properties of atomic and molecular systems. DFT is currently the most popular computational electronic structure method that considers the electron density to describe the chemical reactivity. Thanks to this electronic structure method, DFT-based methods are also known to be very useful and reliable for the studies about prediction of chemical reactivity of molecules, clusters, and solids. From past and current scientific studies based on DFT, it is apparent that this theory provides compatible results with experimental data obtained in the related investigations of organic and inorganic chemistry. The Thomas–Fermi theory that is known as an alternative approach for finding the electronic structure of atoms using one-electron ground state density, $\rho(r)$, proposed in 1926 may be considered as the beginning of the DFT story. In 1950s, Slater presented a relationship between the Thomas–Fermi theorem and Hartree's orbital method. Subsequently, the Hohenberg and Kohn theorem proved that a precise method including the use of $\rho(r)$ exist in principle. It should be noted that modern version of DFT in use today is known as Kohn–Sham (KS) DFT. Kohn-Sham DFT describes self-consistent equations to analyze the chemical systems. In the mentioned equations, the exchange-correlation (XC) energy is given in terms of the $\rho(r)$. In 1990s, Becke and coworkers introduced the B3LYP method that is known as one of the most popular approximation in chemistry and material science, and after these developments, many DFT-based method were also proposed to find a way out for problems related to electronic structures of chemical systems.

Many theoretical studies have been conducted to calculate the ionization energies of atoms and molecules [25, 26]. Some researchers [27] noted that ionization energies of atoms and small molecular systems can be calculated as highly accurate using high level wave function-based approaches such as W1/W2,Gn($n = 1 - 4$) and CBS-QB3. As mentioned above, DFT presents a very useful alternative approach in the determination of chemical properties like ionization energy and electron affinity of a wide range of systems. Early investigations considered new functionals like BLYP, BP86 BPW91, B3LYP, B3P86, and B3PW91 for the calculation of ionization energies of atomic and molecular systems. Until today, many new generations of density functionals have been introduced. Meta-generalized gradient approximation (meta-GGA) functionals, range-separated functionals, dispersion corrected functionals, and doubly hybrid density functionals (DHDFs) are some of them. J. Wu [28] reported that ionization energy and electron affinities calculated by these new generation functionals were very compatible with experimental results.

Pople [29] and coworkers tested the performance of B-LYP exchange correlation functional for the theoretical prediction of some properties like atomization energy, ionization energy, electron affinity, and proton affinity considering the 6-31G(d), 6-31+G(d), 6-311+G(2df ,p), and 6-311+G(3df ,2p) basis sets. In 1996, Proft and Geerlings [30] investigated the performance of B3LYP and B3PW91 exchange correlation functionals in the calculation of ionization energies, electron affinities, electronegativities, and hardness of atoms and some small molecules using Dunning's correlation consistent basis sets and compared the results of the mentioned functionals with local density approximation (LDA) and two gradient corrected density functionals B-P86 and B-LYP. In this study, Proft and Geerlings reported that B3LYP and B3PW91 exchange correlation functionals provide accurate and compatible results with experiment, and these exchange correlation functionals are among the most commonly used Kohn–Sham methods.

Kotochigova [31] analyzed all atoms in the periodic table and noted that ionization energies obtained with the help of local spin-density approximation (LSDA) are in good agreement with experimentally determined results. Then, Kraisler and G. Makov [32] calculated total energies, spin states, and ionization energies of many atoms with the help of the local spin-density approximation (LSDA) and the generalized-gradient approximation (GGA). They found that for many atoms, the ionization energies determined considering GGA are more close to the experimental data than the results obtained using LSDA. In 2004, T. Yanai [33] proposed a new hybrid exchange correlation functional that today is known as CAM-B3LYP. He combined the qualities of the B3LYP method and the long-range correction presented by Tawada. Yanai showed that CAM-B3LYP provides similar results in terms of the atomic ionization energies with the B3LYP method. In brief, many DFT-based methods have been used for the calculation of atomic ionization energies.

The electronic chemical potential, μ, is formally defined [27] as:

$$\mu = \left[\frac{\delta E(\rho)}{\delta \rho}\right]_v \tag{1}$$

where $\rho(r)$ is the electron density function of atom or molecule, $E(\rho)$ is the ground electronic energy functional of that system in equilibrium, and v is the external potential acting on an electron due to the presence of the nucleus.

The differential definition of Eq. (1) for a system of N electrons with ground state energy $E[N,v]$, gives μ as:

$$\mu = -\chi = \left(\frac{\partial E}{\partial N}\right)_{v(r)} \tag{2}$$

Using finite differences approximation, some authors derived the equation: $\mu = -(I+A)/2$ to calculate the chemical potential based on ground-state ionization energy and electron affinity values of chemical systems.

The chemical reactivity theory can provide important insights into the nature of atoms and molecules. A new branch of density functional-based theoretical science known as conceptual density functional theory (CDFT) has been developed to rationalize the chemical behavior of a molecule. The framework and some recent developments of CDFT have been facilitated by the works of conceptual density functional scientists [1–11, 16–26].

Determination of the ionization energies of chemical systems is also a well-studied problem in theoretical domain and is routinely done by invoking Koopmans' theorem within the scope of the closed-shell Hartree–Fock SCF theory. Koopmans' theorem states that the first ionization energy (I) and electron affinity (A) of an atomic or molecular system are equal to the negative values of the orbital energies of the highest occupied molecular orbital (HOMO) and lowest unoccupied molecular orbital (LUMO), respectively.

$$I = -E_{HOMO} \tag{3}$$

$$A = -E_{LUMO} \tag{4}$$

Atomic ionization energy can also be predicted by an analysis using electrostatic potential and the Bohr model of the atom [21]. Spectroscopically, ionization energy is determined by the well-known Rydberg equation that can be derived from Bohr equation [21].

In the present work, semi-empirical and DFT methods were used to evaluate the ionization energy of atoms invoking Rydberg formula for a hydrogen-like element and using spectral frequencies determined by spectroscopic methods. The results of new proposed atomic ionization energy formula were compared with the experimental data and the results of DFT methods. The comparisons showed that the proposed new method is useful and successful for the calculation of ionization energy of many atoms of the periodic table.

14.2　METHOD OF COMPUTATION

14.2.1　SEMI-EMPIRICAL

The Rydberg constant, (R_H), for H atom has the general form:

$$R_H = 2\pi^2 e^4 m / h^3 c \, \text{l} \qquad (5)$$

where c is the velocity of light, e is the electronic charge, and m is the mass of the electron. h is Planck's constant.

For any hydrogen-like element, R_H has the general form-

$$R_H = 2\pi^2 Z^2 e^4 m / h^3 c \qquad (6)$$

where Z is the actual nuclear charge.

The basic tenet of the present method is that we convert a multielectron atom system to a hydrogenic atom and then invoke the Bohr model to determine the ionization energies of multielectron systems. The simple way of doing this is to modify the above expression of R_H by replacing the actual nuclear charge, Z, by the corresponding effective nuclear charge $Z_{eff.}$ of atom. This has virtually converted a multielectron atom to a hydrogenic atom.

The modified Rydberg constant looks

$$R'_H = 2\pi^2 Z_{eff}^2 e^4 m / h^3 c \tag{7}$$

The spectroscopic determination of ionization energies depends on the determination of the series limit–the wave number at which the series terminates and becomes a continuum. We proceed as under:

Let the transition occurs from a higher energy label, E_{Upper}, to a lower energy label, E_{Lower}, and also let for this transition, a photon of wave number, $\bar{v}$, is emitted.

From the Bohr theory [21], we can write the energy of the upper energy state as:

$$E_{Upper} = -hcR_H / n^2 \tag{8}$$

where n is the principal quantum number of the upper energy label.

We know that

$$\Delta E = E_{lower} - E_{Upper} = hc\bar{v} \tag{9}$$

Putting the value of E_{Upper} and rearranging the equation above, we get-

$$\bar{v} = -R_H / n^2 - E_{lower} / hc \tag{10}$$

As the ionization energy, I, is given by the definition

$$I = -E_{lower} \tag{11}$$

it follows that

$$\bar{v} = I / hc - R_H / n \tag{12}$$

Rearranging Eq. (12), we get

$$I = hc \, (\bar{v} + R_H / n^2) \tag{13}$$

Eq. (13) gives the value of ionization energy for a hydrogen atom. In this work, we have converted a multielectron atom to a hydrogenic atom. For this reason, we have used the modified Rydberg constant, R'_H, instead of R_H and replaced the principal quantum number, n, by the effective principal quantum number, n^*, to rewrite Eq. (13), as follows-

$$I = hc\bar{v} + \left(2\pi^2 Z_{eff}^2 e^4 m / h^2 n^{*2}\right) \qquad (14)$$

Because,

$$\xi = Z_{eff} / n^* \qquad (15)$$

where ξ is the orbital exponent of the atoms [22].

Hence, we can rewrite the formula of ionization energy as follows:

$$I = hc\bar{v} + 2\pi^2 \xi^2 e^4 m / h^2 \qquad (16)$$

We have computed the ionization energy of 103 elements of the periodic table using the proposed Eq. (16). The appropriate wave number $(\bar{v})$ values for different elements are taken from reference [23] and are presented in Table 14.1. Furthermore, for the calculation of the ionization energy, we have used the effective nuclear charge, Z_{eff}, published by Ghosh and Biswas [24] and the effective principal quantum number, n^*, for $n=1$ to 6 published by Slater [22]. In case of $n=7$, we have used the n^* value suggested by Ghosh and Biswas [24].

14.2.2   DFT CALCULATIONS

We have mentioned above the commonly used DFT-based models in the calculation of atomic properties like ionization energy and electron affinity. In Tables 14.2 and 14.3, we presented ionization energies calculated by our new method for some selected atoms and compared our results with the results obtained by B3LYP and B3PW91 functionals using various basis sets. Ionization energies calculated using B3LYP and B3PW91 functionals were taken from Ref. [30]. The comparisons showed that the

TABLE 14.1 A Comparison of the Ionization Energy Computed Through the Formula Suggested in the Present Work vis-à-vis the Corresponding Experimental Values

Atom	Wave number (cm⁻¹)	I (theo) in eV	I (exp) in eV	SD (%)	Atom	Wave number (cm⁻¹)	I (theo) in eV	I (exp) in eV	SD (%)
H	109679	13,64	13,598	0,3044	I	84295	10,656	10,451	1,9593
He	198311	24,695	24,587	0,4391	Xe	97834	12,371	12,13	1,988
Li	43487	5,4477	5,3917	1,039	Cs	31406	3,9135	3,8939	0,5022
Be	75193	9,4465	9,3227	1,3279	Ba	42035	5,2434	5,2117	0,6095
B	66928	8,3966	8,298	1,1879	La	44981	5,623	5,5769	0,827
C	90820	11,413	11,26	1,3556	Ce	44672	5,6016	5,5387	1,1356
N	117226	14,753	14,534	1,5034	Pr	44140	5,5554	5,473	1,5051
O	109837	13,911	13,618	2,1478	Nd	44562	5,6304	5,525	1,9077
F	140525	17,805	17,423	2,1911	Pm	45020	5,7128	5,582	2,3427
Ne	173930	22,047	21,565	2,2386	Sm	45520	5,8032	5,6437	2,8257
Na	41449	5,209	5,1391	1,3609	Eu	45735	5,8612	5,6704	3,3644
Mg	61671	7,7628	7,6462	1,5247	Gd	49601	6,3751	6,1498	3,6637
Al	48278	6,1572	5,9858	2,8639	Tb	47294	6,1258	5,8638	4,4686
Si	65748	8,29	8,1517	1,6963	Dy	47900	6,2409	5,9389	5,0857
P	84581	10,671	10,487	1,7608	Ho	48567	6,3665	6,0215	5,7292
S	83559	10,596	10,36	2,2736	Er	49262	6,4984	6,1077	6,3966
Cl	104591	13,263	12,967	2,2776	Tm	49880	6,6236	6,1843	7,1029
Ar	127110	16,121	15,76	2,292	Yb	50443	6,7449	6,2542	7,8465
K	35010	4,3811	4,3407	0,9327	Lu	43763	5,9702	5,4259	10,033

TABLE 14.1 (Continued)

Atom	Wave number (cm⁻¹)	I (theo) in eV	I (exp) in eV	SD (%)	Atom	Wave number (cm⁻¹)	I (theo) in eV	I (exp) in eV	SD (%)
Ca	49306	6,1802	6,1132	1,0962	Hf	55048	7,4277	6,8251	8,8296
Sc	52922	6,6356	6,5615	1,1291	Ta	60891	8,2128	7,5496	8,7851
Ti	55073	6,9095	6,8281	1,1915	W	63428	8,5904	7,864	9,2369
V	54412	6,8349	6,7462	1,3148	Re	63182	8,6257	7,8335	10,112
Cr	54576	6,863	6,7665	1,4256	Os	68079	9,302	8,4382	10,236
Mn	59959	7,5391	7,434	1,4135	Ir	72324	9,9003	8,967	10,407
Fe	63737	8,0163	7,9024	1,441	Pt	72297	9,7059	8,9588	8,3394
Co	63565	8,0037	7,881	1,5563	Au	74409	10,311	9,2255	11,762
Ni	61619	7,8334	7,6398	2,5338	Hg	84184	11,604	10,438	11,174
Cu	62317	7,8675	7,7264	1,8266	Tl	49266	6,1981	6,1082	1,4715
Zn	75769	9,5465	9,3942	1,621	Pb	59819	7,5311	7,4166	1,5435
Ga	48388	6,1954	5,9993	3,2688	Bi	58762	7,3905	7,2855	1,4416
Ge	63713	8,15	7,8994	3,1723	Po	67860	8,541	8,414	1,5088
As	78950	9,9907	9,7886	2,0649	At	77702	9,7858	9,64	1,5126
Se	78658	9,9967	9,7524	2,5054	Rn	86693	10,927	10,749	1,6618
Br	95285	12,106	11,814	2,4743	Fr	32849	4,0881	4,0727	0,3762
Kr	112915	14,344	14	2,4608	Ra	42573	5,3029	5,2784	0,4638
Rb	33691	4,2041	4,1771	0,647	Ac	41700	5,1967	5,17	0,517
Sr	45932	5,7392	5,6949	0,7779	Th	50867	6,3365	6,3067	0,4725
Y	50146	6,2663	6,2173	0,7877	Pa	47500	5,9401	5,89	0,851

Zr	53506	6,6878	6,6339	0,8119	U	49958	6,2602	6,1941	1,0671
Nb	54514	6,8175	6,7589	0,8685	Np	50536	6,349	6,2657	1,3295
Mo	57204	7,1564	7,0924	0,9013	Pu	48603	6,1446	6,026	1,969
Tc	58700	7,3471	7,28	0,922	Am	48182	6,1154	5,9738	2,3703
Ru	59366	7,4352	7,3605	1,015	Cm	48324	6,1386	5,9914	2,4572
Rh	60161	7,5394	7,4589	1,0788	Bk	49989	6,3919	6,198	3,1295
Pd	67242	8,4239	8,3369	1,0435	Cf	50665	6,5051	6,2817	3,5568
Ag	61106	7,6686	7,5762	1,2198	Es	51800	6,6774	6,42	4,0092
Cd	72540	9,0936	8,9938	1,1101	Fm	52392	6,7845	6,5	4,3762
In	46670	5,876	5,7864	1,5489	Md	53037	6,9001	6,58	4,8651
Sn	59233	7,4583	7,3439	1,5579	No	53602	7,0079	6,65	5,382
Sb	69431	8,7502	8,6084	1,6476	Lr	37078	4,9666	4,6	7,9698
Te	72667	9,181	9,0096	1,9025					

TABLE 14.2 Comparison with the Results of B3PW91 Functional of the Results Obtained via New Method for Some Selected Atoms (all values are in eV)

Atom	cc-pVDZ	cc-pVTZ	aug-cc-pVDZ	aug-cc-pVTZ	Exp.	Present work
B	8.73	8.71	8.73	8.71	8.30	8.39
C	11.58	11.58	11.61	11.58	11.26	11.41
N	14.72	14.76	14.81	14.78	14.54	14.75
O	13.79	13.92	13.97	13.96	13.61	13.91
F	17.37	17.53	17.65	17.59	17.42	17.80
Ne	21.32	21.54	21.73	21.62	21.56	22.04
Al	6.12	6.11	6.13	6.12	5.98	6.15
Si	8.25	8.25	8.26	8.25	8.15	8.29
P	10.56	10.58	10.58	10.56	10.49	10.67
S	10.45	10.48	10.48	10.49	10.36	10.59
Cl	13.04	13.03	13.07	13.04	12.97	13.26
Ar	15.83	15.79	15.87	15.79	15.76	16.12

TABLE 14.3 Comparison with the Results of B3LYP Functional of the Results Obtained via New Method for Some Selected Atoms (all values are in eV)

Atom	cc-pVDZ	cc-pVTZ	aug-cc-pVDZ	aug-cc-pVTZ	Exp.	Present work
H	13.64	13.66	13.65	13.67	13.60	13.64
He	24.87	24.94	....	....	24.59	24.69
B	8.76	8.74	8.76	8.74	8.30	8.39
C	11.52	11.53	11.57	11.54	11.26	11.41
N	14.57	14.69	14.69	14.66	14.54	14.75
O	13.92	14.09	14.14	14.14	13.61	13.91
F	17.44	17.66	17.78	17.74	17.42	17.80
Ne	21.32	21.62	21.81	21.73	21.56	22.04
Al	6.03	6.01	6.03	6.02	5.98	6.15
Si	8.11	8.11	8.13	8.12	8.15	8.29
P	10.37	10.62	10.41	10.39	10.49	10.67
S	10.49	10.55	10.55	10.56	10.36	10.59
Cl	13.03	13.06	13.10	13.07	12.97	13.26
Ar	15.79	15.79	15.86	15.80	15.76	16.12

new method proposed by us provides very compatible results with both experimental data and DFT-based methods proposed in the past years.

14.3   RESULTS AND DISCUSSION

Ionization energy is an experimentally determinable quantity, and very accurate experimental ionization energies of elements are available [4] to serve as benchmark. The computed ionization energies for 103 elements of the periodic table using Eq. (16) vis-à-vis their experimental counterparts are presented in Table 14.1. The square of the correlation coefficient (R^2), presented in Table 14.1, is surprisingly near to 1. The standard deviation between theoretical and experimental ionization energy data is also presented in Table 14.1. Looking at the standard deviation values (Table 14.1), we surprisingly note that the two sets of ionization energy data vary between 0.3% and 10%, and in the majority of cases, the standard deviation is in between 1% and 2%.

For a better visualization of the correlation between theoretically evaluated and experimentally determined ionization energies, we have plotted Figure 14.1. From Figure 14.1, it is clear that the theoretical and experimental data are so close that the one curve is just superimposed upon the other curve. The close agreement between the experimental and theoretical ionization potential of so many atoms is not fortuitous. Rather, it is the merit of the model designed with proper philosophy. We have found in many occasions that the effective nuclear charge computed by Ghosh and Biswas [24] is quite reliable.

Because the basic input is the experimentally determined wave number ($\bar{v}$), it is expected that all the factors that control the motion and energy of the electrons of the atoms–the electron correlation, orbital relaxation and relativistic effects–especially for atoms of atomic number greater than 54, are automatically subsumed in the experimental spectral data to make the computed ionization energy values corrected of such discrepancies prevalent in the Hartree–Fock SCF methods.

The strength of any model is its ability to explain experimental observations. One of the fundamental chemical properties of an atom is its tendency to gain, lose, or share electrons. We know that a chemical reac-

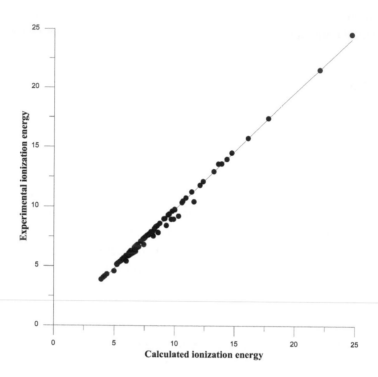

FIGURE 14.1 Graphical comparison of calculated and experimental ionization energies for 103 elements of the periodic table.

tion is based largely upon the interactions between the most loosely bound electrons or the valence electron in atoms. Thus, in order to study the chemical reaction, we need to study the physical property of atoms that influences their chemical behavior. Hence, to justify our computed results presented in Table 14.1 and depicted in Figure 14.1, we have also tried to justify some well-known chemical facts in terms of the computed data below.

I. The ionization potential values of the inert gas elements are highest in each period.

II. Computed ionization energy of Be is greater than that of B. The experimental fact can readily be explained by noting that in the case of boron, the outermost electron occupies a 2p orbital and hence is less strongly bound than if it had entered a 2s orbital.

III. Computed ionization energy of N is greater than that of O. It is an experimental fact that due to the stability of half-filled electronic configuration, the ionization potential of N is greater than that of O.

IV. The computed ionization potential of Cs is the lowest among the 103 elements. The strong chemical reactivity of the element Cs is well documented. The chemical inertness of Hg and its state of aggregation is attributed to its high ionization energy value that supports the least deformability of the element Hg under small perturbation.

14.4 CONCLUSION

In order to adapt the Bohr hydrogenic atom model for the multielectron systems, we have modified Bohr equation for hydrogenic atom to determine the ionization energies by replacing the actual nuclear charge, Z, in the expression of R_H by the corresponding effective nuclear charge, Z_{eff}, of atoms and replacing the principal quantum number, n, by the effective principal quantum number, n^*. The modified expression of R_H is linked to the ionization energy of the atom and the wave number ($\bar{v}$) determined by the spectroscopic method. This assumption converts a multielectron atom to a hydrogenic atom. Because the method relies upon the experimentally determined wave number, it is expected and found that all the factors like the electron correlation, orbital relaxation and relativistic effects are automatically subsumed in the computed ionization energy values. We have computed the ionization energies for the 103 elements of the periodic table using our suggested ansatz. We noted the express periodicity of the ionization potential computed through our proposed equation. The validity tests illustrate that our evaluated ionization potential values have very close agreement with those of experimentally determined ionization potential data correlating known physico-chemical behaviors of some elements.

KEYWORDS

- atomic ionization energy
- Bohr theory
- effective nuclear charge
- effective principal quantum number
- Rydberg formula

REFERENCES

1. Moulder, F., Stickle, W. F., Sobol, P. E., & Bomben, K. D., (1992). Handbook of X-ray Photoelectron Spectroscopy, *Perkin-Elmer Corp*, Eden Prairie, MN, USA.
2. Wagner, C. D., Riggs, W. M., Davis, L. E., Moulder, J. F., & Mullenberg, G. E., (1979). Handbook of X-ray Photoelectron Spectroscopy, *Perkin-Elmer Corp*, Eden Prairie, MN, USA.
3. Atkins, P. W., (1990). QUANTA: A Handbook of Concepts, Oxford University Press.
4. Huheey, J. E., Keiter, E. A., & Keiter, R. L., (2006). Inorganic Chemistry; Principle of Structure and Reactivity, *Pearson Education*, 4th Edition.
5. Reinhardt, W. P., & Doll, J. D., (1969). Direct Calculation of Natural Orbitals by Many-Body Perturbation Theory: *Application to Helium, J. Chem. Phys., 50*, 2767–2768.
6. Ecker, F., & Hohlneicher, G., (1972). Disruption of the oretic calculation of ionization energy and electron affinities with the help of Green's two-point function. *Theor. Chem. Acta., 25*, 289–308.
7. Doll, J. D., & Reinhardt, W. P., (1972). Many-Body Green's Functions for Finite, Nonuniform Systems: Applications to Closed Shell Atoms, *J. Chem. Phys., 57*, 1169–1184.
8. Cederbaum, L. S., Hohlneicher, G., & Peyerimhoff, S., (1971). *Chem. Phys. Lett., 11*, 421.
9. Reinhardt, W. P., & Smith, J. B., (1973). Application of the many-body Green's function formalism to the lithium atom, *J. Chem. Phys., 58*, 2148–2152.
10. Schneider, B., Taylor, H. S., & Yaris, R., (1970). Many-Body Theory of the Elastic Scattering of Electrons from Atom sand Molecules, *Phys. Rev. A., 1*, 855–867.
11. Nerbrant, P. O., (1975). Application of many-body green's functions to the calculation of molecular ionization potentials, *Int. J. Quantum. Chem., 9*, 901–916.
12. Cederbaum, L. S., (1973). Direct calculation of ionization potentials of closed-shell atoms and molecules, *Theor. Chim. Acta., 31*, 239–260.

13. Schirmer, J., Cederbaum, L. S., & Walter, O., (1983). New approach to the one-particle Green's function for finite Fermi systems, *Phys. Rev. A.*, *28*, 1237–1259.

14. Cederbaum, L. S., & Domcke, W., (1977). Theoretical Aspects of Ionization Potentials and Photoelectron Spectroscopy: A Green's Function Approach. *Adv. Chem. Phys.*, *36*, 205–344.

15. Politzer, P. , & Fakher, A. A., (1998). A comparative analysis of Hartree-Fock and Kohn-Sham orbital energies, *Theor. Chem. Acc.*, *99*, 83–87.

16. Hamel, S., Duffy, P., Casida, M. E., & Salahub, D. R., (2002). Kohn–Sham orbitals and orbital energies: fictitious constructs but good approximations all the same, *J. Electron Spectrosc. Relat. Phenom.*, *123*, 345–363.

17. Szabo, A., & Ostlund, N. S., (2004). Modern Quantum Chemistry, Dover Publications, USA, (Chapter 3).

18. Pillar, F. L., (1990). Elementary Quantum Mechanics, 2nd edition, McGraw-Hill.

19. Plakhutina, B. N., Gorelik, E. V., & Breslavskaya, N. N., (2006). Koopmans' theorem in the ROHF method: Canonical form for the Hartree-Fock Hamiltonian, *J. Chem. Phys.*, *125*, 204110(1–10).

20. Isihara, M., (1984). Simple Estimation of Ionization Energies. An Improvement of Koopmans' Approximation by Scaling, Bull. *Chem. Soc. Jpn.*, *57*, 1789–1794.

21. Bohr, N., (1930). On the constitution of atoms and molecules, *Philos. Mag.*, *26*, 1–25.

22. Slater, J. C., (1930). Atomic Shielding Constants, *Phys. Rev.*, *36*, 57–64.

23. Sansonctti, J. E., & Martin, W. C., (2005). *Handbook of Basic Atomic Spectroscopic Data*, American Institute of Physics.

24. Ghosh, D. C., & Biswas, R., (2002). Theoretical Calculation of Absolute Radii of Atoms and Ions. Part 1. The Atomic Radii, *Int. J. Mol. Sci.*, *3*, 87–113.

25. Steenken, S., Telo, J. P., Novais, H. M., & Candeias, L. P., (1992). One-electron-reduction potentials of pyrimidine bases, nucleosides, and nucleotides in aqueous solution. Consequences for DNA redox chemistry, *J. Am. Chem. Soc.*, *114*, 4701–4709.

26. Vijayaraj, R., Subramanian, V., & Chattaraj, P. K., (2009). Comparison of global reactivity descriptors calculated using various density functionals: a QSAR perspective, *J. Chem. Theo. and Comp.*, *10*, 2744–2753.

27. Montgomery, J. A., Frisch, M. J., Ochterski, J. W., & Petersson, G. A., (1999). A complete basis set model chemistry. VI. Use of density functional geometries and frequencies, *J. Chem. Phys.*, *110*, 2822–2827.

28. Su, N. Q., Zhang, I. Y., Wu, J., & Xu, X., (2011). Calculations of ionization energies and electron affinities for atoms and molecules: A comparative study with different methods, *Front. Chem. China.*, *6*, 269–279.

29. Gill, P. M. W., Johnson, B. G., Pople, J. A., & Frisch, M. J., (1992). The performance of the Becke—Lee—Yang—Parr (B—LYP) density functional theory with various basis sets, *Chem. Phys. Lett.*, *197*, 499–505.

30. De Proft, F., & Geerlings, P., (1997). Calculation of ionization energies, electron affinities, electronegativities, and hardnesses using density functional methods, *J. Chem. Phys.*, *106*, 3270–3279.

31. Kotochigova, S., Levine, Z. H., Shirley, E. L., & Stiles, M. D., (1997). Local-density-functional calculations of the energy of atoms, *Phys. Rev. A.*, *55*, 191.

32. Kraisler, E. Makov, G., & Kelson, I., (2010). Ensemble v-representable ab initio-density-functional calculation of energy and spin in atoms: A test of exchange-correlation approximations, *Phys. Rev. A., 82*, 042516.
33. Yanai, T., Tew, D. P., & Handy, N. C., (2004). A new hybrid exchange–correlation functional using the Coulomb-attenuating method (CAM-B3LYP), *Chem. Phys. Lett., 391*, 51–57.

CHAPTER 15

MOLECULAR SIMILARITY FROM MANIFOLD LEARNING ON D2-PROPERTY IMAGES

FARNAZ HEIDAR-ZADEH[1,3] and PAUL W. AYERS[1]

[1]*Department of Chemistry and Chemical Biology, McMaster University, Hamilton, Ontario, Canada*

[2]*Department of Inorganic and Physical Chemistry, Ghent University, Krijgslaan 281 (S3), 9000 Gent, Belgium*

[3]*Center for Molecular Modeling, Ghent University, Technologiepark, Zwijnaarde, Belgium*

CONTENTS

15.1   MOTIVATION

The development of quantitative structure activity relationships (QSAR) is based upon the observation that similar molecules will have similar properties. Therefore, one can predict the properties of an uncharacterized molecule by computing its similarity to the three-dimensional structures of the molecules in a database of reference compounds (3D-QSAR) [1–8]. This strategy is commonly used to predict the properties of biomolecules. In practice, the bottleneck in this approach, both theoretically and computationally, is assessing the similarity of two molecules [9–13]. We are primarily interested in quantum 3D-QSAR, where the similarity between molecules is based on quantum-mechanical indicators [2, 14–20].

Most typically, the similarity of molecules is assessed by aligning the molecules: one molecule is fixed in space, and then, the relative position and orientation of the second molecule is chosen so that the similarity of the molecules is maximized. There are many approaches to this problem in the literature (see Ref. [21] and references cited therein). There are two main problems with this approach. First, for a given measure of molecules' three-dimensional similarity, the problem of aligning molecules is a global optimization problem, and thus NP-hard [19]. Second, different alignments are needed in different contexts. For example, if one is interested primarily in the electrostatic properties of a molecule, then one should align NaCl and KF so that their dipole moments are aligned. However, if one is interested in aligning based on the electron density, then one should align in the reverse direction (NaCl and KF). Every time one wishes to predict a different property using a database, one must solve the NP-hard alignment problem again!

This motivates the development of alignment-free similarity measures [9–13, 19, 22]. It also suggests that we should find a way to represent the distinguishing features of the molecules in the database in a compact way, so that the context-specific molecular similarity can be computed very efficiently. This chapter proposes a method for doing this. The alignment-free method we propose is based on the D2-shape similarity descriptor [23–25], enriched by the values of properties on the molecular surface in a way analogous to methods based on the property encoded surface translator (PEST) [9, 11, 21, 26]. The resulting rectangular "molecular image" inherits the D2-descriptor's invariance to the orientation and the position of the molecule. The molecular images can then be compared directly, but because of the immense information in the image, this is impractical for large databases. However, the distinguishing features between the molecules in the database can be extracted using a manifold learning method, providing a reduced-dimensional description that nonetheless encapsulates the necessary chemical information. Finally, molecular similarity can be computed from the Euclidean distance in the reduced dimensional space.

15.2 PROPERTY ENHANCED INTRAMOLECULAR DISTANCES (PEIDs)

We assume that each molecule, M, is described by a set of $N_{\text{pts}}^{(M)}$ points, $\left\{ \mathbf{r}_i^{(M)} \right\}_{i=1}^{N_{\text{pts}}^{(M)}}$; each point is characterized by a vector of N_{prop} properties, $\left\{ \left[p_1^{(M)}, p_2^{(M)}, \ldots, p_{N_{\text{prop}}}^{(M)} \right] \right\}_{i=1}^{N_{\text{pts}}^{(M)}}$. The points could be the positions of atoms or functional groups in the molecule, in which case they are "condensed" representation of the pointwise molecular properties [27–34]. However, our primary interest is in the properties of the solvent-accessible surface of the molecule; this is where the molecule interacts with other reagents. The particular properties to be evaluated on the surface are not important for the mathematical results in this paper, but we are primarily focused on quantum chemical descriptors and, specifically, the descriptors associated with conceptual density functional theory (DFT) [18, 35–43] Conceptual DFT provides a mathematically well-defined hierarchy of properties that (in principle) captures all the information about molecular interactions and reactivity. [36, 44–51]. Properties like the Fukui function [52–55], local softness [56–58], and local ionization potential (I-bar) [59–61] can be used to capture the covalent reactivity of the molecule; the electrostatic potential can capture information about the ionic interactions and hydrogen bonding [62–69]; more general reactivity indicators like the dual descriptor (ambiphilic reagents) [47, 70–75] and the general purpose reactivity indicator (ambident reagents) [76–78] can be used for more nuanced chemical effects.

There are many ways to describe the inherent shape of a molecule in a way that is independent of the molecules position and orientation. In the 3D-QSAR community, one often examines shape signatures [79–81]. The shape signature of a molecule is constructed by (a) choosing a fixed number of points on the molecular surface, (b) a ray is started from that point (with an arbitrary angle, but typically normal to the surface),

(c) the ray propagates through the molecule (internal properties of the molecule could be accommodated by adding an index of refraction, but this option has not been considered previously) until it intersects the molecular surface, where (d) it reflects according to the law of specular reflection, and (e) this is carried out for a set number of reflections. The lengths of the reflected rays are collected in a histogram, and this histogram is the *shape signature* [81]. A similar idea which is much easier computationally is the D2 descriptor. First, one selects a fixed number of points on a molecular surface; then one computes all the inter-point distances. The original D2 descriptor retains all inter-point distances but, from a chemical standpoint, it makes more sense to retain inter-point distances only if the line segment connecting the points passes exclusively through the interior of the molecule. (This is not problematic when the molecular volume is convex, but when the molecular volume is not convex, some line segments (or portions thereof) will be outside the molecular volume.) The histogram of line segment lengths is the D2 descriptor [23–25]. In the limit where the number of points on the surface is infinite, the list of intermolecular distances in the D2 descriptor contains enough information to reconstruct the molecular surface (but not its chirality).

Both the D2 descriptor and the shape signature contain sufficient information to reconstruct the molecular shape. They are independent of the position and orientation of the molecular (which is favorable), but they are also independent of molecular chirality. In the surface signature approach, this can be easily remedied by making the molecular volume birefringent and by keeping a histogram for both right-circular and left-circular polarized light. Another way to remedy this problem is to keep track of the local chirality [82–91] at the endpoints of the rays (shape signatures) or intramolecular line segments (D2 descriptor). A computationally convenient approach to local chirality is presented in Appendix 1.

The idea of property-enhanced shape signatures was first proposed by Breneman et al.; they store the values of molecular properties at each point of reflection of the shape signature rays and call the resulting histogram of properties and internuclear distances the PEST [26].

A similar idea is proposed here. For each intramolecular line segment (D2 descriptor, eliminating segments that pass outside the surface) and each property, the average property value and the property value difference from the endpoints of the line segment are computed,

$$\mu_{ij}^{(M)} = \frac{1}{2}\left(\mathbf{p}\left(\mathbf{r}_i^{(M)}\right) + \mathbf{p}\left(\mathbf{r}_j^{(M)}\right)\right)$$

$$\Delta_{ij}^{(M)} = \left|\mathbf{p}\left(\mathbf{r}_i^{(M)}\right) - \mathbf{p}\left(\mathbf{r}_j^{(M)}\right)\right| \qquad (1)$$

We prefer these descriptors to the actual property values (as in PEST) because of the analogy to the electrostatic potential, where $\mu_{ij}^{(M)}$ and $\Delta_{ij}^{(M)}$ would contain the essential information about the molecular charge and dipole moment, respectively. If the chirality of the molecule is important for the property one is predicting, then one of the properties should be a local chirality measure [82–91]. We refer to the array of information consisting of the lengths of the intramolecular segments, $d_{ij}^{(M)} = \left|\mathbf{r}_i^{(M)} - \mathbf{r}_j^{(M)}\right|$, along with their vectors of average and difference properties, as the array of property enhanced intramolecular distances (PEIDs).

For a drug-like molecule [92] with ca. 50 atoms, a surface triangulation of modest accuracy will contain 10^4–10^5 points; thus, there will be 10^8–10^{10} line segment lengths, enhanced by the values of 10–100 properties (so both $\mu_{ij}^{(M)}$ and $\Delta_{ij}^{(M)}$ are vectors containing 10–100 elements). This description of the molecule is very rich and should suffice for any practical application. However, it is not practical to use billions of numbers when comparing molecules.

15.3 PEID IMAGES

The PEID information described in the previous section gives a needle-in-a-haystack problem: how can one meaningfully compare molecules using the billions of numbers that PEID includes? We propose to first condense the PEID information into a two-dimensional PEID image. Then, we propose to use techniques from image registration, specifically manifold learning methods [93–99], to identify the key features in the PEID images.

The PEID images that we construct are histograms of the PEID information. To construct these histograms, we first need to bin the data. Recall that our raw data consist of a list of intramolecular distances, $d_{ij}^{(M)} = \left|\mathbf{r}_i^{(M)} - \mathbf{r}_j^{(M)}\right|$, and the averages and differences of the property values at the endpoints of the segments, $\mu_{ij}^{(M)}$ and $\Delta_{ij}^{(M)}$, respectively. That is, each molecule is described by a long list of vectors,

$$\mathbf{v}_{ij}^{(M)} = \left[d_{ij}^{(M)} \quad \mathrm{m}_{ij}^{(M)} \quad \mathrm{D}_{ij}^{(M)} \right] \tag{2}$$

We start by binning the distances. To do this, we first make a master list of *all* the distances from *all* the molecules,

$$\mathcal{D} = \bigcup_{M=1}^{N_{\text{molecules}}} \left\{ d_{ij}^{(M)} \right\} \tag{3}$$

We denote the total number of samples as $n_{\text{total}} = \text{count}(\mathcal{D})$.

We then need to define appropriate bins for these distances, so that we can form a distance histogram. There is a rich literature about how one can find the best bins for data [100], but for the purpose of this paper, we will use the simple procedure described in Appendix 2. That procedure partitions the intermolecular distances into $N_{\text{bins}} = \left\lceil 2\left(n_{\text{total}}/N_{\text{molecules}}\right)^{1/3} \right\rceil$ bins. Each molecule then has, on average, $n_{\text{total}}/(N_{\text{molecules}}N_{\text{bins}})$ values of $d_{ij}^{(M)}$ in each bin. Scaling the number of $d_{ij}^{(M)}$ in each bin by the reciprocal of this factor, $N_{\text{molecules}}N_{\text{bins}}/n_{\text{total}}$, gives us a scaled histogram for the molecule, with a value of $h_k^{(M)}$ for the kth bin. If $h_k^{(M)}$ is greater than (less than) one, this indicates that molecule M is more likely (less likely) than the average value in the dataset to have a value of $d_{ij}^{(M)}$ that falls in the kth bin.

We then subdivide each bin of $\{d_{ij}^{(M)}\}$ values based on the values of the individual properties; the detailed procedure for doing this is explained in Appendix 2. This gives a normalized two-dimensional histogram for each property, $d_{k,l}^{(M)}$, corresponding to the kth bin in intramolecular distance and the lth property value bin. The number of property value bins is

$$N_{bins}^{(2)} = \left\lceil 2\left(n_{total}/\left(N_{molecules}N_{bins}\right)\right)^{1/3} \right\rceil$$

Finally, we combine all the data into a single image. To do this, we stack the histograms of each property. Specifically, we define a PEID image as

$$I_{k,1}^{(M)} = h_k^{(M)}$$

$$I_{k,\ell+1+(m-1)N_{\text{bins}}^{(2)}}^{(M)} = h_{k,\ell}^{(M)} \left(property\ m \right) \quad m = 1, 2, \ldots, N_{\text{properties}} \tag{4}$$

This gives a PEID image with N_{bins} columns and $N_{\text{properties}} \cdot N_{\text{bins}}^{(2)} + 1$ rows. To get an idea of the order of magnitude of these numbers, for a dataset of 10,000 drug-like molecules, with about 10^9 intramolecular distances per molecule, $n_{\text{total}} \sim 10^{13}$, $N_{\text{bins}} \sim 10^3$, and $N_{\text{bins}}^{(2)} \sim 100$. If we store 20 properties per molecule, then the image is about 2000 by 1000, which is about the same size as a single frame of a 1080p screen. At a typical refresh rate of 30 frames/second, our entire dataset has been reduced to a 30 second "movie" on a 1080p screen. Although this seems meager, from a different viewpoint, each molecule is being expressed as a function of two million variables. In order to interpret and use information of such vast dimensionality, a dimensionality reduction technique is essential.

15.4 MANIFOLD LEARNING OF MOLECULAR IMAGES

15.4.1 OVERVIEW

Suppose one wishes to predict a property, Q, from a PEID image. The property can be thought of as function of $D = N_{\text{bins}} \cdot (N_{\text{properties}} \cdot N_{\text{bins}}^{(2)} + 1)$ real variables,

$$Q\left(I_{k,l}^{(M)} \right) : \left(\mathbb{R}^+ \right)^{N_{\text{bins}} \cdot \left(N_{\text{properties}} \cdot N_{\text{bins}}^{(2)} + 1 \right)} \to \mathbb{R} \tag{5}$$

We speculate that in practice, the significant features of the PEID images lie on a low-dimensional manifold, with $d \ll D$. This manifold will generally be a curved hypersurface in the high-dimensional space, if the manifold is smooth then the local distances are well approximated by a Euclidean distance formula. (Consider, for example, that if one wishes to fly from New York to Singapore, then (geodesic) distance one must fly is much longer than the Euclidean distance between the cities. However, the Euclidean distance is accurate for computing the distance between the Empire State Building and the Statue of Liberty.) The goal of manifold learning methods is to estimate the geodesic distance between points without ever explicitly constructing the (probably unknowable) equation by which the manifold of important molecular features is embedded in the high-dimensional space of PEID images.

The first step in many manifold learning methods is to represent the data as a graph. Each molecule M will be represented as a vertex and is connected to other molecules that are "nearby" by edges, e_{LM}. Because the Euclidean distance is reasonable for nearby points, we can use the (scaled) Euclidean distance between nearby molecules,

$$d_{LM} = \frac{\sqrt{\sum_{k=1}^{N_{bins}} \sum_{l=1}^{N_{properties} \cdot N_{bins}^{(2)} + 1} \left(I_{k,l}^{(L)} - I_{k,l}^{(M)} \right)^2}}{N_{bins} \cdot \left(N_{properties} \cdot N_{bins}^{(2)} + 1 \right)} \tag{6}$$

For each molecule, we connect it to the closest $N_{neighbors}$ molecules for which either (a) the value of the target property is similar, $| Q^{(L)} - Q^{(M)} | \le \tau$ or (b) the value of the target property is unknown. The neighbors of molecule M are denoted by the set $\mathcal{N}^{(M)}$. Most manifold learning methods are unsupervised and choose $\tau \to \infty$. However, if we have known values of the target properties for even a subset of the molecules, including that information is beneficial; this type of machine learning method is said to be semi-supervised.

It is important to verify that the graph is connected; if the graph is not connected, then the number of neighbors should be increased until it is connected. It is also important to confirm that the number of neighbors is large enough to describe a manifold of the dimension that one wishes to describe. For example, a simple (hyper)cubic tessellation of a d-dimensional manifold would require $N_{neighbors} = 2d$. The most efficient tessellations (the hexagonal "honeycomb" tiling of a two-dimensional surface, the Weaire–Phelan structure, in three dimensions, etc.) have $N_{neighbors} = d+1$. In practice, determining the right value of $N_{neighbors}$ requires either physical insight or trial-and-error. If $N_{neighbors}$ is too large, one uses the Euclidean distance formula, Eq. (6), between molecules that are not close enough together for the Euclidean distance to be appropriate. If $N_{neighbors}$ is too small, one is trying to model the data with a manifold of insufficient dimensionality.

15.4.2 *PROPERTY ADAPTED LAPLACIAN EIGENMAPS*

One of the simplest manifold-learning methods is Laplacian eigenmaps [101-105]. To provide an intuitive explanation, suppose that you are given a two-dimensional surface embedded in a three-dimensional space. Imagine that the graph of molecular images (vertices) and edges (between nearest-neighbor molecules) provides a wire-frame for this surface. Imagine then, striking the surface with a hammer, so that it rings like a bell. Parts of the surface that are tightly linked to each other will vibrate together, while parts of the surface that are weakly connected will vibrate separately. The low-frequency modes (corresponding to low-pitched sounds) will have few nodes with large portions of the surface vibrating together, while high-frequency modes will have many nodes and have only small portions of the surface vibrating together. Based on the idea that the low-frequency modes capture the main coarse details of the surface, we can describe all but the fine details of the surface with the low-frequency modes. The low-frequency modes, then, provide a dimensional reduction.

The acoustic spectrum we described are the eigenfunctions of the Laplacian. To extend this concept to many dimensions, we define an appropriate graph-theoretic Laplacian. To do this, we first build a weighted adjacency matrix, $\mathbf{W}$, with elements

$$w_{LM} = \begin{cases} \exp\left(-\dfrac{d_{LM}^2}{t^2}\right) & L \in \mathcal{N}^{(M)} \\ 0 & L \notin \mathcal{N}^{(M)} \end{cases} \qquad (7)$$

Notice that in the limit $t \to \infty$, this is just the normal adjacency matrix. Sometimes, it is useful to choose t in an automated, edge-dependent way. For example, it has been suggested to consider the "self-tuning" weights [106, 107]

$$w_{LM} = \begin{cases} \exp\left(-\dfrac{d_{LM}^2}{\sigma_L \sigma_M}\right) & L \in \mathcal{N}^{(M)} \\ 0 & L \notin \mathcal{N}^{(M)} \end{cases} \tag{8}$$

where σ_M is the distance to the furthest-away neighbor of molecule M,

$$\sigma_M = \underbrace{\max}_{L \in \mathcal{N}^{(M)}} d_{LM} \tag{9}$$

The diagonal "mass" matrix, **D**, is defined with elements

$$d_{LM} = \delta_{LM} \sum_{L \in \mathcal{N}^{(M)}} w_{LM} = \delta_{LM} \sum_{L=1}^{N_{\text{molecules}}} w_{LM} \tag{10}$$

and the graph-theoretical Laplacian, **F**, is defined as

$$F = D - W \tag{11}$$

One then solves the generalized eigen problem,

$$Fx_k = \lambda_k D x_k \tag{12}$$

The smallest eigenvalue is zero and corresponds to a constant eigen-vector, $\mathbf{x}_0 = (N_{\text{molecules}})^{-\frac{1}{2}}$. Sort the remaining eigenvalues in the increasing order,

$$\lambda_0 < \lambda_1 \le \lambda_2 \le \lambda_3 \le \cdots \le \lambda_{N_{\text{molecules}}} \tag{13}$$

The first, d nontrivial eigenvectors provide a parameterization of the low-dimensional manifold,

$$\{\mathbf{x}_k\}_{k=1}^d = \left\{ \begin{bmatrix} x_{k1} & x_{k2} & \cdots & x_{kN_{\text{molecules}}} \end{bmatrix}^T \right\}_{k=1}^d \tag{14}$$

In particular, each molecule can be expressed as d-dimensional vector,

$$\mathbf{y}^{(M)} = \begin{bmatrix} x_{1,M} & x_{2,M} & \cdots & x_{d,M} \end{bmatrix} \tag{15}$$

The distance between points in this the new low-dimensional space should be close to the distance along the surface of the manifold (i.e., the geodesic distance).

Often one does not know what the effective dimension is in advance. In such cases, the effective dimension can be ascertained from the spectrum of the Laplacian. Specifically, one looks for an "eigenvalue gap" where there is a large gap in the Laplacian's spectrum. That is, d should be chosen so that $\lambda_d - \lambda_{d-1} << \lambda_{d+1} - \lambda_d$. Note that if $d << 2N_{\text{neighbors}}$, then one should consider repeating the analysis with a smaller number of neighbors. Conversely, if $d > N_{\text{neighbors}}$, then one must repeat the analysis with additional neighbors specified.

Another nuance, which is less important in this application but nonetheless relevant, is that if one has already learned the coordinate representation in Eq. (15), then when one adds new data, it is not necessary to repeat the learning procedure again, from the beginning. Instead, one can use any of the several approaches for out-of-sample extension of Laplacian eigenmaps [98, 108, 109].

As a simpler, and perhaps better, alternative however, one can use the perturbation theory. First, rewrite the generalized eigenvalue problem as an ordinary eigenvalue problem,

$$Hz_k = \lambda_k z_k, \quad H = D^{-\frac{1}{2}} F D^{-\frac{1}{2}} \quad \text{and} \quad x_k = D^{-\frac{1}{2}} z_k \tag{16}$$

and then note that the addition of new molecules to the dataset merely changes the underlying matrices,

$$\begin{aligned} W &\to W + \delta_W \\ D &\to D + \delta_D \\ F &\to F + \delta_F \\ G &\to G + \delta_G \end{aligned} \tag{17}$$

The unperturbed matrices are padded with elements from the identity matrix so that their dimension is the same as the dimension of the larger basis set. Then, the perturbed eigenvalues and eigenvectors are, to first order in the change due to the new data,

$$\lambda_k^{(1)} = \lambda_k + z_k^T \cdot \delta_G \cdot z_k \tag{18}$$

$$z_k^{(1)} = z_k + \sum_{j \neq k} \left(\frac{z_j^T \cdot \delta_G \cdot z_k}{\lambda_k - \lambda_j} \right) z_j \tag{19}$$

From Eq. (19), one uses the definition in Eq. (16) to obtain the parameterization of the manifold. When the number of molecules is large, the sum in Eq. (19) will need to be truncated so that the procedure is practical; this will usually be possible because truncating the sum in Eq. (19) corresponds to restricting the extent to which the equation of the manifold adapts to the presence of the new molecules. If the spectral gap closes after the perturbation (i.e., it is no longer true that $\lambda_d - \lambda_{d-1} << \lambda_{d+1} - \lambda_d$), then one needs to increase the dimensionality of the manifold.

15.4.3   MAXIMUM VARIANCE UNFOLDING

Maximum variance unfolding is a way to learn the reduced-dimensional representation of a manifold directly [110–113]. One starts by requiring that the Euclidean distance between neighboring points in the high-dimensional space is equal to the Euclidean distance between neighboring points on the d-dimensional manifold, i.e.,

$$\left| \mathbf{y}^{(L)} - \mathbf{y}^{(M)} \right|^2 = d_{LM}^2 \qquad L \in \mathcal{N}^{(M)} \tag{20}$$

where $\mathbf{y}^{(M)} \in \mathbf{R}^d$ is the coordinate representation of the molecule in the low-dimensional space. Our goal is to find an unfolded representation of the manifold where the Euclidean metric is valid *globally*, and therefore, we maximize the distance between all the points on the manifold, subject to the constraints in Eq. (20).

$$\underbrace{\max}_{\left\{\begin{array}{l}\left|\mathbf{y}^{(L)}-\mathbf{y}^{(M)}\right|^2=d_{LM}^2\ L\in\mathcal{N}^{(M)}\\0=\sum\limits_{L=1}^{N_{\text{molecules}}}\mathbf{y}^{(L)}\end{array}\right.}\sum_{L=1}^{N_{\text{molecules}}}\sum_{M=1}^{N_{\text{molecules}}}\left|\mathbf{y}^{(L)}-\mathbf{y}^{(M)}\right|^2 \tag{21}$$

The final constraint is used to ensure that there is a unique solution; it removes translational invariance by forcing the manifold to be centered at zero. Note that the existence of this mapping assumes that the manifold is homeomorphic to a hyperplane, i.e., the data are assumed to lie on a deformed hyperplane.

Direct solution of Eq. (21) is impractical; therefore, one instead forms the Gram matrix, $\mathbf{G}$, with elements

$$g_{LM} = \mathbf{y}^{(L)} \cdot \mathbf{y}^{(M)} \tag{22}$$

which has the property

$$g_{LL} + g_{MM} - 2g_{LM} = d_{LM}^2 \tag{23}$$

Then, one can rewrite the optimization as a semidefinite program, where one maximizes the trace of the Gramian over all positive semidefinite symmetric Gram matrices are subjected to the same constraints as in Eq. (21),

$$\underbrace{\max}_{\left\{\mathbf{G}\succeq0\left|\begin{array}{l}g_{LL}+g_{MM}-2g_{LM}=d_{LM}^2\ \ L\in\mathcal{N}^{(M)}\\0=\sum\limits_{L,M}g_{LM}\end{array}\right.\right\}}\text{Tr}\big[\mathbf{G}\big] \tag{24}$$

In this case, the highest eigenvalues of the Gramian are the ones that have the most information. Denote the eigenvectors and eigenvalues of the Gramian as

$$\mathbf{G} = \sum_{L=1}^{N_{\text{molecules}}} \mu_L \mathbf{m}_L \mathbf{m}_L^T \tag{25}$$

$$\mu_1 \geq \mu_2 \geq \cdots \geq \mu_{N_{\text{molecules}}}$$

One wishes to observe a spectral gap at such that $\mu_{d-1}-\mu_d \gg \mu_d-\mu_{d+1}$. In practice, one wants the eigenvalues that are discarded to be approximately zero, $\mu_{k>d} \approx 0$. Note that if $d \ll 2N_{\text{neighbors}}$, then one should consider repeating the analysis with a smaller number of neighbors. Conversely, if $d > 2N_{\text{neighbors}}$, then one must repeat the analysis with additional neighbors specified. The reduced-dimensional representation of the data is then,

$$\mathbf{y}^{(M)} = \begin{bmatrix} m_{1,M}\sqrt{\mu_1} & m_{2,M}\sqrt{\mu_2} & \cdots & m_{d,M}\sqrt{\mu_d} \end{bmatrix}^T \qquad (26)$$

As with Laplacian eigenmaps, there are ways to extend maximum variance unfolding to new molecules without redoing the learning process.

The second "centering" constraint in Eq. (24) can be removed by changing the objective function, obtaining [114],

$$\max_{\left\{ \mathbf{G} \succeq 0 \middle| g_{LL}+g_{MM}-2g_{LM}=d_{LM}^2 \quad L \in \mathcal{N}^{(M)} \right\}} \mathrm{Tr}\left[\mathbf{HGH}\right] \qquad (27)$$

where $\mathbf{H}$ is the centering matrix with elements

$$h_{ij} = \delta_{ij} - \frac{1}{N_{\text{molecules}}} \qquad (28)$$

If one then imposes the constraints with a penalty parameter, one has an unconstrained semidefinite optimization problem [114],

$$\max_{\{\mathbf{G} \succeq 0\}} \left[\mathrm{Tr}\left[\mathbf{HGH}\right] - v \sum_{M=1}^{N_{\text{molecules}}} \sum_{L \in \mathcal{N}^{(M)}} \left(g_{LL} + g_{MM} - 2g_{LM} - d_{LM}^2 \right)^2 \right] \qquad (29)$$

This can be solved as a simple (global) optimization problem for the lower-triangular Cholesky factorization matrices, $\mathbf{G}=\mathbf{LL}^T$. Note that one can easily find low-rank solutions by zeroing all but a few rows of the Cholesky matrix.

One advantage of these algorithm is that it can be extended to include information about the properties we are trying to predict, $Q^{(M)}$, as in colored maximum variance unfolding [114]. To do this, one defines a target Gramian, $\mathbf{J}$, with elements

$$J_{LM} = \begin{cases} Q^{(L)}Q^{(M)} & \textit{properties of molecules L and M are known} \\ \delta_{LM} & \textit{otherwise} \end{cases} \tag{30}$$

and then replaces the objective function in Eq. (27) or (29) with $\mathrm{Tr}[\mathbf{HGHJ}]$.

15.4.4 PROPERTY PREDICTIONS FROM THE PEID MANIFOLD

After performing manifold learning with Laplacian eigenmaps or maximum variance unfolding, each molecule is represented by a d-dimensional vector, $\mathbf{y}^{(M)}$. The distance between molecules can then be computed with the Euclidean distance formula. Covariance between the molecular properties can then be modeled with a Matérn covariance function [115],

$$\Sigma_{LM} = \frac{2^{1-v}}{\Gamma(v)} \left(2v \left(\frac{\left|\mathbf{y}^{(L)} - \mathbf{y}^{(M)}\right|^2 + b^2}{a^2} \right) \right)^{v/2} K_v \left(\sqrt{2v \left(\frac{\left|\mathbf{y}^{(L)} - \mathbf{y}^{(M)}\right|^2 + b^2}{a^2} \right)} \right) \tag{31}$$

where $K_v(x)$ is the modified Bessel function of the second kind. One can then predict the properties of a target molecule whose values for the property, $Q^{(\odot)}$, is not known using the kriging procedure,

$$\begin{bmatrix} \Sigma_{11} & \Sigma_{12} & \cdots & \Sigma_{1M} & 1 \\ \Sigma_{21} & \Sigma_{22} & \cdots & \Sigma_{2M} & 1 \\ \vdots & \vdots & \ddots & \vdots & \vdots \\ \Sigma_{M1} & \Sigma_{M2} & \cdots & \Sigma_{MM} & 1 \\ 1 & 1 & \cdots & 1 & 0 \end{bmatrix} \begin{bmatrix} w_1 \\ w_2 \\ \vdots \\ w_M \\ \mu \end{bmatrix} = \begin{bmatrix} \Sigma_{1\odot} \\ \Sigma_{2\odot} \\ \vdots \\ \Sigma_{M\odot} \\ 1 \end{bmatrix} \tag{32}$$

$$Q^{(\odot)} = \sum_{M=1}^{N_{\text{molecules}}} w_M Q^{(M)} \tag{33}$$

The hyperparameters that define the covariance function, $\{v,a,b\}$, can be determined using cross-validation or other similar methods [115–117].

Certainly, this is not the first time kriging has been used in a (bio)chemical context, but to the best of our knowledge, kriging on a learned manifold has been used for the first time for molecular similarity [116–123].

15.5   SUMMARY

The goal of this paper is to explain how chemical properties, and especially reactivity indicators arising from conceptual density functional theory, can be used to make predictions of molecular properties. Realizing that the chemical reactivity and substrate binding of a molecule is mostly determined by its shapes and its properties (e.g., electrostatic potential, electrophilicity, nucleophilicity, etc.) on its van der Waals surface, we first compute the property enhanced intramolecular distances (Section 15.2). This requires (1) sampling points on the molecular van der Waals surface, (2) finding all the intramolecular line segments between these points (i.e., all segments that are entirely contained within the molecular surface), (3) computing the average of the molecular property values at the two endpoints of the segments, and (4) computing the difference between the molecular property values at the two endpoints of the segments. In the course of formulating the PEID, we derived a new chirality descriptor, which is presented in Appendix 1.

The PEID procedure can easily produce billions of numbers, and therefore, we start by binning the property values and the lengths of the intramolecular segments; the normalized two-dimensional histogram we obtain has the form of an image (Section 15.3). We developed an appropriate binning procedure to form the histogram, as discussed in Appendix B. The amount of data in the PEID images is still vast, and therefore, we propose an additional dimensionality reduction step using manifold learning, specifically Laplacian eigenmaps or maximum variance unfolding. This allows us to define an appropriate distance between molecules (or, more precisely, a distance between molecules' PEIDs); this distance is then used in a covariance function to predict properties using kriging.

The advantage of this method to molecular similarity is that it is rotationally and translationally invariant. Therefore, it is alignment free, and the NP-hard problem of determining an appropriate molecular alignment is thus avoided [124]. In addition, by restating molecular similarity as an image similarity problem, we allow ourselves to use results from the actively developing community of researchers who are developing methods for machine learning from images. The reduced dimensional description one obtains also allows one to identify which features of PEIDs are most closely associated with the properties of interest (e.g., one can visualize the dominant eigenvectors in Laplacian eigenmaps or maximum variance unfolding). This allows one to restate the molecular design problem in an interesting way: can one design a molecule with a desired PEID image? The key features provide a sort of negative image of the binding target. For example, if one finds a certain PEID feature associated with the electrophilicity [125–128] of the surface of a molecule, one knows that the surface of the target must have a geometrically conjugate feature associated with surface nucleophilicity. Similarly, a given signature in the dual descriptor [71, 73–75, 129] of a molecule corresponds to a geometrically conjugate feature in the binding target with an inverted sign.

ACKNOWLEDGMENTS

The authors thank NSERC and Compute Canada for funding.

KEYWORDS

- **chemical reactivity indicators**
- **image recognition**
- **kriging**
- **machine learning**
- **manifold learning**
- **molecular property prediction**
- **quantitative structure property relationships (QSPR)**

REFERENCES

1. Arakawa, M., Hasegawa, K., & Funatsu, K., (2007). The recent trend in QSAR modeling – Variable selection and 3D-QSAR methods. *Current Computer-Aided Drug Design, 3*(4), 254–262.
2. Bultinck, P., Girones, X., & Carbó-Dorca, R., (2005). Molecular quantum similarity: Theory and applications. *Rev. Comput. Chem., 21*, 127–207.
3. Besalú, E., Girones, X., Amat, L., & Carbó-Dorca, R., (2002). Molecular quantum similarity and the fundamentals of QSAR. *Acc. Chem. Res., 35*, 289–295.
4. Carbó-Dorca, R., Amat, L., Besalú, E., Girones, X., & Robert, D., (2000). Quantum mechanical origin of QSAR: theory and applications. *J. Mol. Struct.: THEOCHEM, 504*, 181–228.
5. Carbó-Dorca, R., Amat, L., Besalú, E., & Lobato, M., (1998). Quantum similarity. In *Advances in Molecular Similarity*, Carbó-Dorca, R., Mezey, P. G., Eds., vol. 2, pp. 1–42.
6. Carbó-Dorca, R., & Besalú, E., (1998). A general survey of molecular quantum similarity. *J. Mol. Struct.: THEOCHEM, 451*, 11–23.
7. Carbó, R., Besalú, E., Amat, L., & Fradera, X., (1996). On quantum molecular similarity measures (QMSM) and indices (QMSI). *J. Math. Chem., 19*, 47–56.
8. Carbó, R., Leyda, L., & Arnau, M., (1980). How similar is a molecule to another: an electron-density measure of similarity between two molecular-structures. *Int. J. Quantum Chem., 17*, 1185–1189.
9. Sukumar, N., & Das, S., (2011). Current Trends in Virtual High Throughput Screening Using Ligand-Based and Structure-Based Methods. *Combinatorial Chemistry & High Throughput Screening, 14*(10), 872–888.
10. Ballester, P. J., (2011). Ultrafast shape recognition: method and applications. *Future Medicinal Chemistry, 3*(1), 65–78.
11. Sukumar, N., Krein, M., & Breneman, C. M., (2008). Bioinformatics and cheminformatics: Where do the twain meet? *Current Opinion in Drug Discovery & Development, 11*(3), 311–319.
12. Bender, A., & Glen, R. C., (2004). Molecular similarity: a key technique in molecular informatics. *Organic & Biomolecular Chemistry, 2*(22), 3204–3218.
13. Kitchen, D. B., Decornez, H., Furr, J. R., & Bajorath, J., (2004). Docking and scoring in virtual screening for drug discovery: Methods and applications. *Nature Reviews Drug Discovery, 3*(11), 935–949.
14. Girones, X., Robert, D., & Carbo-Dorca, R., (2001). TGSA: A molecular superposition program based on topo-geometrical considerations. *J. Comput. Chem., 22*(2), 255–263.
15. Bultinck, P., Kuppens, T., Girone, X., & Carbo-Dorca, R., (2003). Quantum similarity superposition algorithm (QSSA): A consistent scheme for molecular alignment and molecular similarity based on quantum chemistry. *Journal of Chemical Information and Computer Sciences, 43*(4), 1143–1150.
16. Bultinck, P., Carbo-Dorca, R., & Van Alsenoy, C., (2003). Quality of approximate electron densities and internal consistency of molecular alignment algorithms in

molecular quantum similarity. *Journal of Chemical Information and Computer Sciences, 43*(4), 1208–1217.

17. Girones, X., & Carbo-Dorca, R., (2004). TGSA-flex: Extending the capabilities of the topo-geometrical superposition algorithm to handle flexible molecules. *J. Comput. Chem., 25*(2), 153–159.

18. Bultinck, P., & Carbó-Dorca, R., (2005). Molecular quantum similarity using conceptual DFT descriptors. *J. Chem. Sci., 117*, 425–435.

19. Heidar Zadeh, F., & Ayers, P. W., (2013). Molecular alignment as a penalized permutation Procrustes problem. *J. Math. Chem., 51*(3), 927–936.

20. Heidar-Zadeh, F., Ayers, P. W., & Carbó-Dorca, R. *A Statistical Perspective on Molecular Similarity*, Chapter 10 (this book).

21. Lemmen, C., & Lengauer, T., (2000). Computational methods for the structural alignment of molecules. *J. Comput. Aided Mol. Des., 14*(3), 215–232.

22. Ballester, P. J., & Richards, W. G., (2007). Ultrafast shape recognition to search compound databases for similar molecular shapes. *J. Comput. Chem., 28*(10), 1711–1723.

23. Osada, R., Funkhouser, T., Chazelle, B., Dobkin, D., Ieee Computer, S., & Ieee Computer, S., (2001). *Matching 3D Models with Shape Distributions.*, pp. 154

24. Osada, R., Funkhouser, T., Chazelle, B., & Dobkin, D., (2002). Shape distributions. *Acm Transactions on Graphics, 21*(4), 807–832.

25. Ohbuchi, R., Minamitani, T., & Takei, T., (2003). *Shape-Similarity Search of 3D Models by Using Enhanced Shape Functions*, pp. 97–104.

26. Breneman, C. M., Sundling, C. M., Sukumar, N., Shen, L. L., Katt, W. P., & Embrechts, M. J., (2003). New developments in PEST shape/property hybrid descriptors. *J. Comput. Aided Mol. Des., 17*(2), 231–240.

27. Yang, W. T., & Mortier, W. J., (1986). The use of global and local molecular parameters for the analysis of the gas-phase basicity of amines. *J. Am. Chem. Soc., 108*, 5708–5711.

28. Bader, R. F. W., (1990). *Atoms in Molecules: A Quantum Theory*. Clarendon: Oxford.

29. Bader, R. F. W., (1985). Atoms in molecules. *Acc. Chem. Res., 18*, 9–15.

30. Cohen, L., (1979). Local kinetic energy in quantum mechanics. *J. Chem. Phys., 70*, 788–789.

31. Bultinck, P., Fias, S., Alsenoy, C. V., Ayers, P. W., & Carbó-Dorca, R., (2007). Critical thoughts on computing atom condensed Fukui functions. *J. Chem. Phys., 127*, 034102.

32. Tiznado, W., Chamorro, E., Contreras, R., & Fuentealba, P., (2005). Comparison among four different ways to condense the Fukui function. *J. Phys. Chem. A., 109*, 3220–3224.

33. Fuentealba, P., Perez, P., & Contreras, R., (2000). On the condensed Fukui function. *J. Chem. Phys., 113*, 2544–2551.

34. Ayers, P. W., Morrison, R. C., & Roy, R. K., (2002). Variational principles for describing chemical reactions: Condensed reactivity indices. *J. Chem. Phys., 116*, 8731–8744.

35. Johnson, P. A., Bartolotti, L. J., Ayers, P. W., Fievez, T., & Geerlings, P., (2012). Charge density and chemical reactivity: A unified view from conceptual DFT". In

Modern Charge Density Analysis, Gatti, C., Macchi, P., Eds. Springer: New York, pp. 715–764.

36. Ayers, P. W., Anderson, J. S. M., & Bartolotti, L. J., (2005). Perturbative perspectives on the chemical reaction prediction problem. *Int. J. Quantum Chem.*, *101*, 520–534.

37. Geerlings, P., De Proft, F., & Langenaeker, W., (2003). Conceptual density functional theory. *Chem. Rev.*, *103*, 1793–1873.

38. Liu, S. B., (2009). Conceptual Density Functional Theory and Some Recent Developments. *Acta Physico-Chimica Sinica.*, *25*, 590–600.

39. Gazquez, J. L., (2008). Perspectives on the density functional theory of chemical reactivity. *Journal of the Mexican Chemical Society*, *52*, 3–10.

40. Van Damme, S., & Bultinck, P., (2010). 3D QSAR based on conceptual DFT molecular fields: Antituberculotic activity. *J. Mol. Struct.: Theochem.*, *943*, 83–89.

41. Van Damme, S., & Bultinck, P., (2009). Conceptual DFT Properties-Based 3D QSAR: Analysis of Inhibitors of the Nicotine Metabolizing CYP2A6 Enzyme. *J. Comput. Chem.*, *30*, 1749–1757.

42. Dolezal, R., Van Damme, S., Bultinck, P., & Waisser, K., (2009). QSAR analysis of salicylamide isosteres with the use of quantum chemical molecular descriptors. *European Journal of Medicinal Chemistry*, *44*, 869–876.

43. Van Damme, S., Langenaeker, W., & Bultinck, P., (2008). Prediction of blood-brain partitioning: A model based on ab initio calculated quantum chemical descriptors. *Journal of Molecular Graphics & Modelling*, *26*, 1223–1236.

44. Liu, S. B., & Parr, R. G., (1997). Second-Order Density-Functional Description of Molecules and Chemical Changes. *J. Chem. Phys.*, *106*(13), 5578–5586.

45. Parr, R. G., (2000). Density-functional theory and chemistry. *Condens.Matter Theor.*, *15*, 297–302.

46. Geerlings, P., & De Proft, F., (2008). Conceptual DFT: the chemical relevance of higher response functions. *PCCP*, *10*, 3028–3042.

47. Fuentealba, P., & Parr, R. G., (1991). Higher-order derivatives in density-functional theory, especially the hardness derivative. *J. Chem. Phys.*, *94*, 5559–5564.

48. Ayers, P. W., & Parr, R. G., (2008). Beyond electronegativity and local hardness: Higher-order equalization criteria for determination of a ground-state electron density. *J. Chem. Phys.*, *129*, 054111.

49. Senet, P., (1996). Nonlinear electronic responses, Fukui functions and hardnesses as functionals of the ground-state electronic density. *J. Chem. Phys.*, *105*, 6471–6489.

50. Cardenas, C., Echegaray, E., Chakraborty, D., Anderson, J. S. M., & Ayers, P. W., (2009). Relationships between third-order reactivity indicators in chemical density-functional theory. *J. Chem. Phys.*, *130*, 244105.

51. Heidar-Zadeh, F., Richer, M., Fias, S., Miranda-Quintana, R. A., Chan, M., Franco-Perez, M., Gonzalez-Espinoza, C. E., Kim, T. D., Lanssens, C., Patel, A. H. G., Yang, X. D., Vohringer-Martinez, E., Cardenas, C., Verstraelen, T., & Ayers, P. W., (2016). An explicit approach to conceptual density functional theory descriptors of arbitrary order. *Chem. Phys. Lett.*, *660*, 307–312.

52. Ayers, P. W., & Levy, M., (2000). Perspective on "Density functional approach to the frontier-electron theory of chemical reactivity" by Parr, R. G., & Yang, W. (1984). *Theor. Chem. Acc.*, *103*, 353–360.

53. Parr, R. G., & Yang, W. T., (1984). Density functional approach to the frontier-electron theory of chemical reactivity. *J. Am. Chem. Soc.*, *106*, 4049–4050.
54. Yang, W. T., Parr, R. G., & Pucci, R., (1984). Electron density, Kohn-Sham frontier orbitals, and Fukui functions. *J. Chem. Phys.*, *81*, 2862–2863.
55. Ayers, P. W., Yang, W. T., & Bartolotti, L. J., (2009). Fukui Function. In *Chemical reactivity theory: A density functional view*, Chattaraj, P. K., Ed. CRC Press: Boca Raton, pp. 255–267.
56. Yang, W. T., & Parr, R. G., (1985). Hardness, softness, and the fukui function in the electron theory of metals and catalysis. *Proc. Natl. Acad. Sci.*, *82*, 6723–6726.
57. Ayers, P. W., & Parr, R. G., (2000). Variational principles for describing chemical reactions: The Fukui function and chemical hardness revisited. *J. Am. Chem. Soc.*, *122*, 2010–2018.
58. Chandra, A. K., & Nguyen, M. T., (2008). Fukui function and local softness. In *Chemical reactivity theory: a density-functional view*, Chattaraj, P. K., Ed. Taylor and Francis: New York, pp. 163–178.
59. Toro-Labbé, A., Jaque, P., Murray, J. S., & Politzer, P., (2005). Connection between the average local ionization energy and the Fukui function. *Chem. Phys. Lett.*, *407*, 143–146.
60. Brinck, T., Murray, J. S., & Politzer, P., (1993). Molecular-surface electrostatic potentials and local ionization energies of Group V-VII hydrides and their anions: relationships for aqueous and gas-phase acidities. *Int. J. Quantum Chem.*, *48*, 73–88.
61. Murray, J. S., Brinck, T., & Politzer, P., (1991). Surface local ionization energies and electrostatic potentials of the conjugate bases of a series of cyclic hydrocarbons in relation to their aqueous acidities. *Int. J. Quantum Chem.*, *18*, 91–98.
62. Politzer, P., & Murray, J. S., (2002). The fundamental nature and role of the electrostatic potential in atoms and molecules *Theor. Chem. Acc.*, *108*, 134 142.
63. Politzer, P., & Truhlar, D., (1981). *Chemical Applications of Atomic and Molecular Electrostatic Potentials*. Plenum: New York.
64. Politzer, P., (1980). Electrostatic Potential-Electronic Density Relationships in Atoms. *J. Chem. Phys.*, *73* (7), 3264–3267.
65. Gadre, S. R., Kulkarni, S. A., & Shrivastava, I. H., (1992). Molecular Electrostatic Potentials - A Topographical Study. *J. Chem. Phys.*, *96*, 5253–5260.
66. Sjoberg, P., & Politzer, P., (1990). Use of the electrostatic potential at the molecular-surface to interpret and predict nucleophilic processes. *J. Phys. Chem.*, *94*, 3959–3961.
67. Shirsat, R. N., Bapat, S. V., & Gadre, S. R., (1992). Molecular Electrostatics - A comprehensive topographical approach. *Chem. Phys. Lett.*, *200*, 373–378.
68. Murray, J. S., Brinck, T., & Politzer, P., (1996). Relationships of molecular surface electrostatic potentials to some macroscopic properties. *Chem. Phys.*, *204*, 289–299.
69. Suresh, C. H., & Gadre, S. R., (1998). A novel electrostatic approach to substituent constants: Doubly substituted benzenes. *J. Am. Chem. Soc.*, *120*, 7049–7055.
70. Morell, C., Grand, A., Gutierrez-Oliva, S., & Toro-Labbé, A., (2007). In *Theoretical Aspects of Chemical Reactivity*, Elsevier: Amsterdam, pp. 31–45.
71. Ayers, P. W., Morell, C., De Proft, F., & Geerlings, P., (2007). Understanding the Woodward-Hoffmann rules using changes in the electron density. *Chem, Eur. J.*, *13*, 8240–8247.

72. Padmanabhan, J., Parthasarathi, R., Elango, M., Subramanian, V., Krishnamoorthy, B. S., Gutierrez-Oliva, S., Toro-Labbé, A., Roy, D. R., & Chattaraj, P. K., (2007). Multiphilic descriptor for chemical reactivity and selectivity. *J. Phys. Chem. A.*, *111*, 9130–9138.

73. Morell, C., Grand, A., & Toro-Labbé, A., (2006). *Chem. Phys. Lett.*, *425*, 342–346.

74. Morell, C., Grand, A., & Toro-Labbé, A., (2005). New dual descriptor for chemical reactivity. *J. Phys. Chem. A.*, *109*, 205–212.

75. Cardenas, C., Rabi, N., Ayers, P. W., Morell, C., Jaramillo, P., & Fuentealba, P., (2009). Chemical Reactivity Descriptors for Ambiphilic Reagents: Dual Descriptor, Local Hypersoftness, and Electrostatic Potential. *J. Phys. Chem. A.*, *113*, 8660–8667.

76. Anderson, J. S. M., Melin, J., & Ayers, P. W., (2007). Conceptual density-functional theory for general chemical reactions, including those that are neither charge nor frontier-orbital controlled. I. Theory and derivation of a general-purpose reactivity indicator. *J. Chem. Theory Comp.*, *3*, 358–374.

77. Anderson, J. S. M., & Ayers, P. W., (2007). Predicting the reactivity of ambidentate nucleophiles and electrophiles using a single, general-purpose, reactivity indicator. *PCCP*, *9*, 2371–2378.

78. Anderson, J. S. M., Melin, J., & Ayers, P. W., (2007). Conceptual density-functional theory for general chemical reactions, including those that are neither charge- nor frontier-orbital-controlled. 2. Application to molecules where frontier molecular orbital theory fails. *J. Chem. Theory Comp.*, *3*, 375–389.

79. Meek, P. J., Liu, Z. W., Tian, L. F., Wang, C. Y., Welsh, W. J., & Zauhar, R. J., (2006). Shape signatures: speeding up computer aided drug discovery. *Drug Discovery Today*, *11*(19–20), 895–904.

80. Chekmarev, D. S., Kholodovych, V., Balakin, K. V., Ivanenkov, Y., Ekins, S., & Welsh, W. J., (2008). Shape signatures: New descriptors for predicting cardiotoxicity in silico. *Chem. Res. Toxicol.*, *21*(6), 1304–1314.

81. Zauhar, R. J., Moyna, G., Tian, L. F., Li, Z. J., & Welsh, W. J., (2003). Shape signatures: A new approach to computer-aided ligand- and receptor-based drug design. *J. Med. Chem.*, *46*(26), 5674–5690.

82. Ruch, E., (1972). Algebraic aspects of the chirality phenomenon in chemistry. *Acc. Chem. Res.*, *5*(2), 49–56.

83. Mezey, P. G., (1997). Chirality measures and graph representations. *Computers & Mathematics with Applications*, *34*(11), 105–112.

84. Mezey, P. G., (1995). Rules on chiral and achiral molecular transformations .2. *J. Math. Chem.*, *18*(2–4), 133–139.

85. Mezey, P. G., (1998). Mislow's label paradox, chirality-preserving conformational changes, and related chirality measures. *Chirality*, *10*(1–2), 173–179.

86. Mezey, P. G., (1995). Rules On Chiral And Achiral Molecular-Transformations. *J. Math. Chem.*, *17*(2–3), 185–202.

87. Mezey, P. G., (1998). The proof of the metric properties of a fuzzy chirality measure of molecular electron density clouds. *J. Mol. Struct.: THEOCHEM*, *455*(2–3), 183–190.

88. Weinberg, N., & Mislow, K., (1993). Distance Functions as Generators of Chirality Measures. *J. Math. Chem.*, *14*(3–4), 427–450.

89. Weinberg, N., & Mislow, K., (1995). A Unification Of Chirality Measures. *J. Math. Chem.*, *17*(1), 35–53.
90. Gilat, G., (1994). On Quantifying Chirality - Obstacles and Problems Towards Unification. *J. Math. Chem.*, *15*(1–2), 197–205.
91. Mislow, K., & Siegel, J., (1984). Stereoisomerism and Local Chirality. *J. Am. Chem. Soc.*, *106*(11), 3319–3328.
92. Lipinski, C. A., Lombardo, F., Dominy, B. W., & Feeney, P. J., (2001). Experimental and computational approaches to estimate solubility and permeability in drug discovery and development settings 1PII of original article: S0169–409X(96)00423–1. The article was originally published in Advanced Drug Delivery Reviews 23 (1997) 3–25. 1. *Advanced Drug Delivery Reviews*, *46*(1–3), 3–26.
93. Szeliski, R., (2006). Image alignment and stitching: a tutorial. *Foundations and Trends in Computer Graphics and Vision*, *2*, 1–104.
94. Wang, J., (2012). *Geometric Structure of High-Dimensional Data and Dimensionality Reduction*. Springer: Berlin.
95. Wachinger, C., & Navab, N., (2010). Manifold learning for multi-modal image registration. In *Proceedings of the British Machine Vision Conference*, Labrosse, F., Zwiggelaar, R., Tiddeman, B., Eds. BMVA Press, pp. 82.
96. Aljabar, P., Wolz, R., & Rueckert, D., (2012). Manifold Learning for Medical Image Registration, Segmentation, and Classification. In *Machine Learning in Computer-Aided Diagnosis: Medical imaging intelligence and analysis*, Suzuki, K., Ed. IGI Global: Hershey, Pennsylvania, USA, pp. 351–372.
97. Mateus, D., Wachinger, C., Atasoy, S., Schwarz, L., & Navab, N., (2012). Learning Manifolds: Design Analysis for Medical Applications. In *Machine Learning in Computer-Aided Diagnosis: Medical Imaging Intelligence and Analysis*, Suzuki, K., Ed. IGI Global: Hershey, Pennsylvania, USA, pp. 374–402.
98. Han, Y., Xu, Z., Ma, Z., & Huang, Z., (2013). Image classification with manifold learning for out-of-sample data. *Signal Processing*, *93*(8), 2169–2177.
99. Sparks, R., & Madabhushi, A., (2016). Out-of-Sample Extrapolation utilizing Semi-Supervised Manifold Learning (OSE-SSL): Content Based Image Retrieval for Histopathology Images. *Scientific Reports*, *6*, 27306–27306.
100. Liu, H., Hussain, F., Tan, C. L., & Dash, M., (2002). Discretization: An enabling technique. *Data Mining and Knowledge Discovery*, *6*(4), 393–423.
101. Belkin, M., & Niyogi, P., (2002). Laplacian eigenmaps and spectral techniques for embedding and clustering. In *Advances in Neural Information Processing Systems 14, Vols 1 and 2*, Dietterich, T. G., Becker, S., Ghahramani, Z., Eds., vol. *14*, pp. 585–591.
102. Belkin, M., & Niyogi, P., (2003). Laplacian eigenmaps for dimensionality reduction and data representation. *Neural Computation*, *15*(6), 1373–1396.
103. Belkin, M., & Niyogi, P., (2004). Semi-supervised learning on Riemannian manifolds. *Machine Learning*, *56*(1–3), 209–239.
104. Belkin, M., & Niyogi, P., (2008). Towards a theoretical foundation for Laplacian-based manifold methods. *Journal of Computer and System Sciences*, *74*(8), 1289–1308.

105. Belkin, M., Niyogi, P., & Sindhwani, V., (2006). Manifold regularization: A geometric framework for learning from labeled and unlabeled examples. *Journal of Machine Learning Research, 7*, 2399–2434.
106. Zelnik-Manor, L., & Perona, P., (2004). Self-tuning spectral clustering. *Advances in Neural Information Processing Systems, 17*, 1601–1608.
107. Ting, D., Huang, L., & Jordan, M. I., 2010, In *An Analysis of the Convergence of Graph Laplacians*, 27th International Conference on Machine Learning, Haifa, Israel, June 21–24, 2010; Haifa, Israel, pp. 1079–1086.
108. Bengio, Y., Paiement, J. F. O., Vincent, P., Delalleau, O., Le Roux, N., & Ouimet, M., (2004). Out-of-sample extensions for LLE, isomap, MDS, eigenmaps, and spectral clustering. In *Advances in Neural Information Processing Systems 16*, Thrun, S., Saul, K., Scholkopf, B., Eds., vol. *16*, pp. 177–184.
109. Bengio, Y., Delalleau, O., Le Roux, N., Paiement, J. F., Vincent, P., & Ouimet, M., (2004). Learning eigenfunctions links spectral embedding and kernel PCA. *Neural Computation, 16*(10), 2197–2219.
110. Weinberger, K. Q., Saul, L. K., & Society, I. C., (2004). Unsupervised learning of image manifolds by semidefinite programming. In *Proceedings of the 2004 IEEE Computer Society Conference on Computer Vision and Pattern Recognition, vol 2*, pp. 988–995.
111. Weinberger, K. Q., & Saul, L. K., (2006). Unsupervised learning of image manifolds by semidefinite programming. *International Journal of Computer Vision, 70*(1), 77–90.
112. Weinberger, K. Q., Sha, F., & Saul, L. K., (2010). Convex Optimizations for Distance Metric Learning and Pattern Classification. *IEEE Signal Process. Mag., 27*(3), 146-+.
113. Weinberger, K. Q., & Saul, L. K., (2009). Distance Metric Learning for Large Margin Nearest Neighbor Classification. *Journal of Machine Learning Research, 10*, 207–244.
114. Song, L., Smola, A., Borgwardt, K., & Gretton, A., (2008). Colored maximum variance unfolding. In *NIPS*, Platt, J. C., Koller, D., Singer, Y., Roweis, S., Eds., vol. 20, pp. 1385–1392.
115. Rasmussen, C. E., & Williams, C. K. I., (2006). *Gaussian Processes for Machine Learning*. MIT Press: Cambridge.
116. Obrezanova, O., Csanyi, G., Gola, J. M. R., & Segall, M. D., (2007). Gaussian processes: A method for automatic QSAR Modeling of ADME properties. *J. Chem Inf. Model., 47*(5), 1847–1857.
117. Obrezanova, O., Gola, J. M. R., Champness, E. J., & Segall, M. D., (2008). Automatic QSAR modeling of ADME properties: blood-brain barrier penetration and aqueous solubility. *J. Comput. Aided Mol. Des., 22*(6–7), 431–440.
118. Hawe, G. I., Alkorta, I., & Popelier, P. L. A., (2010). Prediction of the Basicities of Pyridines in the Gas Phase and in Aqueous Solution. *J. Chem Inf. Model., 50*(1), 87–96.
119. Yin, H., Li, R., Fang, K. T., & Liang, Y. Z., (2007). Empirical Kriging models and their applications to QSAR. *J. Chemom., 21*(1–2), 43–52.

120. Peng, X. L., Yin, H., Li, R., & Fang, K. T., (2006). The application of Kriging and empirical Kriging based on the variables selected by SCAD. *Anal. Chim. Acta.*, *578*(2), 178–185.
121. Fang, K. T., Yin, H., & Liang, Y. Z., (2004). New approach by Kriging models to problems in QSAR. *Journal of Chemical Information and Computer Sciences*, *44*(6), 2106–2113.
122. Heidar-Zadeh, F., Paul W. Ayers, R., & Carbo-Dorca, "A Statistical Perspective on Molecular Similarity" (this book).
123. Heidar-Zadeh, F., & Ayers, P. W., (2018) Spectral Learning for Chemical Prediction. In: *Theoretical and Quantum Chemistry at the Dawn of the 21st Century*. Eds. Ramon Carbo-Dorca and Tanmoy Chakraborty, Apple Academic Press: Waretown NJ, USA.
124. Zadeh, F. H., & Ayers, P. W., (2013). Molecular alignment as a penalized permutation Procrustes problem. *J. Math. Chem.*, *51*(3), 927–936.
125. Parr, R. G., Von Szentpály, L., & Liu, S. B., (1999). Electrophilicity index. *J. Am. Chem. Soc.*, *121*, 1922–1924.
126. Liu, S. B., (2009). Electrophilicity. In *Chemical Reactivity Theory: A Density Functional View*, Chattaraj, P. K., Ed. Taylor and Francis: Boca Raton, pp. 179.
127. Chattaraj, P. K., Sarkar, U., & Roy, D. R., (2006). Electrophilicity index. *Chem. Rev.*, *106*, 2065–2091.
128. Chattaraj, P. K., Maiti, B., & Sarkar, U., (2003). Philicity: A unified treatment of chemical reactivity and selectivity. *J. Phys. Chem. A.*, *107*, 4973–4975.
129. Geerlings, P., Ayers, P. W., Toro-Labbe, A., Chattaraj, P. K., & De Proft, F., (2012). The Woodward-Hoffmann Rules Reinterpreted by Conceptual Density Functional Theory. *Acc. Chem. Res.*, *45*(5), 683–695.
130. Online Statistics Education: A Multimedia Course of Study (http://onlinestatbook.com/). Project Leader: David M. Lane, Rice University (Chapter 2 "Graphing Distributions," section "Histograms").

APPENDIX 1. A MEASURE OF THE LOCAL CHIRALITY OF A PROPERTY ON A SURFACE

One way to enhance descriptor values with information about chirality is to use the projection of a vector curl related to the descriptor. Here we present a less elegant, but simpler, approach. Define a tetrahedron whose center is at a point on the surface $\mathbf{R}_i$; define a vertex of the tetrahedron so that it ε units back towards the "interior" of the molecule along the intramolecular line segment $\mathbf{R}_i - \mathbf{R}_j$, where $\mathbf{R}_j$ is a different point on the molecular surface. For example, let

$$\mathbf{R}_{i,1} = \mathbf{R}_i - \tfrac{\sqrt{6}}{4}\varepsilon\left(\widehat{\mathbf{R}_i - \mathbf{R}_j}\right) \tag{34}$$

We use $\hat{\mathbf{u}} = \mathbf{u}/\sqrt{\mathbf{u} \cdot \mathbf{u}}$ to denote the unit vector corresponding to the vector $\mathbf{u}$

This does not fully define the orientation and position of the tetrahedron as it is still free to rotate about the axis defined by $\mathbf{R}_i - \mathbf{R}_j$. To fix the orientation of the tetrahedron, choose a second vertex in the steepest descent direction of the gradient vector field of the property being studied *along the surface of the molecule*. That is, define

$$\mathbf{a} = \nabla p\left(\mathbf{R}_i\right) - \frac{\left(\nabla p\left(\mathbf{R}_i\right) \cdot \left(\mathbf{R}_i - \mathbf{R}_j\right)\right)\left(\mathbf{R}_i - \mathbf{R}_j\right)}{\left(\mathbf{R}_i - \mathbf{R}_j\right) \cdot \left(\mathbf{R}_i - \mathbf{R}_j\right)} \tag{35}$$

$$\mathbf{R}_{i,2} = \mathbf{R}_i + \tfrac{\sqrt{6}}{12}\varepsilon\left(\widehat{\mathbf{R}_i - \mathbf{R}_j}\right) + \tfrac{1}{\sqrt{3}}\varepsilon\hat{\mathbf{a}} \tag{36}$$

The other two vertices are located at

$$\mathbf{R}_{i,2} = \mathbf{R}_i + \tfrac{\sqrt{6}}{12}\varepsilon\left(\widehat{\mathbf{R}_i - \mathbf{R}_j}\right) - \tfrac{\sqrt{3}}{6}\varepsilon\hat{\mathbf{a}} + \tfrac{1}{2}\varepsilon\left(\mathbf{a} \times \left(\widehat{\mathbf{R}_i - \mathbf{R}_j}\right)\right) \tag{37}$$

This specifies the tetrahedron. Using the property values at the "exterior" vertices of the tetrahedron, we can label the chirality of the tetrahedron based on the sign of $p_{i,4} - p_{i,3}$. (This is analogous to assigning R or S

chirality in elementary organic chemistry.) Then, using the fact that the center of inversion is the point of interest $\mathbf{R}_i$, invert the tetrahedron. This gives a new tetrahedron with points,

$$\mathbf{R}_{i,t^-} = \mathbf{R}_i - \left(\mathbf{R}_{i,t} - \mathbf{R}_i\right) \qquad\qquad t^- = 1,2,3,4 \qquad\qquad (38)$$

property values $\left\{p^-\left(\mathbf{R}_{i,t^-}\right)\right\}_{t^-=1}^{4}$ 12 possible ways of permuting $\left\{\mathbf{R}_{i,t^-}\right\}_{t^-=1}^{4}$ to superpose the inverted tetrahedron with the original tetrahedron. Minimize over all these permutations, $\pi(t^-)$, to find the best match,

$$\delta_i^{(\chi)} = \min_{\left\{\pi\left(t^-\right)\right\}} \frac{1}{4} \sum_{t=1}^{4} \left(\frac{p\left(\mathbf{R}_{i,t}\right) - p^{(-)}\left(\mathbf{R}_{i,\pi\left(t^{(-)}\right)}\right)}{\varepsilon} \right)^2 \qquad\qquad (39)$$

Then, assign the chirality measure as,

$$\chi_i = \delta_i^{(\chi)} \operatorname{sgn}\left(p_{i,4} - p_{i,3}\right) \qquad\qquad (40)$$

In the $\varepsilon \to 0$ limit, this chirality measure could be computed directly from the first and second derivatives of the vector field, $\nabla p(\mathbf{R}_i)$ and $\nabla\nabla^T p(\mathbf{R}_i)$. But the above, more explicit, procedure is easier to understand and implement.

APPENDIX 2. A SIMPLE ALGORITHM FOR ISOFREQUENCY BINNING OF DATA

Consider a molecular dataset where each molecule, M, is associated with a n_M observations, $\left\{x_i^{(M)}\right\}_{j=1}^{n_M}$. The total number of observations is $n_{\text{total}} = n^{(1)} + n^{(2)} + \cdots + n^{(N_{\text{molecules}})}$. Based on Rice's rule [130], it is reasonable to consider

$$N_{\text{bins}} = \left\lceil 2 \left(\frac{n_{\text{total}}}{N_{\text{molecules}}} \right)^{1/3} \right\rceil \qquad\qquad (41)$$

where $\lceil y \rceil$ is the ceiling function, the smallest integer greater than or equal to y.

We wish to consider bins with equal probability of containing data. To achieve this, we first merge the lists of molecular observations,

$$\mathcal{Y} = \bigcup_{M=1}^{N_{molecules}} \left\{ x_n^{(M)} \right\}_{n=1}^{n_M} \tag{42}$$

and then, we sort the elements of this list, $y_i \in \mathcal{Y}$, in the increasing order,

$$y_1 \le y_2 \le \cdots \le y_{n_{total}} \tag{43}$$

The intervals that define the bins are then,

$$\left[y_1, y_{\lceil n_{total}/N_{bins} \rceil} \right)$$

$$\left[y_{\lceil n_{total}/N_{bins} \rceil}, y_{\lceil 2n_{total}/N_{bins} \rceil} \right)$$

$$\vdots \tag{44}$$

$$\left[y_{\lceil (k-1)\cdot n_{total}/N_{bins} \rceil}, y_{\lceil kn_{total}/N_{bins} \rceil} \right)$$

$$\vdots$$

$$\left[y_{\lceil (N_{bins}-1)\cdot n_{total}/N_{bins} \rceil}, y_{n_{total}} \right]$$

We define the general bin interval as,

$$\left[b_{k-1}, b_k \right) \equiv \left[y_{\lceil (k-1)\cdot n_{total}/N_{bins} \rceil}, y_{\lceil kn_{total}/N_{bins} \rceil} \right) \tag{45}$$

(This notation is a bit imprecise because we choose $b_0 = y_1$ and the last interval, with $k = N_{bins}$, is closed on the upper limit.) We then define the normalized molecular histogram of the bins by determining how many samples of each molecule are in the bins and dividing it by the average number of samples in the bin,

$$h_k^{(M)} = \frac{N_{bins} N_{molecules}}{n_{total}} \text{count} \left\{ x_n^{(M)} \in \left[b_{k-1}, b_k \right) \right\}_{n=1}^{n_M} \tag{46}$$

Suppose that each value of $x_i^{(M)}$ is associated with a property value, $p_i^{(M)}$. This means that in each bin, there is a range of property values. We then define a new set that contains all the property values associated with a given bin,

$$\mathcal{Z}_k = \bigcup_{M=1}^{N_{molecules}} \left\{ p_n^{(M)} \text{ such that } x_n^{(M)} \in [b_{k-1}, b_k) \right\} \tag{47}$$

$$M_k = count(\mathcal{Z}_k) \tag{48}$$

and order the properties in this set in the increasing order, $z_{k,1} \le z_{k,2} \le \cdots \le z_{k,M_k}$. On average, the number of properties in each bin is $n_{\text{total}}/N_{\text{bins}}$. This suggests that we should consider

$$N_{\text{bins}}^{(2)} = \left\lceil 2 \left(\frac{n_{\text{total}}}{N_{\text{molecules}} N_{\text{bins}}} \right)^{1/3} \right\rceil \tag{49}$$

secondary bins, which are defined to contain the property values in the intervals,

$$\left[b_{k,\ell-1}^{(2)}, b_{k,\ell}^{(2)} \right) \equiv \left[z_{k, \lceil (\ell-1) \cdot M_k / N_{\text{bins}}^{(2)} \rceil}, z_{k, \lceil \ell \cdot M_k / N_{\text{bins}}^{(2)} \rceil} \right) \tag{50}$$

(This notation is a bit imprecise because we choose $b_0^{(2)} = z_1$ and the last interval, with $\ell = N_{\text{bins}}^{(2)}$, is closed on the upper limit.) We then define the secondary (property) histogram as

$$h_{k,\ell}^{(M)} = \frac{N_{\text{bins}}^{(2)} N_{\text{bins}} N_{\text{molecules}}}{n_{\text{total}}} count \left\{ p_i^{(M)} \in \left[b_{k,\ell-1}^{(2)}, b_{k,\ell}^{(2)} \right) \text{ and } x_i^{(M)} \in [b_{k-1}, b_k) \right\}_{n=1}^{n_M} \tag{51}$$

INDEX

Milton Keynes UK
Ingram Content Group UK Ltd.
UKHW050300161024
449569UK00048B/799

9 781774 635322